Safe, Secure and Sustainable Oil and Gas Drilling, Exploitation and Pipeline Transport Offshore

Safe, Secure and Sustainable Oil and Gas Drilling, Exploitation and Pipeline Transport Offshore

Editors

Dejan Brkić
Pavel Praks

MDPI • Basel • Beijing • Wuhan • Barcelona • Belgrade • Manchester • Tokyo • Cluj • Tianjin

Editors
Dejan Brkić   Pavel Praks
VSB—Technical University of Ostrava   VSB—Technical University of Ostrava
Czech Republic   Czech Republic

Editorial Office
MDPI
St. Alban-Anlage 66
4052 Basel, Switzerland

This is a reprint of articles from the Special Issue published online in the open access journal *Journal of Marine Science and Engineering* (ISSN 2077-1312) (available at: https://www.mdpi.com/journal/jmse/special_issues/safe_oil_gas_drilling).

For citation purposes, cite each article independently as indicated on the article page online and as indicated below:

LastName, A.A.; LastName, B.B.; LastName, C.C. Article Title. *Journal Name* **Year**, *Volume Number*, Page Range.

ISBN 978-3-0365-2171-8 (Hbk)
ISBN 978-3-0365-2172-5 (PDF)

Contents

About the Editors

Dejan Brkić holds a Ph.D., M.Sc., and B.Sc. from the University of Belgrade, in Belgrade, Serbia, Department of Petroleum and Natural Gas Engineering, and M.Sc. from the University of Niš, in Niš, Serbia, Department of Mechanical Engineering. He had worked for the European Commission as a policy advisor for offshore oil and gas safety affiliated with the Energy Security, Distribution and Markets Unit of the Joint Research Centre (JRC), in Ispra (Italy), before he joined IT4Innovations of the VSB-Technical of Ostrava (VSB-TU), in Ostrava (the Czech Republic), as a researcher affiliated with waste to energy projects and Department of Electronic engineering of the University of Niš, in Niš (Serbia), as an independent researcher. His research interest includes hydraulics, transport and distribution of natural gas, energy efficiency, etc. To date, he has published around 50 scientific papers in international journals indexed by Scopus and Clarivate (h-index: 21 in Google Scholar).

Pavel Praks holds a Ph.D. in Computer Science and Applied Mathematics from the VSB-Technical University of Ostrava (VSB-TU), Ostrava, Czech Republic. Since 2018, he has been working as a researcher at IT4Innovations National Supercomputing Center, VSB-TU Ostrava. His research includes artificial intelligence in the energy sector, uncertainty analysis, probabilistic modeling and optimization of energy networks. During the period 2012–2018, he was a research fellow at the European Commission, Joint Research Centre (JRC), in Ispra (Italy), at the Energy Security, Distribution and Markets Unit. In JRC, he contributed to probabilistic modeling of the EU natural gas infrastructure from the reliability, vulnerability and security perspectives. During the period 1999–2012, he worked at VSB-TU as a lecturer and researcher of numerical methods and statistics. He published more than 65 papers indexed in Scopus (h-index: 11).

Preface to "Safe, Secure and Sustainable Oil and Gas Drilling, Exploitation and Pipeline Transport Offshore"

The Special Issue "Safe, Secure and Sustainable Oil and Gas Drilling, Exploitation and Pipeline Transport Offshore" was focused on regulations, including technical and operational standards, safety technologies, and organizational factors, which can greatly contribute to the occurrence of accidents in the offshore oil and gas sector. All methods, computational procedures, innovations, and technologies, which can increase the production rate, safety of pipelines, usability, and efficiency, were also of interest, as were all aspects related to the production of oil and gas and drilling, both offshore and onshore, and those related to an increased degree of utilization and efficiency of drilling, all safety aspects, and all aspects of security of supply. Contributions from academia, standardization and regulatory bodies, manufacturers of equipment, service and exploitation companies, and from all other types of industry were welcome. This Special Issue contains 13 papers and one editorial paper.

Dejan Brkić, Pavel Praks
Editors

Journal of
*Marine Science
and Engineering*

Safe, Secure and Sustainable Oil and Gas Drilling, Exploitation and Pipeline Transport Offshore

Dejan Brkić [1,2,*] and Pavel Praks [1,*]

1   IT4Innovations, VŠB—Technical University of Ostrava, 708 00 Ostrava, Czech Republic
2   Faculty of Electronic Engineering, University of Niš, 18000 Niš, Serbia
*   Correspondence: dejan.brkic@elfak.ni.ac.rs or dejanbrkic0611@gmail.com (D.B.); pavel.praks@vsb.cz (P.P.)

Citation: Brkić, D.; Praks, P. Safe, Secure and Sustainable Oil and Gas Drilling, Exploitation and Pipeline Transport Offshore. *J. Mar. Sci. Eng.* **2021**, *9*, 404. https://doi.org/10.3390/jmse9040404

Received: 31 March 2021
Accepted: 31 March 2021
Published: 10 April 2021

Publisher's Note: MDPI stays neutral with regard to jurisdictional claims in published maps and institutional affiliations.

The Special Issue "Safe, Secure and Sustainable Oil and Gas Drilling, Exploitation and Pipeline Transport Offshore" was focused on regulations, including technical and operational standards, safety technologies, and organizational factors, which can greatly contribute to the occurrence of accidents in the offshore oil and gas sector.

All methods, computational procedures, innovations and technologies which can increase the production rate, safety of pipelines, usability and efficiency were also of interest, as well as all aspects related to the production of oil and gas and drilling, both offshore and onshore, and those related to an increased degree of utilization and efficiency of drilling, all safety aspects, and all aspects of security of supply. Contributions from academia, standardization and regulatory bodies, manufacturers of equipment, service and exploitation companies, and from all other types of industry were welcome.

This Special Issue contains 13 papers:

1. The purpose of [1] is to demonstrate the possibility of using a mathematical model of a k-out-of-n system to support decision-making in the preventive maintenance of an unmanned underwater vehicle to monitor the condition of a subsea pipeline;
2. To improve oil spill detection, combined with underwater image processing technology, an unsupervised detection algorithm for oil spill in underwater pipelines is proposed for the first time in [2];
3. The purpose of [3] is to demonstrate the possibilities of assessing the reliability of oil and gas industry structures with the help of mathematical models of k-out-of-n systems;
4. Tuned Mass Damper could effectively reduce the vibration response of the Top Tensioned Risers, as described in [4], where the modal superposition method is used to calculate the model while current loading in the South China Sea was then applied to the riser;
5. The strength characteristic of blast wall on drillship based on the blast load profile from fire and explosion risk analysis results, as well as the ability of the current design scantling of the blast wall to endure the blast pressure during the well test are examined in [5];
6. To improve the anti-explosion performance of blast wall in offshore platforms, an auxetic re-entrant blast wall is proposed and designed based on the indentation resistance effect of an auxetic structure, as described in [6];
7. The objective of [7] is to reveal the freak wave effects on dynamic behaviours of offshore pipelines for deepwater installations;
8. The aim of [8] is to study the variation in drag force, vertical offset angle, resistance, and attitude for towing operations with a view towards optimizing these operations;
9. In [9], the characteristic of natural convection under yawing motion is studied systematically to clarify the interaction between yawing motion and thermal-dynamic behaviour;

10. The thermal and hydraulic characteristics of the liquid hold were investigated under different combinations of dimensionless parameters, and the combined effect of rolling and fluid non-Newtonian behavior is investigated in [10];
11. The purpose of [11] is to provide a structural review of the progress made on the detection and localization of leaks in pipelines by using approaches based on the Kalman filter;
12. The "International Electrotechnical Commission System for Certification to Standards Relating to Equipment for Use in Explosive Atmospheres" (IECEx) and European "Atmosphere Explosible" (ATEX) schemes are compared in [12], and a recommendation about their use in offshore oil and gas offshore industry is made;
13. In [13], how technical standards and procedures, which are recognized worldwide by the petroleum industry, can be accepted by various standardization bodies is discussed, as well as how to select the most appropriate technical standards that can increase the overall level of safety and environmental protection whilst avoiding the costs of additional certifications.

Author Contributions: Both editors contributed equally to this special issue. All authors have read and agreed to the published version of the manuscript.

Institutional Review Board Statement: Not applicable.

Acknowledgments: The academic editors of this special issue acknowledge the support of the staff of the MDPI, other academic editors who made decisions for certain papers, the reviewers who thoroughly checked the submitted manuscripts and last but not least contributions of the authors who submitted their manuscripts. The work of the academic editors of this special issue was supported by the Ministry of Education, Science, and Technological Development of Serbia and partially also by the Technology Agency of the Czech Republic through the project "CEET—Center of Energy and Environmental Technologies" TK03020027.

Conflicts of Interest: The authors declare no conflict of interest.

References

1. Rykov, V.; Kochueva, O.; Farkhadov, M. Preventive Maintenance of a k-out-of-n System with Applications in Subsea Pipeline Monitoring. *J. Mar. Sci. Eng.* **2021**, *9*, 85. [CrossRef]
2. Song, H.; Song, J.; Ren, P. Underwater Pipeline Oil Spill Detection Based on Structure of Root and Branch Cells. *J. Mar. Sci. Eng.* **2020**, *8*, 1016. [CrossRef]
3. Rykov, V.V.; Sukharev, M.G.; Itkin, V.Y. Investigations of the Potential Application of k-out-of-n Systems in Oil and Gas Industry Objects. *J. Mar. Sci. Eng.* **2020**, *8*, 928. [CrossRef]
4. Song, J.; Wang, T.; Chen, W.; Guo, S.; Yan, D. Vibration Control of Marine Top Tensioned Riser with a Single Tuned Mass Damper. *J. Mar. Sci. Eng.* **2020**, *8*, 785. [CrossRef]
5. Jung, B.; Kim, J.H.; Seo, J.K. Investigation of the Structural Strength of Existing Blast Walls in Well-Test Areas on Drillships. *J. Mar. Sci. Eng.* **2020**, *8*, 583. [CrossRef]
6. Luo, F.; Zhang, S.; Yang, D. Anti-Explosion Performance of Composite Blast Wall with an Auxetic Re-Entrant Honeycomb Core for Offshore Platforms. *J. Mar. Sci. Eng.* **2020**, *8*, 182. [CrossRef]
7. Xu, P.; Du, Z.; Gong, S. Numerical Investigation into Freak Wave Effects on Deepwater Pipeline Installation. *J. Mar. Sci. Eng.* **2020**, *8*, 119. [CrossRef]
8. Zan, Y.; Guo, R.; Yuan, L.; Wu, Z. Experimental and Numerical Model Investigations of the Underwater Towing of a Subsea Module. *J. Mar. Sci. Eng.* **2019**, *7*, 384. [CrossRef]
9. Yu, G.; Zhang, L.; Jia, S.; Geng, Y.; Liu, J. Numerical Study on the Natural Convection of Air in a Cubic Cavity Subjected to a Yawing Motion. *J. Mar. Sci. Eng.* **2019**, *7*, 204. [CrossRef]
10. Yu, G.; Jia, S.; Geng, Y. Numerical Investigation into the Two-Phase Convective Heat Transfer within the Hold of an Oil Tanker Subjected to a Rolling Motion. *J. Mar. Sci. Eng.* **2019**, *7*, 94. [CrossRef]
11. Torres, L.; Jiménez-Cabas, J.; González, O.; Molina, L.; López-Estrada, F.-R. Kalman Filters for Leak Diagnosis in Pipelines: Brief History and Future Research. *J. Mar. Sci. Eng.* **2020**, *8*, 173. [CrossRef]
12. Brkić, D.; Stajić, Z. Offshore Oil and Gas Safety: Protection against Explosions. *J. Mar. Sci. Eng.* **2021**, *9*, 331. [CrossRef]
13. Brkić, D.; Praks, P. Proper Use of Technical Standards in Offshore Petroleum Industry. *J. Mar. Sci. Eng.* **2020**, *8*, 555. [CrossRef]

Journal of
Marine Science and Engineering

Technical Note

Offshore Oil and Gas Safety: Protection against Explosions

Dejan Brkić [1,2,*] and **Zoran Stajić** [1,3]

1 Faculty of Electronic Engineering, University of Niš, 18000 Niš, Serbia; zoran.stajic@elfak.ni.ac.rs
2 IT4Innovations, VŠB—Technical University of Ostrava, 708 00 Ostrava, Czech Republic
3 Research and Development Centre IRC Alfatec, 18000 Niš, Serbia
* Correspondence: dejan.brkic@elfak.ni.ac.rs or dejanbrkic0611@gmail.com

Abstract: Offshore oil and gas operations carry a high risk of explosions, which can be efficiently prevented in many cases. The two most used approaches for prevention are: (1) the "International Electrotechnical Commission System for Certification to Standards Relating to Equipment for Use in Explosive Atmospheres" (IECEx) and (2) European "Atmosphere Explosible" (ATEX) schemes. The main shortcoming for the IECEx scheme is in the fact that it does not cover nonelectrical equipment, while for the ATEX scheme, it is due to the allowed self-certification for a certain category of equipment in areas with a low probability of explosions, as well as the fact that it explicitly excludes mobile offshore drilling units from its scope. An advantage of the IECEx scheme is that it is prescribed by the US Coast Guard for protection against explosions on foreign mobile offshore drilling units, which intend to work on the US continental shelf but have never operated there before, with an additional requirement that the certificates should be obtained through a US-based Certified Body (ExCB). Therefore, to avoid bureaucratic obstacles and to be allowed to operate with minimized additional costs both in the US and the EU/EEA's offshore jurisdictions (and very possibly worldwide), all mobile offshore drilling units should be certified preferably as required by the US Coast Guard.

Keywords: safety legislation; market access; explosions; ATEX; IECEx; offshore oil and gas; certified equipment; hazardous area classification; gas atmospheres; international standards

Citation: Brkić, D.; Stajić, Z. Offshore Oil and Gas Safety: Protection against Explosions. *J. Mar. Sci. Eng.* **2021**, *9*, 331. https://doi.org/10.3390/jmse9030331

Academic Editor: José A. F. O. Correia

Received: 24 February 2021
Accepted: 15 March 2021
Published: 16 March 2021

Publisher's Note: MDPI stays neutral with regard to jurisdictional claims in published maps and institutional affiliations.

1. Introduction

The offshore oil and gas industry involves many risks of explosions for personnel and installations. Explosions can occur under certain conditions in the presence of a mixture of air and released hydrocarbon gases due to a number of different reasons, including blowouts, the use of inadequate equipment or its malfunction, negligence, lack of training, poor or incomplete maintenance, etc., while sources of ignition can include naked flames, electrical sparks, static electricity, hot surfaces, friction, ionizing radiation, ultrasound, hot gases, etc. (e.g., methane is capable to cause an explosion in concentration between 4.4% and 16.5% in its mixture with air, while the ignition temperature for methane should be $\geq$595 °C).

Protection against explosions on mobile offshore drilling units needs to be arranged through worldwide accepted regulations at the lowest possible cost, which will be accepted by most host countries, while on fixed platforms, it needs only to fulfill legislative requirements of the host country. Two main approaches for protection against explosions are the "International Electrotechnical Commission System for Certification to Standards Relating to Equipment for Use in Explosive Atmospheres" (IECEx) [1] and the European "ATmosphere EXplosible" (ATEX) schemes (the ATEX approach can be treated as a regional European version of the IECEx approach, while certain differences can be noted) [2–4].

Offshore oil and gas operations typically start with a relatively short one-off drilling phase performed by mobile offshore drilling units [5], which is shortly after replaced with a long-lasting exploitation phase that can take several years and even decades and is performed from fixed production platforms. Legislative requirements for protection against

3

explosions on fixed platforms and offshore drilling ships are very often different (also, the legislatures treats a certain type of ships as fixed platforms, e.g., floating production storage and offloading units are typical ships in terms of shape, navigation, and propulsion, but they are treated as fixed facilities in terms of legislative requirements for protection against explosions because, as a rule, they are settled for a long time or even permanently at one place; although anchored in a similar way as ships).

The EU/EEA and the USA require different levels of protection against explosions for oil and gas mobile offshore drilling units compared with their related requirements imposed for fixed offshore platforms and offshore (Table 1). Both the ATEX and IECEx schemes provide a high level of protection against explosions but with some differences that are described in this note [6,7].

Table 1. Protection scheme against explosion on the US and the EU/EEA continental shelf [1].

ATEX	IECEx
Fixed platforms (including equipment onboard) on the EU/EEA continental shelf	Mobile units (including equipment onboard) on the EU/EEA continental shelf [2]
Facilities and equipment onshore the EU/EEA	Foreign mobile units (including equipment onboard) on the US continental shelf [3]

[1] US onshore installations and US domestic mobile units (including equipment onboard) operating on the US continental shelf are covered through the US National Electrical Code (NEC) and need to be certified through a Nationally Recognized Testing Laboratory (NRTL). [2] Practically prescribed by the IMO MODU Code. "Code for the Construction and Equipment of Mobile Offshore Drilling Units" by the International Maritime Organization. [3] Need to be certified through a US-based "International Electrotechnical Commission System for Certification to Standards Relating to Equipment for Use in Explosive Atmospheres" (IECEx) body if they intend to operate on the US continental shelf but had never operated before there (foreign mobile units with equipment onboard that frequently operate on the US continental shelf need to follow the same rules as US domestic vessels).

2. Main Certification Schemes for Protection against Explosions

The principles for the classification of equipment proofed for use in areas with a higher risk of explosions are similar in most countries, and they are based on requirements introduced by the International Electrotechnical Commission (IEC) [8]. Despite this fact, many local modifications and variants exist.

ATEX protection is described in Section 2.1 and IECEx in Section 2.2. Although based on the almost same principles but with some modifications, as described in Section 2.3, various certification schemes for protection against explosions are available and required for use in different countries. Section 2.3.1 describes the situation in the USA, Section 2.3.2 in Canada, Section 2.3.3 in the Eurasian Union (Armenia, Belarus, Russia, Kazakhstan, and Kyrgyzstan), Section 2.3.4 in China, Section 2.3.5 in Brazil, and Section 2.3.6 in Australia, together with New Zealand.

2.1. European "ATmosphere Explosible" (ATEX)

Two European ATEX directives are available for protection in explosive atmospheres (ATEX is a French acronym for "ATmosphere EXplosible"):

(1)　1999/92/EC ATEX directive for the protection of the user (personnel and workers)—"Minimum requirements for improving the safety and health protection of workers potentially at risk from explosive atmospheres";

(2)　2014/34/EU ATEX directive for product safety—"Equipment and protective systems intended for use in potentially explosive atmospheres," a new version of 94/9/EC.

Using the ATEX scheme, equipment shall be certified through one of the European notified bodies listed in the "New Approach Notified and Designated Organizations" (NANDO) database. The obligatory certification has been in force since 1 July 2003 through the 94/9/EC directive and since 20 April 2016 through the related recast 2014/34/EU of the previous directive. Both ATEX directives for product safety (2014/34/EU) and the protection of the user (personnel and workers) (1999/92/EC) are mandatory in the EU/EEA.

The new goal-oriented European approach [2], which has been in force since 2000, allows the use of any available technical standard that can assure application of the obligatory directives, such as ATEX. Instead of referring to a certain list of harmonized standards (the European Standards Committee (CEN) and the European Committee for Electrotechnical Standardization (CENELEC) define the harmonized standards), the directives prescribe Essential Health and Safety Requirements (EHSRs) that shall be fulfilled. The use of harmonized technical standards gives only a presumption of conformity to the directives, while compliance only to the Essential Health and Safety Requirements (EHSRs) is considered full conformity (to achieve that, any appropriate available technical standard can be used).

The ATEX directive for product safety covers electrical equipment, nonelectrical equipment such as mechanical, hydraulic (pumps), and pneumatic equipment, including assemblies, protective systems, components (push button, relays, valves), controlling and regulating devices, etc. The ATEX directive for product safety deals with both electrical and nonelectrical equipment. Such equipment is divided into (i) Groups and (ii) Categories and can be placed in (iii) Zones, as shown in Table 2.:

Table 2. Groups, Categories, and Zones prescribed by the "ATmosphere EXplosible" (ATEX) directive.

(i) Groups	
Group I	Relevant to underground coal mines, where the occurrence of dust and methane firedamp is frequent [9–14]
Group II	Relevant to the oil and gas industry and refers to potentially gaseous explosive atmospheres
(ii) Categories	
Category 1	Most strict and intended for use in areas with the highest risk of explosions
Category 2	For use in areas in which explosive atmospheres are likely to occur
Category 3	For areas with a low probability for explosions
(iii) Hazardous Zones	
Zone 0	An explosive atmosphere is present continuously, for long periods, or frequently
Zone 1	An explosive atmosphere is likely to occur occasionally in normal operation
Zone 2	An explosive atmosphere is not likely to occur during normal operations (if it occurs, it only persists for a short period)

For Categories 1 and 2, the procedure to affix the CE marking following appropriate type-examination procedures shall be issued by a notified body, while for Category 3, the manufacturer can ensure full conformity. The European auxiliary "Ex" mark, together with the European general mark for conformity "CE," shall be affixed on the certified equipment.

Group II of equipment for use in offshore oil and gas installations is divided into Categories, which can be used or installed in hazardous Zones, as given in Table 3.

Table 3. Use of Group II of equipment in Hazardous Zones depending on its Category.

Zone	Category
Zone 0	Category 1
Zone 1	Category 1 or 2
Zone 3	Category 1, 2, or 3

Categories of equipment and Hazardous Zones [15–19] can be determined by using any of the available technical standards. For electrical equipment, IEC 60079: Explosive atmospheres—CENELEC 60079 series can be used, and for nonelectrical equipment, EN 13463 series (EN ISO 80079) can be used. The types of protection are given in Table 4.

Table 4. Examples of types of protection for a certain Zone[1].

Zone	Types of Protection	Technical Standard
Zone 0	Ex ia—Intrinsic safety (higher grade)	IEC EN 60079-11
	Ex s—Specifically designed for Zone 0	IEC 60079-33
Zone 0 and 1	Ex ib—Intrinsic safety (lower grade)	IEC EN 60079-11
	Ex d—Flameproof enclosures	EN 13463-3 and IEC EN 60079-1
	Ex e—Increased safety	IEC EN 60079-7
	Ex p—Pressurizations	IEC EN 60079-2 and 13
Zone 0, 1, and 2	Ex n—Type of protection	IEC EN 60079-15
	Ex o—Oil immersed	IEC EN 60079-6
	Ex q—Powder filled	IEC EN 60079-5

[1] Certain technical standards such as EN 13463-3 cover nonelectrical equipment, and such standards do not have an IEC equivalent, while certain such as IEC 60079-33 are not harmonized with the ATEX directive and hence not marked as EN (not developed by the CEN/CENELEC) but can be used if they meet the Essential Health and Safety Requirements of the ATEX directive.

Zones are determined by the grade of release: (1) continuous grade, (2) periodically or occasionally, and (3) normally not to expect; by the degree of ventilation: (1) high, (2) medium, and (3) low; and by availability of ventilation: (1) good, (2) fair, and (3) poor. The potentially explosive atmosphere can be prevented from igniting if the surface temperature of the item of equipment is lower than the ignition temperature of the surrounding gas, while the temperature classes for Europe and the USA are maximal (T_1 = 450 °C, T_2 = 300 °C, T_3 = 200 °C, T_4 = 135 °C, T_5 = 100 °C, and T_6 = 85 °C (subcategories exist in the USA)). The gas groups are I—methane, IIA—propane, IIB—ethylene, IIC—acetylene, and hydrogen, etc. (ignition energy is I—280 μJ, IIA—160 μJ, IIB—80 μJ, and IIC—20 μJ) [20,21].

Use of the 2014/34/EU directive for product safety is obligatory in the EU/EEA, both onshore and offshore, with the exception of mobile drilling units operated on the EU/EEA continental shelf. On the other hand, the 2013/30/EU directive for overall oil and gas offshore safety requires the use of all the best worldwide available technical standards [22] and prescribes the use of the IMO MODU Code, "Code for the Construction and Equipment of Mobile Offshore Drilling Units" by the International Maritime Organization, which, in practice, requires the use of the IECEx scheme for protection on mobile drilling units.

2.2. "International Electrotechnical Commission System for Certification to Standards Relating to Equipment for Use in Explosive Atmospheres" (IECEx)

The "International Electrotechnical Commission System for Certification to Standards Relating to Equipment for Use in Explosive Atmospheres" (IECEx) scheme is not mandatory [1], but it is widely used in the offshore oil and gas industry and especially on mobile offshore drilling units in the US and EU/EEA's offshore jurisdictions.

The IECEx certification scheme is operated by International Electrotechnical Commission (IEC), and it is based on its technical standards [23–26]. Around 30 nations participate in the IECEx scheme for protection against explosions, and many are based on it, such as the already described European ATEX, as well as many others described in Section 2.3. A nation without its own national scheme for protection against explosions can adopt the IECEx scheme to fill its legislative gaps using a pretailored model for regulation in the sector of equipment used in environments with an explosive atmosphere through the United Nations Economic Commission for Europe (UNECE) [27].

The offshore oil and gas industry under the IECEx scheme uses the same classification of equipment as under the ATEX protective scheme: Groups, Categories of equipment, and Hazardous Zones (Technical Committee 31 of the IEC is responsible for equipment for explosive atmospheres). The IEC 60079 series of Technical Standards can be used, but in the offshore industry, the Recommended Practice by the American Petroleum Institute API505 "Classification of Locations for Electrical Installations at Petroleum Facilities Classified as Class I, Zone 0, Zone 1, and Zone 2" is more used (a similar approach is used in the US onshore industry through the NEC505, as described in Section 2.3.1) [16]. The requirements between these two approaches for Hazardous Zones are similar; the IEC 60079 is more analytical, while the RP API505 is more prescriptive [6].

Certificates issued by the IECEx scheme are listed on its website; therefore, their validity can be verified in a very convenient way. Under the IECEx scheme, four types of certificates can be issued:

(1) Equipment—Only technical standards issued by Technical Committee 31 "Equipment for explosive atmospheres" of the International Electrotechnical Commission (IEC) can be used for the certification of equipment (only electrical equipment can be certified), while the certification through an IECEX Certification Body (ExCB) is required for all Categories of equipment (Categories 1, 2, and 3 in IECEx Ga, Gb, and Gc, respectively) (ATEX allows for the self-certification of equipment by the manufacturer in Category 3);

(2) Service Facilities—In addition to the certification of equipment, the IECEx scheme allows the certification of the repair and overhaul of equipment (nothing similar is available in the ATEX scheme);

(3) Persons—Certification for qualified persons who were properly trained and meet the prerequisites to implement safety requirement required by the IECEx scheme (nothing similar is available in the ATEX scheme);

(4) Conformity Mark—It is issued by approved IECEx Certification Bodies (ExCBs) for the equipment manufactured and tested using appropriate IEC technical standards.

2.3. Other Main Certification Schemes for Protection against Explosions

Possession of the IECEx or ATEX certifications cannot replace the obligatory certification imposed by most countries but can facilitate and support a procedure to apply for it. The IECEx Test Report is likely to be accepted as a base for national certification in the USA, Canada, Russia, Ukraine, Belarus, China, South Korea, India, Brazil, Argentina, Chile, South Africa, etc. Some countries, such as China, also accept the ATEX certificate as a base for its national certification. On the other hand, the ATEX Test Report can be used as a base for a national certificate in Hong Kong (special administrative region of China), Taiwan (de facto different legislation apply, but de jure part of China), Vietnam, Indonesia, etc. Finally, in some countries, such as in Japan, many domestic tests are required, and the IECEx or the ATEX Test Reports cannot facilitate the procedure.

2.3.1. USA

The Occupational Safety and Health Administration (OSHA) requires certification against explosions through one of its Nationally Recognized Testing Laboratories (NRTLs). In the National Electrical Code and National Fire Protection Association prescript NEC/ NFPA 70 for hazardous areas, NEC500 is based on a two-division classification of hazardous locations, while NEC505 is based on a three-zone classification. The three-zone area classification system is compatible with the IEC 60079 series of technical standards for explosive atmospheres; such an approach is compatible with the IECEx scheme and can be used as a basis for obtaining certification for the protection of equipment installed/or used on mobile offshore drilling units on the US offshore continental shelf. Certified equipment for use in areas protected against explosions shall be marked with a label of a Nationally Recognized Testing Laboratory (NRTL). In the offshore sector in the USA, the Bureau of Safety and Environmental Enforcement (BSEE) has authority over fixed platforms and the United States Coast Guard (USCG) over mobile units (ruling published in the Federal Register (Vol. 80, No. 61) on 31 March, 2015, and came into force on 2 April 2018). The IECEx certificates are not directly accepted under the National Electrical Code (NEC) in the USA, but the IECEx Test Report can facilitate appropriate US certification (only electrical equipment can be certified against explosions).

2.3.2. Canada

Similar rules for protection against explosions as in the USA are in force in Canada. The Standards Council of Canada (SCC) requires certification through one of its recognized certification bodies, while product approval related to electrical safety is under the jurisdiction of Provincial Governments. Classification of hazardous areas is based on the three-zone classification, as prescribed by the Canadian Electrical Code (CEC) (two-division classification is also accepted) and is compatible with the IEC 60079 series of technical standards for explosive atmospheres. Certified equipment shall be marked with a label of a recognized Canadian certification body. The IECEx certificate can be used as a basis for obtaining Canadian certification.

2.3.3. Eurasian Union

The Eurasian Union (ex. Customs Union; Armenia, Belarus, Russian Federation, Kazakhstan, Kyrgyzstan) requires technical regulations GOST/CU-TR on the safety of equipment in explosion hazardous environments issued by the Euro-Asian Council for Standardization, Metrology, and Certification (EASC). Only electrical equipment can be certified (in practice, only equipment that satisfies at least the ATEX level of protection can be certified). The appropriate product conformity mark of the Eurasian Union shall be affixed on the certified equipment.

2.3.4. China

The Chinese State Administration for Market Regulation (SAMR) requires a China Compulsory Certificate (CCC) Ex certification for protection against explosions (ATEX and IECEx certificates are not accepted, but if they exist, they can significantly reduce the work and preparation effort for the CCC certification). The China Compulsory Certificate mark, commonly known as a CCC mark, shall be affixed on the certified equipment.

2.3.5. Brazil

The Brazilian Ministry of Development, Industry and Foreign Trade, requires a certification against explosions by the National Institute of Metrology, Standardization and Industrial Quality (INMETRO). Brazilian certification bodies accept IECEx tests as a base for the certificate [28]. The INMETRO mark shall be affixed on the certified equipment.

2.3.6. Australia and New Zealand

In Australia and New Zealand, an ANZEx certification against explosions is required. To obtain it, an IECEx certificate is sufficient, and only certain national differences need to be tested additionally (the IECEx certification will be directly accepted for Group II equipment).

3. Differences between the ATEX and IECEx Certification Schemes

Historically, the National Electrical Code (NEC), established in 1947 as a concept of areas with two Divisions, followed by the American Petroleum Institute (API), which started related work in 1951 on its area classification document API500 based on Divisions (API500A for electrical installations in petroleum refineries was finalized in 1955, initially without the suffix "A;" API500B for electrical installations at drilling rigs and production facilities on land and marine platforms was finalized in 1961; and API500C for electrical installations at pipelines was finalized in 1966. These merged in 1991 into a single API 500 document [29]). API 500 is based on a hazardous area classification based on two Divisions, while API 505, introduced in 1997 [30], is based on three Zones (both the ATEX and IECEx schemes are based on Zone classification) [31] (Table 5 and Figure 1). After these early efforts, which were made in the USA and followed by many countries, the first international system for protection against explosions is dated in 1996 when IECEx was established and in 2003 when the first certificates were issued (IEC-60079-10-1 is compatible with US API 505 for the classification of hazardous areas). In Europe, the history of ATEX directives started also in the 1990s.

Table 5. Zone and Division systems: a comparison.

API 505, IEC-60079-10-1	Zone 0	Zone 1	Zone 2
Presence of explosive gases	>1000 h/year	10 to 1000 h/year	<10 h/year
API 500 Presence of explosive gases	Division 1 >10 h/year	Division 1 >10 h/year	Division 2 <10 h/year

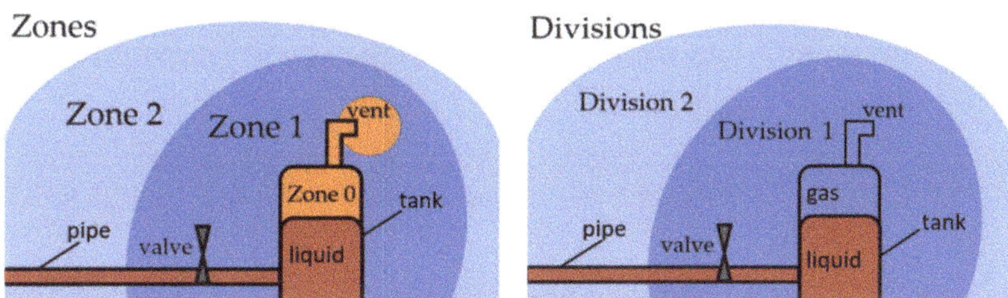

Figure 1. Zone and Division Systems: an illustrative example.

Categories of equipment in the ATEX and IECEx schemes are equivalent are given in Table 6, where G refers to gaseous potentially explosive atmospheres.

Table 6. Categories of equipment in the ATEX and IECEx schemes.

ATEX	Group 1G	Group 2G	Group 3G
IECEx	Group Ga	Group Gb	Group Gc

The main differences between the ATEX and IECEx protection schemes are:

- In the EU/EEA, the ATEX scheme is mandatory, and it is developed by the European Commission (government), while the IECEx scheme is operated by the International Electrotechnical Commission (IEC), i.e., regulated through an industry representative at international level.
- Safety of personnel is regulated under the ATEX scheme through the mandatory use of a separate directive, 1999/92/EC, while nothing similar is available through the IECEx.
- A certificate through an ATEX Notified Body is mandatory for equipment in Categories 1 and 2, while self-certification by the manufacturer is allowed for equipment in Category 3 (on the other hand, the IECEx scheme requires attestation of all categories of equipment through its Certified Bodies (ExCBs)).
- The ATEX scheme requires certification through any available technical standards that can meet Essential Health and Safety Requirements of the ATEX directives, while the IECEx requires use of the technical standards developed only by the International Electrotechnical Commission (IEC).
- Although the ATEX scheme uses terms "equipment" and "protective systems," it covers both electrical and nonelectrical equipment (EN 13463 and EN ISO 80079 "Non-electrical equipment for explosive atmospheres"), while the IECEx covers only electrical equipment.
- Protection of certain Categories of equipment is sometimes different in the ATEX and IECEx schemes (every individual case should be studied if different protections are required) [32–34].
- The ATEX scheme explicitly does not include mobile offshore drilling units in its scope.

Different marking of equipment is required using the ATEX and IECEx schemes (Figure 2).

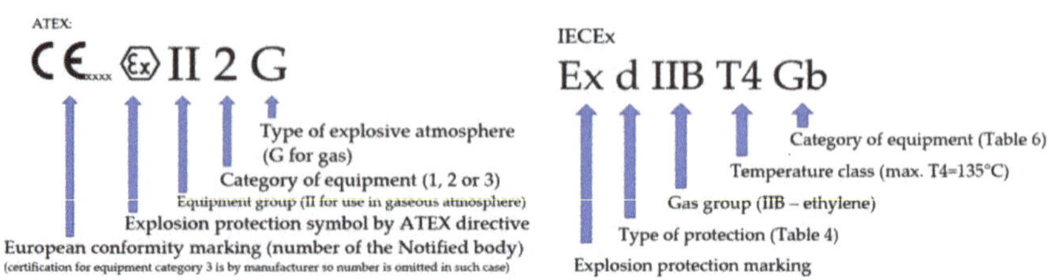

Figure 2. Marking according to the ATEX and the IECEx schemes for protection against explosions.

4. Conclusions

All European Directives are designed to protect the European single market in the first place, but in respect to protection against explosion on mobile offshore units for oil and gas drilling in the EU/EEA offshore continental shelf, some concerns exist if the European market is fully protected. Furthermore, mobile units in the EU/EEA offshore jurisdiction are excluded out of the scope of the ATEX Directive 2014/34/EU-ATEX by Article 1, Section 2(e): "This Directive shall not apply to: seagoing vessels and mobile offshore units together with equipment on board such vessels or units." On the other hand, the US market is fully protected due to the fact that the US Coast Guard requires the use of domestic protection scheme on the US offshore continental shelf for foreign mobile drilling units, while foreign-flagged vessels that have never operated, but intend to operate under the US offshore jurisdiction, should be certified through the IECEx scheme against explosions by a domestic US-based Certified Body (ExCB). In all other cases, full compliance with US regulations is obligatory.

No main differences between the ATEX and the IECEx systems for protection against explosions are detected. A shortcoming for the ATEX scheme is due to the allowed self-certification for the equipment Category 3, which should be placed in Hazardous Zone 2, while for the IECEx in the fact that it does not cover nonelectrical equipment.

Mobile offshore drilling units [35–38] should be preferably certified through a US-based Certified Bodies (ExCB)) using the IECEx protective scheme against explosions because such units will be allowed to operate with minimized additional costs, both in US and EU/EEA's offshore jurisdictions (and very possible worldwide). On the other hand, fixed offshore platforms should be certified strictly following the rules of host countries.

Author Contributions: D.B. has extensive experience with offshore oil and gas operations and Z.S. in design and practical use of electrical equipment. D.B. and Z.S. in particular contributed to this article with: Conceptualization, D.B.; methodology, D.B.; validation, D.B. and Z.S.; formal analysis, D.B.; investigation, D.B. and Z.S.; writing—original draft preparation, D.B.; writing—review and editing, D.B.; supervision, Z.S.; project administration, D.B.; funding acquisition. Both authors have read and agreed to the published version of the manuscript.

Funding: The work of Z.S. and D.B. has been supported by the Ministry of Education, Science, and Technological Development of Serbia, and the work of D.B. has been supported by the Technology Agency of Czechia through the project "Center of Energy and Environmental Technologies" (CEET) TK03020027.

Institutional Review Board Statement: Not applicable.

Informed Consent Statement: Not applicable.

Conflicts of Interest: The authors declare no conflict of interest.

The following are used in this paper:

1999/92/EC ATEX European directive: Minimum requirements for improving the safety and health protection of workers potentially at risk from explosive atmospheres
2013/30/EU: European directive for safety of offshore oil and gas operations

2014/34/EU ATEX European directive (a new version of 94/9/EC): Equipment and protective systems intended for use in potentially explosive atmospheres
ANZEx: certification scheme against explosions in Australia and New Zealand
API: American Petroleum Industry
API500: "Classification of Locations for Electrical Installations at Petroleum Facilities Classified as Class I, Division 1, and Division 2;" API505a, b, and c merged in 1991 in a single API 500 document
API500A for electrical installations in petroleum refineries, finalized in 1955, initially without suffix A
API500B for electrical installations at drilling rigs and production facilities on land and on marine platforms, finalized in 1961
API500C for electrical installations at pipelines, finalized in 1966
API505: "Classification of Locations for Electrical Installations at Petroleum Facilities Classified as Class I, Zone 0, Zone 1, and Zone 2;" issued in 1997
ATEX: in French: ATmosphere Explosible
Brazilian Ministry of Development, Industry and Foreign Trade
BSEE: Bureau of Safety and Environmental Enforcement
Categories: Categories of equipment (1 or Ga—very high protection, 2 or Gb—moderate, 3 or Gc—lowest protection against explosions)
CCC Ex: Chinese mark for protection against explosions
CE: EU/EEA conformity mark
CEC: Canadian Electrical Code
CEN: European Standards Committee
CENELEC: European Committee for Electrotechnical Standardization
CENELEC IEC 60079 "Explosive atmospheres;" a series of technical standards
Coast Guard: deal with protections against explosion on mobile units in US waters; Federal Register (Vol. 80, No. 61) on 31 March 2015 and in force from 2 April 2018
EASC: Euro-Asian Council for Standardization, Metrology, and Certification
EEA: European Economic Area
EHSRs: Essential Health and Safety Requirements of a European directive
EN 13463 and EN ISO 80079 "Non-electrical equipment for explosive atmospheres": Technical standards
EU: European Union
Eurasian Union (ex. Customs union): Armenia, Belarus, Russia, Kazakh-stan, Kyrgyzstan
Ex: EU/EEA conformity mark for equipment certified for use in potentially explosive atmospheres
ExCB: IECEX Certification Body
GOST/CU-TR: Eurasian Union Technical Standards
Group: Group of equipment (I—dust and II—gas)
Hazardous Areas: probability of explosions; based on Zones or Divisions
IEC: International Electrotechnical Commission
IECEx: International Electrotechnical Commission System for Certification to Standards Relating to Equipment for Use in Explosive Atmospheres
IMO MODU code: Code for the Construction and Equipment of Mobile Offshore Drilling Units by the International Maritime Organization
INMETRO: Brazilian National Institute of Metrology, Standardization, and Industrial Quality
ISO: International Technical Standards
NANDO database: New Approach Notified and Designated Organizations
NEC/NFPA 70: US National Electrical Code for hazardous areas
NEC500: related to the API500
NEC505: related to the API505
New Approach: goal-oriented European approach for the application of obligatory directives
NFPA: National Fire Protection Association
Notify Body: an ATEX certification body
NRTL: US Nationally Recognized Testing Laboratory

OSHA: US Occupational Safety and Health Administration
SAMR: Chinese State Administration for Market Regulation
UNECE: United Nations Economic Commission for Europe

References

1. Towle, C. The IECEX Certification Scheme. *Meas. Control* **2003**, *36*, 154–155. [CrossRef]
2. Marangon, A.; Carcassi, M. ATEX Directives: The New Approach. In Proceedings of the Conference Valutazione e Gestione del Rischio Negli Insediamenti Civili ed Industriali, Pisa, Italy, 17–19 October 2006; Available online: http://conference.ing.unipi.it/vgr2006/archivio/Archivio/2008/Archivio/2006/Articoli/700213.pdf (accessed on 19 January 2021).
3. Leroux, P. New Regulations and Rules for ATEX Directives. *IEEE Ind. Appl. Mag.* **2007**, *13*, 43–51. [CrossRef]
4. Kaiser, M.J.; Snyder, B. The five offshore drilling rig markets. *Mar. Policy* **2013**, *39*, 201–214. [CrossRef]
5. Geng, J.; Murè, S.; Demichela, M.; Baldissone, G. ATEX-HOF Methodology: Innovation Driven by Human and Organizational Factors (HOF) in Explosive Atmosphere Risk Assessment. *Safety* **2020**, *6*, 5. [CrossRef]
6. Tommasini, R.; Pons, E.; Palamara, F. Area Classification for Explosive Atmospheres: Comparison between European and North American Approaches. In Proceedings of the 60th Annual IEEE Petroleum and Chemical Industry Technical Conference, Chicago, IL, USA, 23–25 September 2013. [CrossRef]
7. Pommé, R.; Sijrier, H.J. IECEX Certification Schemes versus ATEX Directives. In Proceedings of the PCIC Europe 2010, Oslo, Norway, 15–17 June 2010; IEEE: Piscataway, NJ, USA, 2010; pp. 1–8. Available online: https://ieeexplore.ieee.org/document/5525731 (accessed on 19 January 2021).
8. Johnson, B.; House, P.; Mulder, R.; Pijpker, T.; Pomme, R. Role of IEC Standards. *IEEE Ind. Appl. Mag.* **2008**, *14*, 14–20. [CrossRef]
9. Novák, P.; Kot, T.; Babjak, J.; Konečný, Z.; Moczulski, W.; Rodriguez López, Á. Implementation of Explosion Safety Regulations in Design of a Mobile Robot for Coal Mines. *Appl. Sci.* **2018**, *8*, 2300. [CrossRef]
10. Palazzi, E.; Currò, F.; Fabiano, B. Accidental Continuous Releases from Coal Processing in Semi-Confined Environment. *Energies* **2013**, *6*, 5003–5022. [CrossRef]
11. Hughes, A. Why Dust Explodes. *Meas. Control* **2012**, *45*, 271–272. [CrossRef]
12. Abbasi, T.; Abbasi, S.A. Dust explosions–Cases, causes, consequences, and control. *J. Hazard. Mater.* **2007**, *140*, 7–44. [CrossRef] [PubMed]
13. Eckhoff, R.K.; Li, G. Industrial Dust Explosions. A Brief Review. *Appl. Sci.* **2021**, *11*, 1669. [CrossRef]
14. Eckhoff, R.K. Differences and similarities of gas and dust explosions: A critical evaluation of the European 'ATEX' directives in relation to dusts. *J. Loss Prev. Process Ind.* **2006**, *19*, 553–560. [CrossRef]
15. Alves, J.J.; Neto, A.T.; Araújo, A.C.; Silva, H.B.; Silva, S.K.; Nascimento, C.A.; Luiz, A.M. Overview and Experimental Verification of Models to Classify Hazardous Areas. *Process Saf. Environ. Prot.* **2019**, *122*, 102–117. [CrossRef]
16. Rangel, E.; Luiz, A.M.; de P. M. Filho, H.L. Area Classification is Not a Copy-and-Paste Process: Performing Reliable Hazardous-Area-Classification Studies. *IEEE Ind. Appl. Mag.* **2015**, *22*, 28–39. [CrossRef]
17. Propst, J.E.; Barrios, L.A.; Lobitz, B. Applying the API Alternate Method for Area Classification. *IEEE Trans. Ind. Appl.* **2007**, *43*, 162–171. [CrossRef]
18. Stefanowicz, D. An Overview of Hazardous Areas. *Meas. Control* **2003**, *36*, 142–144. [CrossRef]
19. McMillan, A.J. Hazardous Areas—An Overview. *Meas. Control* **1986**, *19*, 6–9. [CrossRef]
20. Zhang, Q.; Li, W.; Zhang, S. Effects of spark duration on minimum ignition energy for methane/air mixture. *Process Saf. Prog.* **2011**, *30*, 154–156. [CrossRef]
21. Cui, G.; Li, Z.; Yang, C.; Zhou, Z.; Li, J. Experimental study of minimum ignition energy of methane–air mixtures at low temperatures and elevated pressures. *Energy Fuels* **2016**, *30*, 6738–6744. [CrossRef]
22. Brkić, D.; Praks, P. Proper Use of Technical Standards in Offshore Petroleum Industry. *J. Mar. Sci. Eng.* **2020**, *8*, 555. [CrossRef]
23. Thomas, G.A.N.; Reeve, P.T.N.; Spanhel, J.M. Sharing Information to Promote Oil and Gas Industry Standardisation in ISO. In Proceedings of the Offshore Technology Conference, Offshore Technology Conference, Houston, TX, USA, 1–4 May 1995. [CrossRef]
24. Reeve, P.T.N.; Johansen, A.R.; Lautier, J. International Standards for the Oil and Natural Gas Industries: A Review Paper. In Proceedings of the Offshore Technology Conference, Offshore Technology Conference, Houston, TX, USA, 2–5 May 1994. [CrossRef]
25. Arney, C.E. Toward One Set of International Standards for the Petroleum Industry Worldwide. In Proceedings of the Offshore Technology Conference, Houston, TX, USA, 4–7 May 1992. [CrossRef]
26. Wilson, D.E. Internationalization of Oil Industry Standards. In Offshore Technology Conference. In Proceedings of the Offshore Technology Conference, Houston, TX, USA, 4–7 May 1992. [CrossRef]
27. Agius, C.; Jachia, L. International IECEx System Evolution and Role of the United Nations, UNECE. In Proceedings of the 2012 Petroleum and Chemical Industry Conference Europe Conference Proceedings (PCIC EUROPE), Prague, Czech Republic, 19–21 June 2012; IEEE: Piscataway, NJ, USA, 2012; pp. 1–15. Available online: https://ieeexplore.ieee.org/document/6243265 (accessed on 19 January 2021).
28. Rangel, E. Brazil moves from divisions to zones. In Proceedings of the Annual Petroleum and Chemical Industry Conference, New Orleans, LA, USA, 23–25 September 2002; IEEE: Piscataway, NJ, USA, 2002; Volume 49, pp. 23–30. [CrossRef]
29. Hamer, P.S.; Bishop, D.N.; Bried, F. API RP 500-Recommended Practice for Classification of Locations for Electrical Installations at Petroleum Facilities. In Proceedings of the 1992 Conference of Industry Applications Society 39th Annual Petroleum and Chemical Industry Conference, San Antonio, TX, USA, 28–30 September 1992; IEEE: Piscataway, NJ, USA, 1992; pp. 37–42. [CrossRef]

30. Bishop, D.N.; Jagger, D.M.; Propst, J.E. New area classification guidelines. In Proceedings of the Industry Applications Society 45th Annual Petroleum and Chemical Industry Conference, Indianapolis, IN, USA, 28–30 September 1998; IEEE: Piscataway, NJ, USA, 1998; pp. 9–19. [CrossRef]
31. Carlson, R.; Flanders, P.; Roussel, B. Conversion from Division to Zone electrical classification-why and how the world's largest oil company made the change. In Proceedings of the Industry Applications Society Forty-Seventh Annual Petroleum and Chemical Industry Technical Conference, San Antonio, TX, USA, 13 September 2000; IEEE: Piscataway, NJ, USA, 2000; pp. 1–9. [CrossRef]
32. Hamer, P.S.; Wood, B.M.; Doughty, R.L.; Gravell, R.L.; Hasty, R.C.; Wallace, S.E.; Tsao, J.P. Flammable Vapor Ignition Initiated by Hot Rotor Surfaces within an Induction Motor: Reality or Not? *IEEE Trans. Ind. Appl.* **1999**, *35*, 100–111. [CrossRef]
33. Muhammad Niazi, U.; Shakir Nasif, M.; Muhammad, M.; Khan, F. Investigating Vapour Cloud Explosion Dynamic Fatality Risk on Offshore Platforms by Using a Grid-Based Framework. *Processes* **2020**, *8*, 685. [CrossRef]
34. Zhu, R.; Li, X.; Hu, X.; Hu, D. Risk Analysis of Chemical Plant Explosion Accidents Based on Bayesian Network. *Sustainability* **2020**, *12*, 137. [CrossRef]
35. Jung, B.; Kim, J.H.; Seo, J.K. Investigation of the Structural Strength of Existing Blast Walls in Well-Test Areas on Drillships. *J. Mar. Sci. Eng.* **2020**, *8*, 583. [CrossRef]
36. Luo, F.; Zhang, S.; Yang, D. Anti-Explosion Performance of Composite Blast Wall with an Auxetic Re-Entrant Honeycomb Core for Offshore Platforms. *J. Mar. Sci. Eng.* **2020**, *8*, 182. [CrossRef]
37. Yu, G.; Zhang, L.; Jia, S.; Geng, Y.; Liu, J. Numerical Study on the Natural Convection of Air in a Cubic Cavity Subjected to a Yawing Motion. *J. Mar. Sci. Eng.* **2019**, *7*, 204. [CrossRef]
38. Yu, G.; Jia, S.; Geng, Y. Numerical Investigation into the Two-Phase Convective Heat Transfer within the Hold of an Oil Tanker Subjected to a Rolling Motion. *J. Mar. Sci. Eng.* **2019**, *7*, 94. [CrossRef]

Technical Note

Proper Use of Technical Standards in Offshore Petroleum Industry

Dejan Brkić * and Pavel Praks *

IT4Innovations National Supercomputing Centre, VŠB—Technical University of Ostrava,
70800 Ostrava, Czech Republic
* Correspondence: dejanbrkic0611@gmail.com or dejan.brkic@vsb.cz (D.B.); pavel.praks@vsb.cz (P.P.)

Received: 9 June 2020; Accepted: 16 July 2020; Published: 24 July 2020

Abstract: Ships for drilling need to operate in the territorial waters of many different countries which can have different technical standards and procedures. For example, the European Union and European Economic Area EU/EEA product safety directives exclude from their scope drilling ships and related equipment onboard. On the other hand, the EU/EEA offshore safety directive requires the application of all the best technical standards that are used worldwide in the oil and gas industry. Consequently, it is not easy to select the most appropriate technical standards that increase the overall level of safety and environmental protection whilst avoiding the costs of additional certifications. We will show how some technical standards and procedures, which are recognized worldwide by the petroleum industry, can be accepted by various standardization bodies, and how they can fulfil the essential health and safety requirements of certain directives. Emphasis will be placed on the prevention of fire and explosion, on the safe use of equipment under pressure, and on the protection of personnel who work with machinery. Additionally considered is how the proper use of adequate procedures available at the time would have prevented three large scale offshore petroleum accidents: the Macondo Deepwater Horizon in the Gulf of Mexico in 2010; the Montara in the Timor Sea in 2009; the Piper Alpha in the North Sea in 1988.

Keywords: equipment certification; maritime safety; offshore accidents; oil and gas; technical standards; petroleum engineering

---

1. Introduction

Today, when the protection of the environment is a top priority, it is also imperative to protect oceans from pollution. Oceans know no national borders, and hence one oil and gas accident can have a large impact on many states. In the current world, we witness huge amounts of gas being burnt, unused on the flares of offshore oil fields or on distant onshore oil fields [1,2]. On the other hand, even the trade of quotas for the gasses with greenhouse effects exists, and some unused sources of natural gas are being investigated by energy companies for possible future exploration. In such a controversial world, it is important to re-evaluate current practices in the application of legislation for oil and gas drilling that are used worldwide [3,4]. For example, every region has its own characteristics, where, for example, drilling in cold regions [5–8] has additional requirements compared with drilling in moderate climatic regions. Similar to nuclear accidents, the probability that a major offshore oil and gas accident will occur is relatively low, but if it happens the consequences can be catastrophic. The offshore petroleum industry is complex, with different hazards, such as from leakage [9], pressure [10], moving objects, fatigue [11], fire, explosions, and in different various shocking events [12–15].

Devices and equipment which are used in the offshore petroleum industry are highly sophisticated, complex and they are developed, designed, and manufactured using certain technical standards. Some technical standards are developed on the national level, but in practice, if they are widely accepted as a good example of the best practice by the worldwide offshore petroleum industry, informally they are treated as international. Obviously, the recommended best procedures ought to be accepted and used around the world, but this should be done carefully to avoid damages to the domestic industry of a host country [16–18].

Offshore petroleum industries are based on a variety of overlapping philosophies with very different approaches to achieving high levels of safety. A very short overview of three offshore accidents to describe the complexity of this industry follows (two accidents from drilling mobile units and one from a production permanent platform): (1) the Macondo Deepwater Horizon oil spill in the Gulf of Mexico; (2) the Montara in the Timor Sea; (3) the Piper Alpha accident in the North Sea:

(1)   The Macondo Deepwater Horizon oil spill occurred in the Gulf of Mexico on 20 April 2010 on the Macondo prospect, around 60 km seaward from the Louisiana coast, where British Petroleum (BP) was the oil company operator while Transocean was the rig contractor. It is to date the largest offshore petroleum accident ever in terms of hydrocarbon release [19–24]. It should be noted that the mentioned spill was caused by the Macondo Deepwater Horizon semi-submersible rig, operated by Transocean Ltd., the world's second largest offshore drilling contractor who is based in Switzerland but also with offices in many other countries. The rig was registered by the American Bureau of Shipping (ABS) and operated under a Marshall Islands flag of convenience and was built by South Korean company Hyundai Heavy Industries. The rig was chartered to British Petroleum, headquartered in London, the United Kingdom, which was the developer of the Macondo prospect with 65% stake, while the United States-based now defunct Anadarko Petroleum Corporation had 25% stake. Last but not least, United States-based MOEX Offshore 2007, a subsidiary of large company, Japanese Mitsui had 10% stake. The cause of the accident was a failure of the cement to form a proper shield of concrete at the base of the Macondo well at the level at which it was supposed to contain oil and gas.

(2)   The Montara oil spill occurred on 21 August 2009 in the Montara offshore oil field northwest of the Australian coast in the Timor Sea. Although much smaller in terms of hydrocarbon release compared with the Macondo Deepwater Horizon accident, the Montara oil spill was the first such accident in this part of the world for 25 years [25]. The West Atlas jack-up drilling rig at the Montara prospect was built by Keppel Fels at the Keppel Shipyard in Singapore in 2007 and was one of the largest in the world. The West Atlas rig was owned by the Norwegian–Bermudan Seadrill and operated by PTTEP Australasia (PTTEPAA), a subsidiary of PTT Exploration and Production (PTTEP), an oil and gas exploration company from Thailand. The cause of the accident was a failure of the cement to form a barrier of concrete (failure to install a pressure containment cap on the well).

(3)   The Piper Alpha accident occurred on 6 July 1988, offshore in the North Sea about 190 km northeast of Aberdeen, Scotland [26]. It is, to date, the deadliest offshore petroleum accident. The Piper Alpha was a large fixed production platform located at the offshore Piper oilfield, which was operated by Occidental Petroleum (Caledonia) Limited-OXY, a US company. The platform, which had four modules separated by firewalls, was constructed partly by McDermott Engineering in Scotland and partly by the Union Industrielle d'Entreprise in France. The cause of the accident has been unclear to date, but most probably it was due to bad maintenance management—the release of as little as 30 kg of condensate (mainly propane) over thirty seconds through an unsecured blind flange, where a pressure safety relief valve had been removed as part of maintenance on the standby condensate pump [27].

From these three examples, it is obvious that devices and equipment for the offshore petroleum industry are manufactured in different countries and most probably using many different technical standards. Besides, it can be foreseen that such equipment and devices installed and used on drilling units will, by default, often be transferred during their lifetime from one country to another (sometimes possibly avoiding custom duties and related technical controls). This can cause problems if the technical standards in the different host countries are incompatible. Therefore, selecting the appropriate and the best technical standards is complex, as some countries prescribe exactly which standards should be used, while in others the best worldwide available appropriate technical standards are expected to be used (sometimes without going into too much detail about how to choose such standards).

Here we will describe the specific situation in the European Union and European Economic Area (EU/EEA), where the product safety directives explicitly exclude from their scope offshore mobile units for drilling, and in some cases, related equipment on board (with some extensions to the well-control equipment under pressure, which is used onshore). Details about these exclusions will be explained especially in the spotlight of the EU/EEA offshore safety directive, which explicitly requires the use of the best worldwide available technical standards. Some notes about the prevention of major or other types of accidents in the offshore petroleum industry will be examined, mostly in respect to the safe use of equipment under pressure, protection against fire and explosions, the protection of staff who work with machinery, etc.

2. Legislation and Technical Standards

European Union directives are legally binding pieces of legislation in the European Union (EU) and also in the European Economic Area (EEA), which includes Norway, Iceland, and Lichtenstein (and in the UK, after leaving the EU during the transitional period, at least until the end of 2020, and in some other countries through bilateral agreements, such as Switzerland, which is out of both the EU and the EEA). All EU/EEA directives are principally designed to protect the single market of the EU/EEA [28].

The directives relevant for the offshore petroleum industry are (i) Directive 2013/30/EU on safety of offshore oil and gas operations (used as an umbrella Directive for offshore safety) and (ii) the product safety directives (these directives are relevant but, as explained in Section 2.1, they explicitly exclude from their scope drilling ships and related equipment)—(ii-1) the ATEX directive 2014/34/EU, for prevention against explosions; (ii-2) the pressure equipment directive 2014/68/EU; (ii-3) the machinery directive 2006/42/EC. Some others, such as the EU/EEA marine equipment directive 2014/90/EU, apply to equipment installed and used on-board ships that are registered in the EU and the EEA, but does not interfere with oil and gas drilling (therefore the "Wheelmark" sign; i.e., certification according to this directive does not apply to oil and gas equipment intended for drilling).

Conformity with the EU/EEA directives can be achieved through the harmonized technical standards, which are nonbinding pieces of regulation. To prevent technical barriers and to assure freedom of trade in the EU/EEA where various standards existed or still exist, even the use of harmonized technical standards are sometimes not sufficient to assure full compliance with the essential health and safety requirements prescribed by the directives (some directives refer only to the essential safety requirements). In general, all other available technical standards can be used in addition to the harmonized standards if they can assure full conformity with the prescribed essential requirements. In Section 2.2. we will analyze the role of harmonized and other technical standards in the EU/EEA offshore petroleum industry and interactions among different standardization bodies. Technical standards should be used in the EU/EEA to provide easier compliance with the provisions of the essential health and safety requirements of the relevant directives (consequently, the Conformité Européenne "CE" sign could be affixed to a product. Certifications should be done in general by the notified bodies listed in the EU/EEA NANDO database (the New Approach Notified and Designated Organizations).

2.1. Explicit Exclusions from EU/EEA Product Safety Directives

In general, all equipment and machinery in the EU/EEA need to fulfil the essential health and safety requirements of the relevant directives, or only safety requirements in the case of directives which primarily protect equipment and only secondarily humans (e.g., the machinery directive 2006/42/EC deals with essential health and safety requirements, but the pressure equipment directive 2014/68/EU only deals with essential safety requirements). Additionally, as a rule in the EU/EEA, required by the "New Legislative Framework" [29] from 2008, which is an EU/EEA package of measures that aim to improve market surveillance and boost the quality of conformity assessments, the mandatory Conformité Européenne "CE" mark must be affixed on equipment and machinery that complies with the directives. This ensures fair competition by holding all companies accountable to the same rules (combining equipment manufactured using technical standards of different countries in an assembly is strongly discouraged—it is also not permitted to then affix the CE marking to components).

The importance of offshore environmental protection [30,31] and safety is noted in the European Parliament resolution, "Facing the challenge of the safety of offshore oil and gas activities". EU/EEA oil and gas offshore safety is based on the 2013/30/EU Directive, which deals with major accidents, while in general the product safety directives should be used for protection against all other types of accidents. However, the three mentioned product safety directives that are relevant for offshore safety in the EU/EEA—(1) the ATEX directive 2014/34/EU for prevention against explosions; (2) the pressure equipment directive 2014/68/EU; (3) the machinery directive 2006/42/EC—explicitly do not cover mobile offshore drilling units and, under certain conditions, related equipment on board. The exclusions are as follows:

(1)   From the ATEX directive 2014/34/EU: In article 1, section 2(e) "seagoing vessels and mobile offshore units together with equipment on board such vessels or units";

(2)   From the pressure equipment directive 2014/68/EU (essential safety requirements are designed to protect the equipment against hazards caused by pressure): in article 1, section 2(a): "pipelines comprising piping or a system of piping designed for the conveyance of any fluid or substance to or from an installation (onshore or offshore) … "; section 2(i): "well-control equipment used in the petroleum, gas or geothermal exploration and extraction industry and in underground storage which is intended to contain and/or control well pressure; this shall comprise the wellhead (Christmas tree), the blow out preventers (BOP), the piping manifolds and all their equipment upstream"; section 2(j–ii) " … compressors, pumps … "; section 2(n): "ships, rockets, aircraft and mobile off-shore units, as well as equipment specifically intended for installation onboard or the propulsion thereof";

(3)   From the machinery directive 2006/42/EC (the essential health and safety requirements of this directive protect workers and personnel primarily): in article 1, section 2(f): "seagoing vessels and mobile offshore units and machinery installed onboard such vessels and/or units".

In general, so-called other types of accidents that should be covered by the product safety directives, in the worst-case scenario, can lead to major accidents with catastrophic consequences through a chain of unpredicted events. In addition to climatic changes or significant damage to the environment, the term "major accident" is defined in the directive 2013/30/EU as an incident with an explosion, fire, loss of well control, release of dangerous fluids or any other cause that has significant potential to cause fatalities or severe injuries to five or more people, and can lead to significant damage to an installation. Instead, through the product safety directives, drilling ships and oil and gas equipment onboard has been covered through the International Convention for the Safety of Life at Sea (SOLAS) and more specifically through the Code for the construction and equipment of Mobile Offshore Drilling Units of the International Maritime Organization (the MODU code). The MODU code has been developed to protect any ship as a vessel, and its integrity and stability in the first place in terms of navigation [32–35], emphasizing the additional requirements caused by drilling activities [36], but does not include requirements for the drilling of subsea wells or the procedures for their control (such drilling operations are subject to control by the coastal state). The MODU code is used in relation to the EU/EEA product safety directives as follows: (1) Protection against explosions; (2) Protection of pressure equipment; (3) Protection of personnel who works with machinery:

(1)   Protection against explosions: The ATEX directive 2014/34/EU protects equipment onshore and on offshore fixed platforms and the ATEX directive 1999/92/EC protects personnel [37].

However, as prescribed by the MODU code, protection against explosions onboard mobile drilling units in the EU/EEA waters goes through the International Electrotechnical Commission System for Certification to Standards Relating to Equipment for Use in Explosive Atmospheres (IECEx). The IECEx was developed mostly for electrical devices [38] and hence explosions caused by mechanical equipment is not extensively covered through this scheme. For the many decades during which mechanical equipment has been used in hazardous areas without causing damage, some accidents have undoubtedly occurred, mostly due to friction and overheating [39]. Following that reasoning, the EU/EEA also included anti-explosive protection for machinery through the ATEX scheme [40]. On the other hand, the main shortcoming under the ATEX scheme is that it allows self-certification in some zones (manufacturer can demonstrate the safety), which is not allowed in the IECEx scheme [41–46]. Additionally, the ATEX scheme refers only to equipment and protective systems and does not make any distinction between machinery and electrical equipment [47–50]. Some other requirements are prescribed in other countries, such as in Russia [51,52].

(2)   Protection of pressure equipment: Some equipment under pressure used offshore in the EU/EEA which is not specifically modified by adding, for example, moving compensators [53] (to annul the effect of waves at sea) are under the scope of the pressure equipment directive 2014/68/EU. The pressure equipment directive 2014/68/EU also excludes from its scope almost all equipment under pressure for drilling, exploitation or transport used in petroleum industries, including well-control equipment [54–57], both offshore on fixed or mobile units, or onshore. Besides, pumps are in the scope of the pressure equipment directive 2014/68/EU only if technical analyses show that the hazard, due to pressure, is the dominant and main factor of risk. Additionally, all assemblies under pressure need to be certified through the notified bodies listed in the EU/EEA NANDO database (the New Approach Notified and Designated Organizations), even when all parts of such assemblies have already been individually attested.

(3)   Protection of personnel who works with machinery: Machinery which is used in the EU/EEA onshore or on fixed offshore platforms are under the scope of the machinery directive 2006/42/EC. The transfer of such machinery from mobile units to fixed platforms or onshore without the proper EU/EEA certification, or avoiding payments of custom duties is problematic.

2.2. Certification Using Technical Standards

Technical standards are not legally binding documents, but they are widely accepted in engineering practice. In the EU/EEA, they should be used to support the application of the relevant directives. In the EU, conformity with the appropriate directives can be achieved by using the appropriate harmonized standards. With the "New approach" of the Council of the European Union from 1985, there is an obligation of results (i.e., to meet the essential health and safety requirements, and not of means). For the application of imposed standards, any standard may be used, provided that it meets the essential health and safety requirements [38,58].

The most relevant standards for the offshore petroleum industry are by the Norwegian NORSOK (Norsk Sokkels Konkurranseposisjon—the Norwegian shelf's competitive position), by the United States' American Petroleum Institute (API), and by the International Organization for Standardization (ISO) Technical Committee TC 67. On the other hand, standards developed by classification societies and private institutions can be used in the offshore petroleum industry, but only in addition to the abovementioned technical standards. The role of classification societies in the shipping industry is to establish and maintain technical standards for the construction, operation, and classification of ships and offshore structures.

Harmonized standards are developed by the recognized European Standards Organizations, the Comité Européen de Normalisation (CEN), and the Comité Européen de Normalisation Électrotechnique (CENELEC). Although adherence to the standards is ultimately voluntary, these standards are created to support manufacturers, other economic operators, and conformity assessment bodies to demonstrate that products, services, and processes comply with the relevant directives. These two European Standardization Organizations—CEN and CENELEC—are officially recognized as competent in the area of voluntary technical standardization in Regulation 1025/2012. Besides, in the EU/EEA, ultimately any standard may be used provided that it meets the essential health and safety requirements of the relevant directives. For certain directives, only essential safety requirements are required, where the application of the harmonized CEN/CENELEC standards gives only a presumption of conformity to these directives.

The use of all the best worldwide available technical standards is also prescribed by the umbrella directive for offshore oil and gas safety in the EU/EEA: Directive 2013/30/EU (harmonized standards with this directive are not available). Here it should be noted that harmonized standards, which can support the product safety directives exist; they are extensively developed for the machinery directive 2006/42/EC and for the pressure equipment directive 2014/68/EU, but only partially for the ATEX directive 2014/34/EU. Due to the lack of harmonized standards, all other available technical standards should be used if they can assure full compliance with the essential health and safety requirements of the ATEX directive 2014/34/EU.

The International Organization for Standardization (ISO) develops technical standards for the offshore petroleum industry through its Technical Committee TC 67 "Materials, equipment and offshore structures for petroleum, petrochemical and natural gas industries". Some of the ISO standards are developed jointly with the American Petroleum Institute (API) [59]. Sometimes, the API standards are recognized by the ISO through TC 67, where they can be harmonized with appropriate EU/EEA directives through the Vienna agreement. The Vienna Agreement on technical co-operation between ISO and CEN, signed in 1991, was drawn up to prevent duplication of efforts of CEN and ISO, and reducing time spent preparing standards (a similar agreement exists between IEC and CENELEC—the Frankfurt agreement). The path for the harmonization of API standards with the relevant EU/EEA directives through these two agreements is depicted with a red arrow in Figure 1.

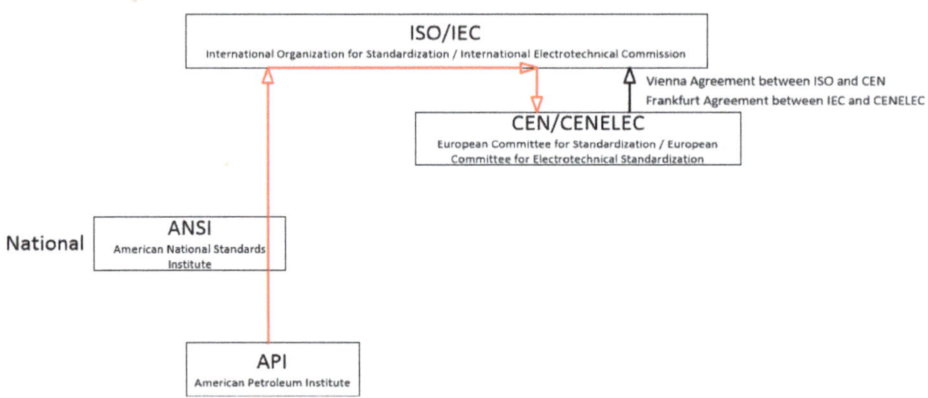

Figure 1. CEN/CENELEC harmonization of API standards with the appropriate EU/EEA directives through ISO/IEC following the Vienna and Frankfurt Agreements.

Some concerns exist regarding the combination of devices to form an assembly where some of the devices in the assembly are certified using different technical standards. Prior to being used, such assemblies must again be certified, but this time as a whole.

Even the best standards, such as ISO and API, can have shortcomings, and therefore additional re-evaluations or additional checks can be useful.

3. Conclusions

The technical standards used in the offshore petroleum industry are high level, but sometimes they are inadequately applied. For example, the three accidents mentioned in this technical note could have been prevented: (1) The Macondo Deepwater Horizon accident—that the concrete at the bottom of the Macondo borehole did not create a seal would easily be detected by performing a cement bond log; a measurement to verify the presence and quality of the cement used to seal the oil and gas-bearing rock formations; (2) The Montara accident:—there is no doubt that the Montara oil spill in Australia would not have happened in other countries, such as the UK and Norway, where the regulatory framework requires the mandatory inspection of wells during and at the completion of construction; (3) The Piper Alpha accident—proper maintenance management would most probably have been sufficient to prevent the disaster at the Piper Alpha platform.

Offshore drilling ships, including equipment onboard, which by definition operate in various countries, are very modern and therefore can usually be certified easily following many different legislative and technical norms. However, these parallel certifications are costly and should be simplified through international collaboration to avoid overlapping by signing agreements for mutual recognition among various countries (e.g., such as through Mutual Recognition Agreements (MRAs) and the designated Conformity Assessment Bodies (CABs)).

In the current situation, most drilling ships and equipment onboard are certified through various certification rules, which assure high levels of safety. These certifications are recognized or tolerated worldwide and they are complementarily used with domestic standards (in some countries, such domestic standards for offshore oil and gas drilling are not fully developed). All re-evaluations should be done carefully, and only if the additional cost would increase the overall level of safety and environmental protection significantly.

Related to the offshore petroleum activities in the EU/EEA waters, special care should be taken to the following:

- In the past, harmonized standards developed by CEN or CENELEC assured full conformity, while today to remove barriers for trade in the EU/EEA, any other available technical standards can be used if they can assure compliance to the health and safety requirements of the relevant directives. The umbrella directive 2013/30/EU for oil and gas offshore safety in the EU/EEA allows, and even requires, the use of the best available standards, which are recognized internationally.
- Due to exemption from the scope of the product safety directives, the machinery directive 2006/42/EC, the pressure equipment directive 2014/68/EU, and the ATEX directive 2014/34/EU, drilling ships and, in certain cases, oil and gas equipment used offshore are out of their scope (therefore are not required to fulfil the essential health and safety requirements of these directives and to be certified according to them). However, the use of some of their harmonized standards can be recommended if such practice can increase the overall level of safety. Additionally, it should be double-checked if the observed equipment is really excluded from the scope in every observed particular case (in case of exclusion, the international MODU code applies).
- Well-control equipment is also out of the scope of the pressure equipment directive 2014/68/EU onshore, and it is not required to follow its essential safety requirements (this directive does not refer to essential health requirements). Besides, well-control equipment onshore and offshore on fixed platforms is under the scope of the machinery directive 2006/42/EC and the relevant essential health and safety requirements of this directive should be satisfied.
- Protection against explosions on fixed offshore platforms goes through the European ATEX scheme, while on mobile offshore units through the IECEx scheme.

Author Contributions: Conceptualization, D.B.; methodology, D.B.; formal analysis, D.B.; investigation, D.B.; resources, P.P.; data curation, D.B. and P.P.; writing—original draft preparation, D.B.; writing—review and editing, D.B. and P.P.; visualization, D.B.; supervision, P.P.; project administration, D.B. and P.P.; funding acquisition, P.P. Both authors have read and agreed to the published version of the manuscript.

Funding: This work was supported by the Ministry of Education, Youth and Sports from the National Programme of Sustainability (NPS II) project "IT4Innovations excellence in science—LQ1602".

Acknowledgments: We thank John Cawley from IT4Innovations who, as a native speaker, kindly checked the correctness of English expressions throughout the paper.

Conflicts of Interest: The authors declare no conflict of interest. The opinions expressed in this paper are solely those of the authors.

Notations: The following are used in this paper:

European Union directives:

- Offshore safety directive 2013/30/EU-Directive on safety of offshore oil and gas operations (used as umbrella Directive for offshore oil and gas safety)
- Product safety directives: (1) ATEX directive 2014/34/EU; (2) pressure equipment directive 2014/68/EU; (3) machinery directive 2006/42/EC
- Marine equipment directive 2014/90/EU
- ATEX directive 1999/92/EC for protection of personnel

Documents, Regulations, and Databases:

- New Legislative Framework package, that reinforces the application and enforcement of internal market legislation in the EU/EEA
- Facing the challenge of the safety of offshore oil and gas activities: 2011/2072(INI), Motion for a European parliament resolution from 26 July 2011
- IECEx: International Electrotechnical Commission System for Certification to Standards Relating to Equipment for Use in Explosive Atmospheres
- International Convention for the Safety of Life at Sea (SOLAS) from 1974 and in force since 25 May 1980
- MODU code: "Code for the construction and equipment of Mobile Offshore Drilling Units of the International Maritime Organization" from 2009
- NANDO: New Approach Notified and Designated Organizations Information System (Conformity Assessment and Acceptance of Industrial Products Notification Bodies)
- Regulation (EU) No 1025/2012, related to European standardization
- "New approach": Council Resolution of 7 May 1985 on a new approach to technical harmonization and standards

Standardization and other bodies:

- NORSOK: Norwegian Technical standards
- API: American Petroleum Institute
- ISO: International Organization for Standardization
- TC 67: Technical committee of the ISO "Materials, equipment and offshore structures for petroleum, petrochemical and natural gas industries"
- CEN: Comité Européen de Normalisation
- CENELEC: Comité Européen de Normalisation Électrotechnique
- IEC: International Electrotechnical Commission
- ANSI: American National Standardization Institute
- CABs: Conformity Assessment Bodies

Agreements:

- Vienna agreement—technical cooperation between ISO and CEN
- Frankfurt agreement—technical cooperation between IEC and CENELEC
- MRAs—Mutual Recognition Agreements promote trade in goods between the EU and third countries and facilitate market access

Offshore accidents:

- Macondo Deepwater Horizon: the Gulf of Mexico in 2010 (offshore drilling rig, to date the biggest in terms of hydrocarbon release)
- Montara: the Timor Sea in 2009 (offshore drilling rig)
- Piper Alpha: the North Sea in 1988 (fixed production platform, to date the deadliest offshore petroleum accident)

Companies:

- British Petroleum (BP)
- MOEX Offshore-subsidiary of Mitsui

- Occidental Petroleum (Caledonia) Limited-OXY
- Anadarko Petroleum Corporation—Now defunct
- PTTEP Australasia (PTTEPAA)—Subsidiary of PTT Exploration and Production (PTTEP)
- Transocean: rig contractor
- Norwegian–Bermudan Seadrill: rig contractor
- Keppel Fels: rig manufacturer
- Hyundai Heavy Industries: rig manufacturer
- McDermott Engineering: platform manufacturer
- Union Industrielle d'Entreprise: platform manufacturer

Oil fields and Wells:

- Macondo (the Gulf of Mexico)
- Montara (the Timor Sea)
- Piper (the North Sea)

Rigs and platforms:

- Deepwater Horizon: semi-submersible rig
- West Atlas: jack-up drilling rig
- Piper Alpha: fixed production platform

Ship registers:

- American Bureau of Shipping (ABS)

Product marking:

- Conformité Européenne "CE" sign: general conformity with the relevant EU/EEA Directives
- "Wheelmark" sign: conformity with marine equipment Directive 2014/90/EU

Equipment:

- BOP—Blow out preventers
- "Christmas tree"—Wellhead

References

1. Elvidge, C.D.; Ziskin, D.; Baugh, K.E.; Tuttle, B.T.; Ghosh, T.; Pack, D.W.; Erwin, E.H.; Zhizhin, M. A fifteen-year record of global natural gas flaring derived from satellite data. *Energies* **2009**, *2*, 595–622. [CrossRef]
2. Rabe, B.; Kaliban, C.; Englehart, I. Taxing flaring and the politics of state methane release policy. *Rev. Policy Res.* **2020**, *37*, 6–38. [CrossRef]
3. Baalisampang, T.; Abbassi, R.; Khan, F. Overview of marine and offshore safety. *Met. Chem. Proc. Saf.* **2018**, *2*, 1–97. [CrossRef]
4. Fattakhova, E.Z.; Barakhnina, V. Accident rate analysis on the offshore oil and gas production installations and platforms. *Int. J. Appl. Fund. Res.* **2015**, *1*. Available online: http://www.science-sd.com/460-24767 (accessed on 14 July 2020).
5. Zolotukhin, A.; Gavrilov, V. Russian Arctic petroleum resources. *Oil Gas Sci. Technol.* **2011**, *66*, 899–910. [CrossRef]
6. Aalto, P. Modernisation of the Russian energy sector: Constraints on utilizing Arctic offshore oil resources. *Eur. Asia Stud.* **2016**, *68*, 38–63. [CrossRef]
7. Helle, I.; Mäkinen, J.; Nevalainen, M.; Afenyo, M.; Vanhatalo, J. Impacts of oil spills on Arctic marine ecosystems: A quantitative and probabilistic risk assessment perspective. *Environ. Sci. Technol.* **2020**, *54*, 2112–2121. [CrossRef]
8. Lindholt, L.; Glomsrød, S. The Arctic: No big bonanza for the global petroleum industry. *Energy Econ.* **2012**, *34*, 1465–1474. [CrossRef]
9. Torres, L.; Jiménez-Cabas, J.; González, O.; Molina, L.; López-Estrada, F.-R. Kalman Filters for leak diagnosis in pipelines: Brief history and future research. *J. Mar. Sci. Eng.* **2020**, *8*, 173. [CrossRef]
10. McBain, G. A user's view of the pressure equipment directive. *Meas. Control* **2003**, *36*, 300–304. [CrossRef]
11. Ciavarella, M.; Carbone, G.; Vinogradov, V. A critical assessment of Kassapoglou's statistical model for composites fatigue. *Facta Univ. Mech. Eng.* **2018**, *16*, 115–126. [CrossRef]

12. Zhao, X.; Wang, S.; Wang, X.; Fan, Y. Multi-state balanced systems in a shock environment. *Reliab. Eng. Syst. Saf.* **2020**, *193*, 106592. [CrossRef]
13. Wang, X.; Zhao, X.; Wang, S.; Sun, L. Reliability and maintenance for performance-balanced systems operating in a shock environment. *Reliab. Eng. Syst. Saf.* **2020**, *195*, 106705. [CrossRef]
14. Zhao, X.; Wu, C.; Wang, X.; Sun, J. Reliability analysis of k-out-of-n: F balanced systems with multiple functional sectors. *Appl. Math. Model.* **2020**, *82*, 108–124. [CrossRef]
15. Luo, F.; Zhang, S.; Yang, D. Anti-explosion performance of composite blast wall with an auxetic re-entrant honeycomb core for offshore platforms. *J. Mar. Sci. Eng.* **2020**, *8*, 182. [CrossRef]
16. Lindøe, P.H.; Baram, M.; Renn, O. *Risk Governance of Offshore Oil and Gas Operations*; Cambridge University Press: Cambridge, UK, 2013. [CrossRef]
17. Lindøe, P.H.; Baram, M.S. The role of standards in hard and soft approaches to safety regulation. In *Standardization and Risk Governance*; Olsen, O.E., Juhl, K.V., Lindøe, P.H., Engen, O.A., Eds.; Routledge: Abingdon, UK, 2020; pp. 235–254. [CrossRef]
18. Yang, Y. Reforming health, safety, and environmental regulation for offshore operations in China: Risk and resilience approaches? *Sustainability* **2019**, *11*, 2608. [CrossRef]
19. Crone, T.J.; Tolstoy, M. Magnitude of the 2010 Gulf of Mexico oil leak. *Science* **2010**, *330*, 634. [CrossRef]
20. Croisant, S.A.; Lin, Y.L.; Shearer, J.J.; Prochaska, J.; Phillips-Savoy, A.; Gee, J.; Jackson, D.; Panettieri, R.A.; Howarth, M.; Sullivan, J.; et al. The Gulf coast health alliance: Health risks related to the Macondo spill (GC-HARMS) study: Self-reported health effects. *Int. J. Environ. Res. Public Health* **2017**, *14*, 1328. [CrossRef]
21. Zilversmit, L.; Wickliffe, J.; Shankar, A.; Taylor, R.J.; Harville, E.W. Correlations of biomarkers and self-reported seafood consumption among pregnant and non-pregnant women in southeastern Louisiana after the Gulf oil spill: The GROWH study. *Int. J. Environ. Res. Public Health* **2017**, *14*, 784. [CrossRef]
22. Nunziata, F.; Buono, A.; Migliaccio, M. COSMO–SkyMed Synthetic aperture radar data to observe the deep-water horizon oil spill. *Sustainability* **2018**, *10*, 3599. [CrossRef]
23. Zhang, Y.; Li, S.; Guo, Z. The Evolution of the coastal economy: The role of working waterfronts in the Alabama Gulf Coast. *Sustainability* **2015**, *7*, 4310–4322. [CrossRef]
24. Moerschbaecher, M.; Day, J.W., Jr. Ultra-deep water Gulf of Mexico oil and gas: Energy return on financial investment and a preliminary assessment of energy return on energy investment. *Sustainability* **2011**, *3*, 2009–2026. [CrossRef]
25. Hayes, J. Operator competence and capacity–Lessons from the Montara blowout. *Saf. Sci.* **2012**, *50*, 563–574. [CrossRef]
26. Paté-Cornell, M.E. Learning from the piper alpha accident: A postmortem analysis of technical and organizational factors. *Risk Anal.* **1993**, *13*, 215–232. [CrossRef]
27. Cullen, L.W. The public inquiry into the Piper Alpha disaster. *Drilling Contractor* **1993**, *49*. Available online: https://www.osti.gov/biblio/6208554 (accessed on 27 June 2020).
28. Messerlin, P.A. The European Union single market in goods: Between mutual recognition and harmonisation. *Aust. J. Int. Aff.* **2011**, *65*, 410–435. [CrossRef]
29. Pollack, M.A. Theorizing the European Union: International organization, domestic polity, or experiment in new governance? *Annu. Rev. Polit. Sci.* **2005**, *8*, 357–398. [CrossRef]
30. Fiorino, D.J. *The New Environmental Regulation*; MIT Press: Cambridge, MA, USA, 2006.
31. Manheim, F.T. *The Conflict Over Environmental Regulation in the United States: Origins, Outcomes, and Comparisons with the EU and Other Regions*; Springer Science and Business Media: Berlin/Heidelberg, Germany, 2008. [CrossRef]
32. Xu, P.; Du, Z.; Gong, S. Numerical investigation into freak wave effects on deep water pipeline installation. *J. Mar. Sci. Eng.* **2020**, *8*, 119. [CrossRef]
33. Zan, Y.; Guo, R.; Yuan, L.; Wu, Z. Experimental and numerical model investigations of the underwater towing of a subsea module. *J. Mar. Sci. Eng.* **2019**, *7*, 384. [CrossRef]
34. Yu, G.; Zhang, L.; Jia, S.; Geng, Y.; Liu, J. Numerical study on the natural convection of air in a cubic cavity subjected to a yawing motion. *J. Mar. Sci. Eng.* **2019**, *7*, 204. [CrossRef]
35. Yu, G.; Jia, S.; Geng, Y. Numerical investigation into the two-phase convective heat transfer within the hold of an oil tanker subjected to a rolling motion. *J. Mar. Sci. Eng.* **2019**, *7*, 94. [CrossRef]
36. Kaiser, M.J.; Snyder, B. The five offshore drilling rig markets. *Mar. Policy* **2013**, *39*, 201–214. [CrossRef]
37. Marx, I. Combining the best of both worlds. *IEEE Ind. Appl. Mag.* **2010**, *16*, 30–34. [CrossRef]

38. Petitfrere, C.; Proust, C. Analysis of ignition risk on mechanical equipment in ATEX. Proceeding of the 4th IEEE European Conference on Electrical and Instrumentation Applications in the Petroleum and Chemical Industry, Paris, France, 13–15 June 2007; pp. 1–9. [CrossRef]

39. Leroux, P. New regulations and rules for ATEX directives. *IEEE Ind. Appl. Mag.* **2006**, *13*, 43–51. [CrossRef]

40. Parise, G.; Sutherland, P.E.; Moylan, W.J. Electrical safety for employee workplaces in Europe and in the USA. *IEEE Trans. Ind. Appl.* **2005**, *41*, 1091–1098. [CrossRef]

41. Campbell, J.D.; Chudleigh, J.P. Problems encountered in designing electrical systems for hazardous areas. In *IEEE Transactions on Industry and General Applications*; IEEE: Piscataway, NJ, USA, 1970; Volume IGA-6, pp. 326–329. [CrossRef]

42. Rodrigues, A.M.T.G. A Software Application to Define and Rank ATEX Zones. Ph. D. Thesis, Faculdade de Engenharia da Universidade do Porto, Porto, Portugal, 2016. Available online: https://repositorio-aberto.up.pt/bitstream/10216/88507/2/156964.pdf (accessed on 11 June 2020).

43. Riikonen, J. ATEX certification for hazardous areas. *World Pumps* **2010**, *2010*, 22–24. [CrossRef]

44. Pommé, R.; Sijrier, H.J. IECEX certification schemes versus ATEX directives. Proceeding of the Petroleum and Chemical Industry Conference Europe—Electrical and Instrumentation Applications (PCIC Europe), Oslo, Norway, 15–17 June 2010; pp. 1–8. Available online: https://ieeexplore.ieee.org/xpl/conhome/5512515/proceeding (accessed on 29 June 2020).

45. Nicols, R. ATEX directives for the UK sector. *Meas. Control* **2003**, *36*, 147–155. [CrossRef]

46. Propst, J.E.; Barrios, L.A.; Lobitz, B. Applying the API alternate method for area classification. *IEEE Trans. Ind. Appl.* **2007**, *43*, 162–171. [CrossRef]

47. Khaymedinova, Z. Explosion-Proof Requirements for Electrical Machines in Chemical, Oil and Gas Industry in Russia and CIS Countries. Ph.D. Thesis, Lappeenranta University of Technology, Lappeenranta, Finland, 2009. Available online: https://lutpub.lut.fi/bitstream/handle/10024/46840/nbnfi-fe200908031986.pdf (accessed on 11 June 2020).

48. Sinclair, R. ATEX good and bad/A notified body's perspective. *Meas. Control* **2003**, *36*, 140–141. [CrossRef]

49. Dearden, H.T. Who's afraid of ATEX? *Meas. Control* **2006**, *39*, 17–18. [CrossRef]

50. Towle, C. ATEX directives/A manufacturer's viewpoint. *Meas. Control* **2003**, *36*, 152–154. [CrossRef]

51. Hitchen, I.R. Vibration monitoring for rotating machinery. *Meas. Control* **1980**, *13*, 97–102. [CrossRef]

52. Shu-Chao, L. The main differences of general requirements and intrinsic safety between IEC and Russian standards. electric explosion protection. *China Nat. Knowl. Infrastruct. J.* **2004**, *3*. Available online: http://en.cnki.com.cn/Article_en/CJFDTotal-DQFB200403001.htm (accessed on 11 June 2020).

53. Stefanowicz, D. ATEX for installers and maintainers. *Meas. Control* **2003**, *36*, 144–158. [CrossRef]

54. Harvey, S.; Elashvili, I.; Valdes, J.J.; Kamely, D.; Chakrabarty, A.M. Enhanced removal of Exxon Valdez spilled oil from Alaskan gravel by a microbial surfactant. *Biotechnology* **1990**, *8*, 228–230. [CrossRef] [PubMed]

55. Khan, F.I.; Sadiq, R.; Husain, T. Risk-based process safety assessment and control measures design for offshore process facilities. *J. Hazard. Mater.* **2002**, *94*, 1–36. [CrossRef]

56. Mutlu, M.; Tang, Y.; Franchek, M.A.; Turlak, R.; Gutierrez, J.A. Dynamic performance of annular blowout preventer hydraulic seals in deepwater environments. *J. Offshore Mech. Arct. Eng.* **2018**, *140*, 061301. [CrossRef]

57. Crumpton, H. *Well Control for Completions and Interventions*; Gulf Professional Publishing: Houston, TX, USA, 2018. [CrossRef]

58. Marangon, A.; Carcassi, M. ATEX Directives: The New Approach. Valutazione e Gestione del Rischio negli Insediamenti Civili ed Industriali, 2006. Available online: http://conference.ing.unipi.it/vgr2006/archivio/Archivio/2006/Articoli/700213.pdf (accessed on 29 June 2020).

59. Weightman, R.T.; Warnack, M.F. API and ISO standards can be combined. *Oil Gas J.* **1992**, *90*, 50–52. Available online: https://www.ogj.com/home/article/17218930/api-and-iso-standards-can-be-combined (accessed on 10 March 2020).

 Journal of
Marine Science and Engineering

Article

Numerical Investigation into the Two-Phase Convective Heat Transfer within the Hold of an Oil Tanker Subjected to a Rolling Motion

Guojun Yu [1,2,*], Sheng Jia [1] and Yanting Geng [1]

[1] Merchant Marine College, Shanghai Maritime University, Shanghai 201306, China; 201730110052@stu.shmtu.edu.cn (S.J.); 201610129071@stu.shmtu.edu.cn (Y.G.)
[2] School of Naval Architecture, Ocean & Civil Engineering, Shanghai Jiao Tong University, Shanghai 200240, China
* Correspondence: gjyu@shmtu.edu.cn

Received: 20 March 2019; Accepted: 1 April 2019; Published: 3 April 2019

Abstract: A crude oil tanker usually encounters a rolling motion during sea transportation, which leads to rotational movement and sometimes a sloshing of the liquid hold. This rolling-induced body motion seriously affects the thermal and hydraulic behavior of the liquid hold, which then affects the heating process and heat preservation of the tanker. Clarification of the involved thermal and hydraulic characteristics is the basic requirement for establishment of a scientific heating scheme and heat preservation method. A two-phase 3D model considering the free liquid surface and non-Newtonian behavior of the fluid was established for the thermal calculation of the liquid holds in oil tankers. The thermal and hydraulic characteristics of the liquid hold were investigated under different combinations of dimensionless parameters, and the combined effect of rolling and fluid non-Newtonian behavior was investigated. It was found that rolling intensifies the heat transfer based on the combination of the Richardson number (Ri) and the rotation-strength number (ω^*), and non-Newtonian behavior of the fluid effectively affects the heat transfer in a rolling motion. This research is expected to provide a reference for design and optimization of the heating and heat preservation method for oil tanker operation.

Keywords: oil tanker; rolling motion; mixed convection

---

1. Introduction

Tanker transportation is one of the most important ways for crude oil transportation, especially during international crude oil trade. Some high-pour point or high-viscosity crude oil requires heating during transportation to ensure good flow ability for flexible offloading. This heating process consumes large amounts of energy, and a scientific heating scheme bears great potential to save energy. However, such a scientific heating scheme is based on the full understanding of the thermal and hydraulic behavior of the crude oil in the ocean condition. The thermal-hydraulic behavior of crude oil in ocean conditions is much more complicated than that in static conditions, since the flow and heat transfer is influenced by the motion-induced additional forces, such as centrifugal force, Euler force, Coriolis force, and buoyancy forces induced by these forces, in addition to the forces for static cases. The motion of the tanker during sea transportation includes oscillations of six degrees of freedom, such as linear motions, including heaving, swaying, and surging, and rotational motions, including rolling, pitching, and yawing. Among these movements, the rolling motion has a more important influence on the thermal-hydraulic behavior [1]. Therefore, only the rolling motion is investigated in this research.

There are many studies regarding the effect of a rolling motion on flow and heat transfer within oil tankers in the open literature. Hiroharu [2] studied the heat transfer characteristic of oil in a 1/50 scale model of a tanker in a rolling condition, finding that oil temperature in the tanker becomes uniform with rolling motion, and the heat transfer coefficient increases with increasing of amplitude and frequency of rolling. Yu et al. [3] studied the temperature drop of an oil tanker with a free liquid surface subjected to a rolling motion, indicating that the heat transfer of oil is dominated by mixed convection rather than natural convection under a rolling condition, and the temperature drop rate increases with the oscillation frequency. Similarly, only a small-size model tanker was investigated. Doerffer et al. [4] studied the influence of low-frequency harmonic oscillation on the natural convection occurring at the vertical walls of a tanker, pointing out that the dimensionless group $Gr/(Re^2Pr^{1/3})$ (Gr, Re, Pr are Grashof number, Reynolds number and Prandtl Number respectively) can be regarded as the principal criterion for this kind of mixed convection and it decreases corresponding to the increasing participation of forced convection in mixed convection. Akagi et al. [5] studied the mixed convection heat transfer in an oil tanker subjected to a rolling motion, taking the inertia forces, including the centrifugal force and Coriolis force into consideration, indicating that forced convection plays an important role during a rolling motion.

Similar research can also be found in the area of convection heat transfer in pipe flow or channel flow. Most of these studies also indicated that the rolling affected the thermal-hydraulic behavior. Tan et al. [6,7] studied single-phase natural circulation heat transfer in a pipe system with a rolling motion. They found that the fluid flow fluctuates periodically due to the rolling motion, and the heat transfer coefficient increases with the rolling amplitude and frequency. Yu et al. [8] studied the fluid temperature fluctuation at the outlet of the test section in a single-phase natural circulation loop with a heat narrow rectangular channel under rolling motion conditions, indicating that the average temperature of fluid at the outlet of the test section under rolling motion conditions is higher than that under static conditions because of the decreasing cycle averaged flow rate. In addition, Liu et al. [9] studied the safety boundary of flow instability and critical heat flux for parallel channels in static and ship motion. Wu et al. [10] studied the thermal-hydraulic behavior of a nuclear reactor core in an ocean environment. Xing et al. [11] studied the effects of ocean conditions on coolant flow in nuclear power systems in sea transportation. Besides, Hu et al. [12] studied the hydraulic dynamics behavior in floating LNG (Liquefied Natural Gas) systems under ocean conditions. Grotle et al. [13] studied the thermal response in marine LNG fuel tanks by experiments and modelling. Jiang et al. [14] analyzed the effect between ship motion response and internal sloshing flow. Cercos-Pita [15] investigated the coupled nonlinear dynamics of a vessel with a free surface tank onboard. All this research indicates rolling has obvious effects on flow and heat transfer.

It can be found from the above literature review that there are many researchers focusing on the heat and flow characteristic of fluid in a rolling system, but their models are restricted to either small-size geometry or a single combination of influencing factors, and the general flow and heat transfer characteristics are not systematically investigated. In addition, another important factor—the non-Newtonian behavior of the crude oil—is not taken into consideration. In this paper, the convection heat transfer under different combinations of dimensionless numbers as well as the flow-behavior index n for characterizing the non-Newtonian behavior will be investigated.

2. Physical and Mathematical Model

2.1. Physical Model

The full physical model of a real tanker should be a three-dimensional one, as shown in Figure 1 (top). However, the calculation of a 3D model is too much time-consuming, since the geometry of the oil tanker is usually very large. Considering the longitude temperature gradient is very weak compared to the transverse one, the physical model can be simplified to a two-dimensional one shown in Figure 1 (bottom), in which the red part is crude oil, the green part is ballast water, and the blue part is air (inert gas).

For the convenience of establishment and computation of the mathematical model, the following assumptions were made:

(1) The upper part of the oil tank is a single layer wall without any building above, which directly touches the air;
(2) The oil tank contains two thirds of the oil; the initial temperature of the oil is 323.15 K;
(3) The four wing spaces at the bottom are filled with ballast water, while the other wing tanks are filled with air. The initial temperature of water and air is 293.15 K;
(4) The tanker's draught is two thirds of its height;
(5) The outside boundary of the tanker is subjected to the third boundary conditions. The temperature of the external sea water and air are both 291.15 K, and the convective heat transfer coefficient between walls and sea water was set to be 1250 W/(m²·K), and that between walls and air was 50 W/(m²·K), respectively [16].

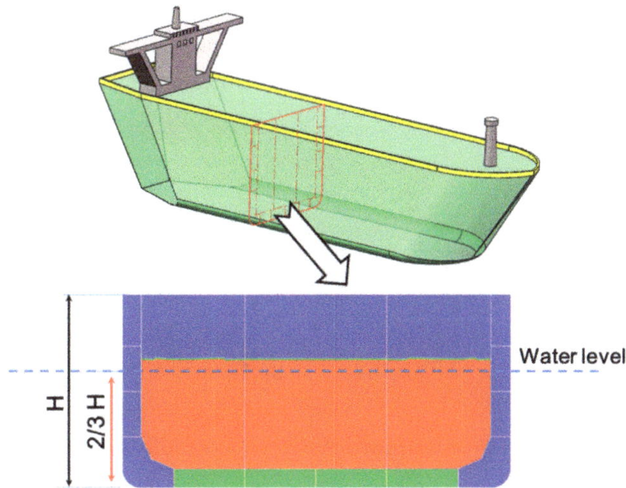

Figure 1. Physical model of the thermal system of an oil tanker.

2.2. Mathematical Model

The mathematical model was established in a non-inertial coordinate system shown in Figure 2. The origin of the coordinates was located at the center of the bottom wall. The *xoy* coordinate system is a non-inertial coordinate system at rest relative to the model; the *x'oy'* coordinate system is a static coordinate system. At $t = 0$, the model is stationary, and the model starts to move since $t > 0$ with the following parameters.

$$\begin{cases} \theta = \theta_m \sin\left(\frac{2\pi}{T_c}t\right) \\ \omega = \frac{2\pi}{T_c}\theta_m \cos\left(\frac{2\pi}{T_c}t\right) \\ \frac{d\omega}{dt} = -\frac{4\pi^2}{T_c^2}\theta_m \sin\left(\frac{2\pi}{T_c}t\right) \end{cases} \quad (1)$$

Here, θ_m and T_c are the amplitude and period of the rolling motion, respectively.

During the rotation, the crude oil in the tanker is subjected to the inertial force caused by the rotation, including centrifugal force, Euler force, and Coriolis force. All the inertial forces are listed below:

(1) Gravity force, ρg;

(2)　Centrifugal force, $-\rho\omega \times (\omega \times r)$;

(3)　Euler force, $-\rho\frac{d\omega}{dt} \times r$;

(4)　Coriolis force, $-2\rho\omega \times u$.

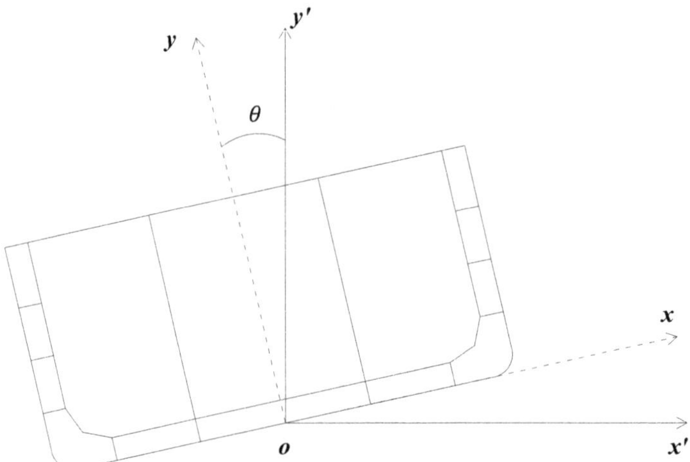

Figure 2. Non-inertial coordinate system.

In the non-inertial coordinate system, the direction of gravity acceleration changes with time, and Rodrigues' rotation formula [17] is used to determine the gradational acceleration in the non-inertial coordinate system, that is

$$g^* = \begin{pmatrix} -gsin\theta \\ -gcos\theta \\ 0 \end{pmatrix}. \tag{2}$$

With the Boussinesq assumption, the governing equations are given as follows.

The continuous equation:

$$\frac{\partial u_x}{\partial x} + \frac{\partial u_y}{\partial y} = 0. \tag{3}$$

The momentum equation:

$$\frac{\partial(\rho_0 u_x)}{\partial t} + \frac{\partial(\rho_0 u_x u_x)}{\partial x} + \frac{\partial(\rho_0 u_x u_y)}{\partial y} = -\frac{\partial p_{eff}}{\partial x} + \frac{\partial}{\partial x}\left(\mu\frac{\partial u_x}{\partial x}\right) + \frac{\partial}{\partial y}\left(\mu\frac{\partial u_x}{\partial y}\right)$$
$$+\rho_0\beta(T - T_0)gsin\theta - \rho_0\beta(T - T_0)\omega^2 x + \rho_0\frac{d\omega}{dt}y + 2\rho_0\omega u_y \tag{4}$$

$$\frac{\partial(\rho_0 u_y)}{\partial t} + \frac{\partial(\rho_0 u_y u_x)}{\partial x} + \frac{\partial(\rho_0 u_y u_y)}{\partial y} = -\frac{\partial p_{eff}}{\partial y} + \frac{\partial}{\partial x}\left(\mu\frac{\partial u_y}{\partial x}\right) + \frac{\partial}{\partial y}\left(\mu\frac{\partial u_y}{\partial y}\right)$$
$$+\rho_0\beta(T - T_0)gcos\theta - \rho_0\beta(T - T_0)\omega^2 y - \rho_0\frac{d\omega}{dt}x - 2\rho_0\omega u_x \tag{5}$$

From Equations (3)–(5), u_x and u_y are velocities in x and y direction, respectively; ρ_0 is the reference density of the crude oil, and the subscript "0" here indicates the reference value; β is the thermal expansion coefficient; ω is the rotation angular velocity. p_{eff} is the effective pressure defined as:

$$-\frac{\partial p_{eff}}{\partial x} = -\frac{\partial p}{\partial x} - \rho_0 gsin\theta + \rho_0\omega^2 x \tag{5a}$$

$$-\frac{\partial p_{eff}}{\partial y} = -\frac{\partial p}{\partial y} - \rho_0 gcos\theta + \rho_0\omega^2 y. \tag{5b}$$

The energy equation:

$$\frac{\partial(\rho_0 T)}{\partial t} + \frac{\partial(\rho_0 u_x T)}{\partial x} + \frac{\partial(\rho_0 u_y T)}{\partial y} = \frac{\partial}{\partial x}\left(\frac{\lambda}{c_p}\frac{\partial T}{\partial x}\right) + \frac{\partial}{\partial y}\left(\frac{\lambda}{c_p}\frac{\partial T}{\partial y}\right). \tag{6}$$

Here, λ and c_p are conductivity and specific heat capacity of the crude oil.
Gas-liquid phase distribution function:

$$\frac{\partial \alpha_q}{\partial t} + \frac{\partial\left(u_x \alpha_q\right)}{\partial x} + \frac{\partial\left(u_y \alpha_q\right)}{\partial y} = 0. \tag{7}$$

Here, α_q is the volume fraction of the qth phase.

Surface tension is considered in the calculation, based on the continuum surface force model (CSF) proposed by Brackbill et al. [18]; surface tension is added as a source term to the momentum equation

$$F_{csf} = \sum_{i<j} \sigma_{ij} \frac{\alpha_i \rho_i k_j \nabla a_j + \alpha_j \rho_j k_i \nabla a_i}{\frac{1}{2}(\rho_i + \rho_j)}. \tag{8}$$

In this model, because there are only two phases, air and crude oil, in a cell, the equation can be simplified as follows.

$$F_{csf} = \sigma_{12} \frac{\rho k_1 \nabla a_1}{\frac{1}{2}(\rho_1 + \rho_2)} \tag{9}$$

Here, ρ is the volume averaged density and determined by $\rho = \alpha_1 \rho_1 + (1 - \alpha_1)\rho_2$.

In addition, the Nu number used in this paper is calculated by the following formulas [19].

$$Nu_X = -\frac{\partial \Theta}{\partial X}, \; Nu_Y = -\frac{\partial \Theta}{\partial Y}, \; \overline{Nu_X} = \int_{min}^{max} Nu_X dX, \; \overline{Nu_Y} = \int_{min}^{max} Nu_Y dY \tag{10}$$

Governing Equations (3)–(7) are transformed into dimensionless form using the following formulas, and the dimensionless equations are given by Equations (13)–(17).

$$X = \frac{x}{l}, Y = \frac{y}{l}, U = \frac{u_x}{\overline{\omega}l}, V = \frac{u_y}{\overline{\omega}l}, \Theta = \frac{T - T_0}{T_h - T_0}, P = \frac{P_{eff}}{\rho \overline{\omega}^2 l^2}$$

$$\tau = \overline{\omega}t, \omega^* = \frac{\overline{\omega}l^2}{\alpha}, \mu^* = \frac{\mu}{\mu_0}, Pr = \frac{v_0}{\alpha}, Ra = \frac{g\beta\Delta Tl^3}{v_0\alpha}, Ra_w = \frac{\overline{\omega}l\beta\Delta Tl^3}{v_0\alpha} \tag{11}$$

$$Re = \frac{\overline{\omega}l^2}{v}, Ri = \frac{Gr}{Re^2} = \frac{1}{\omega^{*2}}PrRa$$

In Equation (11), Ra_w is the rotational Rayleigh number, ω^* is the rotation-strength number characterizing the importance of centrifugal force relative to diffusivity, Ri is the Richardson number, l is the reference length which equates the width of the largest tank, and $\overline{\omega}$ is the average angular velocity calculated by Equation (12).

$$\overline{\omega} = \frac{2\int_{-\frac{T}{4}}^{\frac{T}{4}} \omega dt}{T_c} = \frac{4\theta_m}{T_c} \tag{12}$$

Dimensionless governing equations are written as follows.

$$\frac{\partial U}{\partial X} + \frac{\partial V}{\partial Y} = 0 \tag{13}$$

$$\frac{\partial U}{\partial \tau} + \frac{\partial (UU)}{\partial X} + \frac{\partial (UV)}{\partial Y} = -\frac{\partial P}{\partial X} + \frac{Pr}{\omega^*}\frac{\partial}{\partial X}\left(\mu^*\frac{\partial U}{\partial X}\right) + \frac{Pr}{\omega^*}\frac{\partial}{\partial Y}\left(\mu^*\frac{\partial V}{\partial Y}\right)$$
$$+ Ri\Theta \sin(\theta_m \sin \tau) - Ri\Theta g^* X \theta_m^2 \cos^2 \tau - \theta_m Y \sin \tau + 2\theta_m \cos \tau V \tag{14}$$

$$\frac{\partial V}{\partial \tau} + \frac{\partial (UV)}{\partial X} + \frac{\partial (VV)}{\partial Y} = -\frac{\partial P}{\partial Y} + \frac{Pr}{\omega^*}\frac{\partial}{\partial X}\left(\mu^*\frac{\partial V}{\partial X}\right) + \frac{Pr}{\omega^*}\frac{\partial}{\partial Y}\left(\mu^*\frac{\partial V}{\partial Y}\right)$$
$$+ Ri\Theta \cos(\theta_m \sin \tau) + Ri\Theta g^* Y \theta_m^2 \cos^2 \tau + \theta_m X \sin \tau - 2\theta_m \cos \tau U \tag{15}$$

$$\frac{\partial (\Theta)}{\partial \tau} + \frac{\partial (U\Theta)}{\partial X} + \frac{\partial (V\Theta)}{\partial Y} = \frac{1}{\omega^*}\left(\frac{\partial^2 \Theta}{\partial X^2} + \frac{\partial^2 \Theta}{\partial Y^2}\right) \tag{16}$$

$$\frac{\partial \alpha_q}{\partial \tau} + \frac{\partial (U\alpha_q)}{\partial X} + \frac{\partial (V\alpha_q)}{\partial Y} = 0 \tag{17}$$

In Equations (14) and (15), g^* is the dimensionless centrifugal acceleration and is determined by $g^* = \overline{\omega}^2 l / g$.

It can be found from Equations (13)–(17) that there are four independent dimensionless parameters, Pr, ω^*, Ri, and θ_m. In this research, the properties and amplitude of rolling are fixed, so ω^* and Ri are the independent variables that regulate the flow and heat transfer behavior. Besides, in the non-Newtonian behavior cases, which will be introduced in detail below, flow-behavior index n, ω^*, and Ri are the three independent dimensionless variables.

Boundary conditions:

$$\begin{aligned}
L/2 < x < L/2, y = H \quad &: \quad T = 293.15\,\text{K}, h = 50\,\text{W}/(\text{m}^2{\cdot}\text{K}) \\
x = -L/2, 0 < y < 2L/3 \quad &: \quad T = 291.15\,\text{K}, h = 1250\,\text{W}/(\text{m}^2{\cdot}\text{K}) \\
x = -L/2, 2L/3 < y < L \quad &: \quad T = 293.15\,\text{K}, h = 50\,\text{W}/(\text{m}^2{\cdot}\text{K}) \\
x = L/2, 0 < y < 2L/3 \quad &: \quad T = 291.15\,\text{K}, h = 1250\,\text{W}/(\text{m}^2{\cdot}\text{K}) \\
x = L/2, 2L/3 < y < L \quad &: \quad T = 293.15\,\text{K}, h = 50\,\text{W}/(\text{m}^2{\cdot}\text{K}) \\
-L/2 < x < L/2, y = 0 \quad &: \quad T = 291.15\,\text{K}, h = 1250\,\text{W}/(\text{m}^2{\cdot}\text{K})
\end{aligned} \tag{18}$$

For easy reference, boundary conditions are also illustrated in Figure 3.

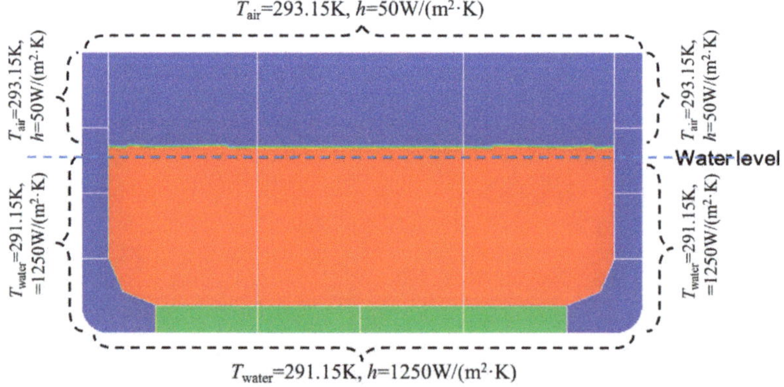

Figure 3. Boundary conditions of the mathematical model.

3. Numerical Methods and Their Validation

3.1. Numerical Methods

The basic equations for the present three-dimensional unsteady rotating cavity flow were solved numerically in the framework of the control-volume method by Ansys-Fluent 17.0 solver. In particular, the power-law scheme was used to discretize Equations (3)–(7) on a non-staggered grid system with the pressure and temperature defined at the mesh centers. The convection terms were discretized by a second-order upstream scheme, and the diffusion terms were discretized by the central differencing scheme. The resulting finite difference equations were solved by the SIMPLE algorithm [20] with a *k-e* model for turbulence effect. To resolve the steep velocity and temperature gradients in the near-wall region, a non-uniform grid shown in Figure 4 was used.

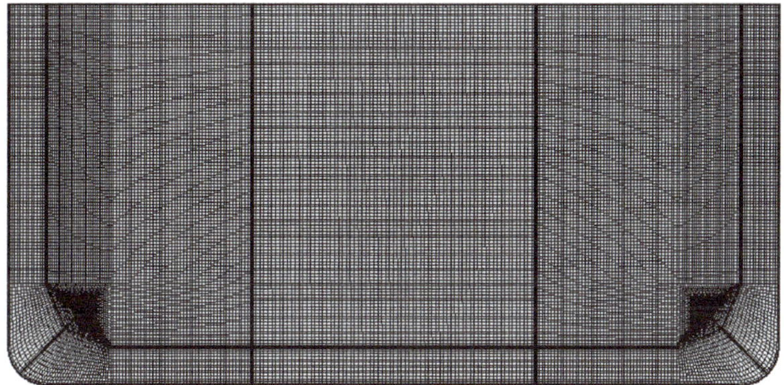

Figure 4. The structural grids generated by ICEM-CFD (Integrated Computer Engineering and Manufacturing code for Computational Fluid Dynamics).

3.2. Grid-Independence Test

A grid-independence test was carried out for a rolling case with $Ri = 144.5$ and $\omega^* = 4.86 \times 10^5$, on three different grids—16,000 cells, 32,000 cells, and 64,000 cells—and the average temperatures are compared in Figure 5. It was found that the results calculated from the grid of 32,000 cells were very close to that calculated from the grid of 64,000 cells. Thus, all the calculations in this paper were performed on the grid of 32,000 cells.

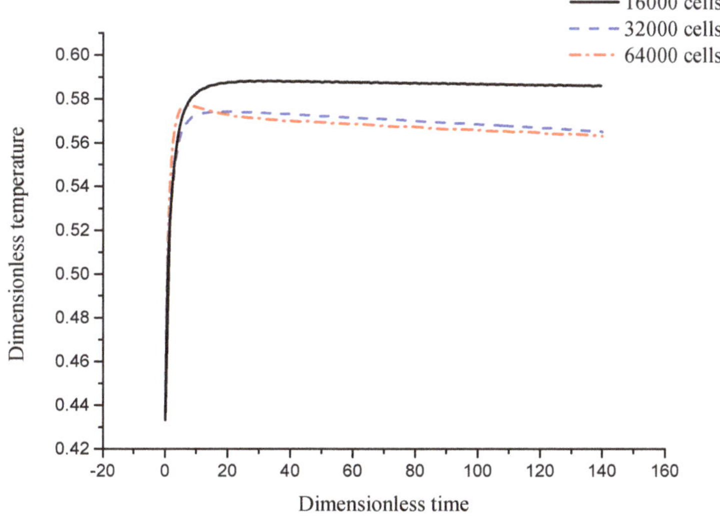

Figure 5. Average temperature calculated on different grid systems.

3.3. Model Validation

The numerical validations were performed with two cases—a partially-filled cavity subjected to a linear acceleration motion for validating the phase interface, and a constant-velocity rotational natural convection case for validation of the heat transfer and flow.

In the first validation case, a partially-filled cavity was accelerated at 3 m/s^2 on a horizontal flat surface. The free surface of the liquid can be determined by an analytical correlation given by Equation (19). The two-phase flow problem was solved by the proposed model at the same time and compared with the analytical correlation, which is shown in Figure 6. In the figure, the black dashed line indicates the analytical results of the free surface, while the contour shows the distribution of the liquid and gas phase. It can be found that the numerical results agree well with the analytical results.

$$y = \arctan\frac{3}{9.8}x + 0.5 = 0.3x + 0.5 \tag{19}$$

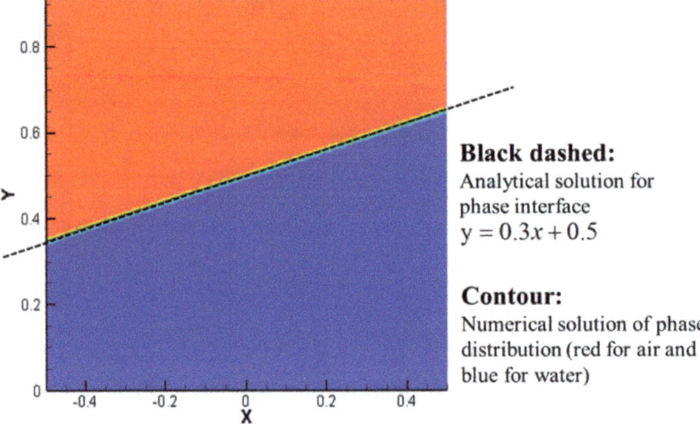

Figure 6. Comparison of phase interface between numerical results and analytical results.

In addition, a natural convection in a rotation system with respect to a horizontal axis, which was an ideal case for the rolling motion in this research, was employed to validate the flow and heat transfer. The numerical results were compared with those calculated by Tso et al. [21] (middle) and the experimental data performed by Saleh et al. [22] (right), as shown in Figure 7. It was found that the numerical results agree well with the results in the literature.

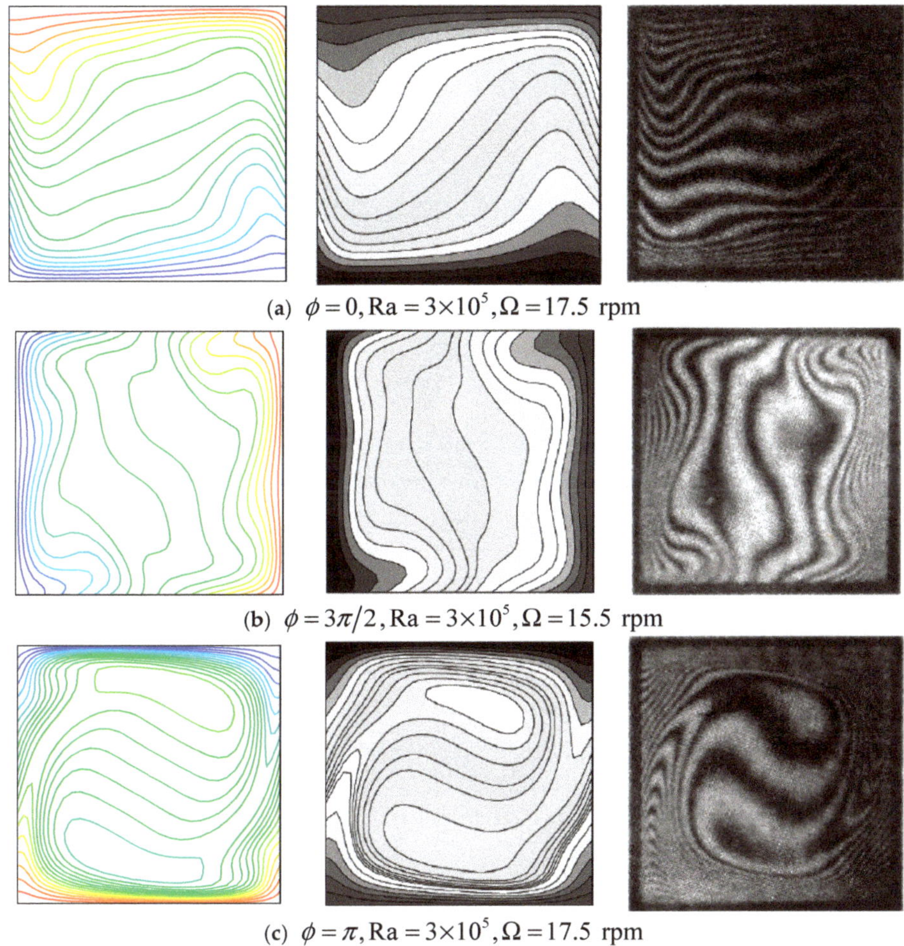

(a) $\phi = 0, \mathrm{Ra} = 3 \times 10^5, \Omega = 17.5$ rpm

(b) $\phi = 3\pi/2, \mathrm{Ra} = 3 \times 10^5, \Omega = 15.5$ rpm

(c) $\phi = \pi, \mathrm{Ra} = 3 \times 10^5, \Omega = 17.5$ rpm

Figure 7. Comparison of computed isotherms between present work (left) with Tso et al. [21] (middle) and Saleh et al. [22] (right) results under (a) $\phi = 0, \mathrm{Ra} = 3 \times 10^5, \Omega = 17.5\,\mathrm{rpm}$, (b) $\phi = 3\pi/2, \mathrm{Ra} = 3 \times 10^5, \Omega = 15.5\,\mathrm{rpm}$, and (c) $\phi = \pi, \mathrm{Ra} = 3 \times 10^5, \Omega = 17.5\,\mathrm{rpm}$.

4. Results and Discussion

It was discussed in Section 2.2 that Pr, ω^*, Ri, and θ_m are four dimensionless parameters regulating the flow and heat transfer behavior. In this research, the properties and amplitude (10 degrees) of rolling were fixed (a medium value with the normal range [4] was selected), so ω^* and Ri were the only two dimensionless independent variables.

When the tanker was subjected to a rolling motion, the surface of the crude oil oscillated accordingly, as shown in Figure 8. Thus, the mechanism of the heat transfer in this research was a mixed convection heat transfer, which can be described by the Richardson number (Ri)—the forced convection dominates the heat transfer when $Ri < 0.1$, natural convection dominates when $Ri > 10$, and both dominate when $0.1 < Ri < 10$. Thus, the selection of Ri ensures the coverage of different ranges. The selection of rotation-strength number ω^* considers the popular angular velocity and tank size in real practice. In this section, 12 cases were designed to investigate effects of different dimensionless parameters as well as non-Newtonian behavior in terms of the flow-behavior index on flow and heat transfer.

The combinations of the dimensionless parameters for case 1 to 12 are listed in Table 1, in which cases 7–12 consider the non-Newtonian behavior, while cases 1–6 do not consider this behavior. In this research, non-Newtonian behavior is described by a power law with a flow-behavior index n; different n indicates different non-Newtonian behavior. It should be noted that case 1 was the reference case that was not subjected to a rolling motion. The designed cases were calculated with the combinations of the dimensionless parameters listed in Table 1.

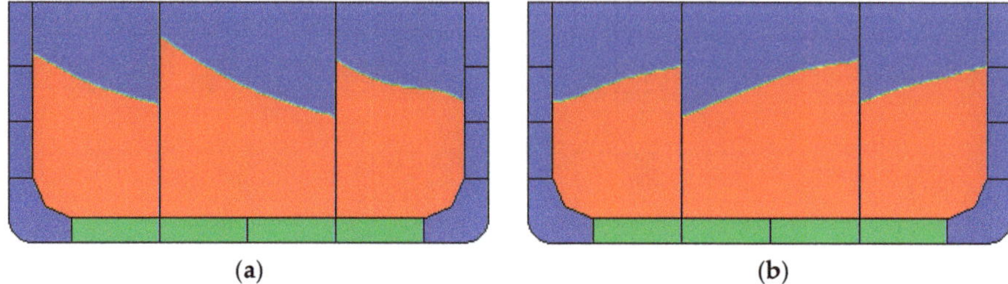

(a)   (b)

Figure 8. The gas-liquid phase distribution images at two time instants in case 2: (**a**) $\tau = 27$, (**b**) $\tau = 54$.

Table 1. The combinations of the dimensionless parameters for cases 1 to 12.

Case	Ri	ω^*	n
Case 1	∞	0	1
Case 2	1	1.67×10^8	1
Case 3	5	1.67×10^8	1
Case 4	10	1.67×10^8	1
Case 5	5	1.67×10^7	1
Case 6	5	1.67×10^6	1
Case 7	5	1.67×10^8	0.5
Case 8	5	1.67×10^8	0.7
Case 9	5	1.67×10^8	0.9
Case 10	∞	0	0.5
Case 11	∞	0	0.7
Case 12	∞	0	0.9

4.1. The Effect of Ri with Fixed ω^*

As discussed at the beginning of Section 4, the Richardson number Ri represents the importance of the natural convection relative to the forced convection, and ω^* characterizes the importance of centrifugal forces due to the rolling motion, relative to diffusivity. It can be concluded that Ri and ω^* have effects on the thermal-hydraulic behavior of the crude oil. In this part, three different Ri, lying in different ranges of mixed convection, were investigated with fixed ω^* in cases 2 to 4. The results were compared with each other and with those of the reference case (case 1).

First, the temperature and the velocity fields at $\tau = 11,500$ are compared in Figure 9. It can be found that the temperature field presents a forced convection dominated distribution when Ri was low. When Ri increases, the flow and heat transfer was gradually dominated by natural convection. Figure 10 compares the dimensionless temperature drop of different Ri. Together with Figure 9, it was found that the temperature drops faster at lower Ri, since forced convection enhances the heat exchange.

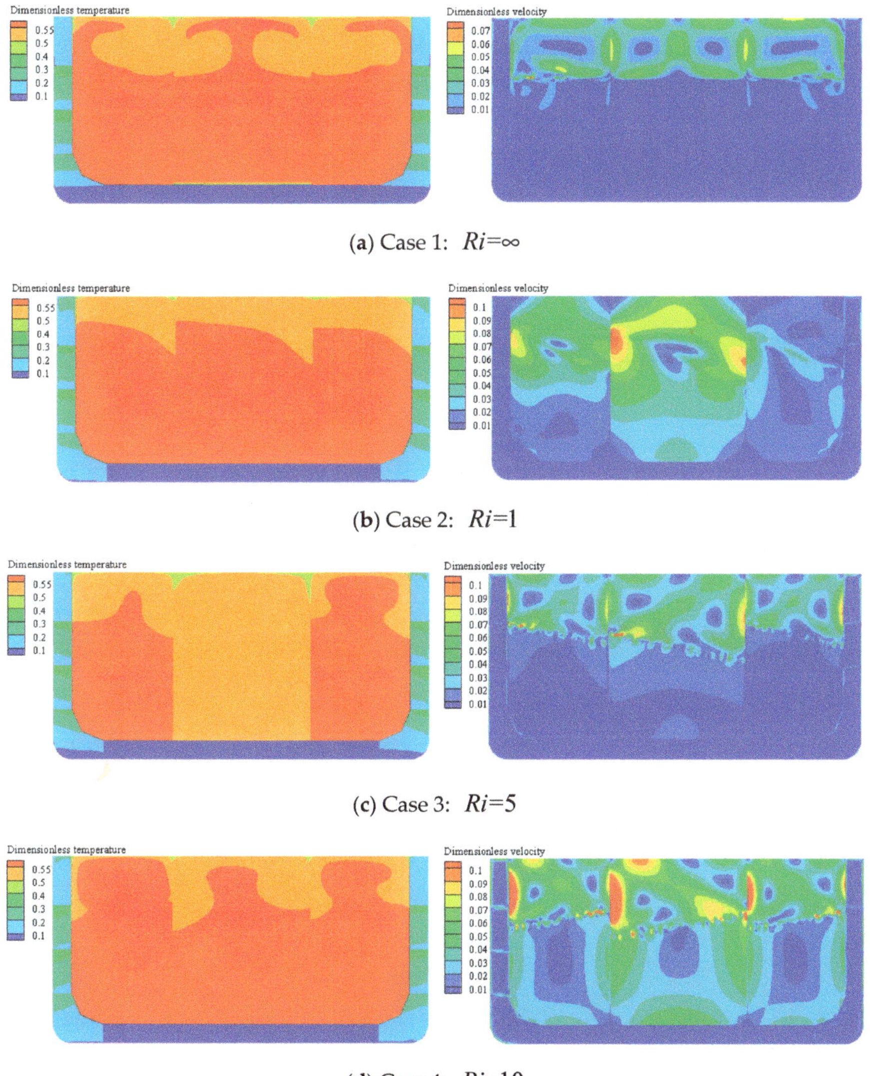

(**a**) Case 1: $Ri=\infty$

(**b**) Case 2: $Ri=1$

(**c**) Case 3: $Ri=5$

(**d**) Case 4: $Ri=10$

Figure 9. The temperature and the velocity fields for cases 1–4 at $\tau = 11,500$: (**a**) Case 1, (**b**) Case 2, (**c**) Case 3, (**d**) Case 4.

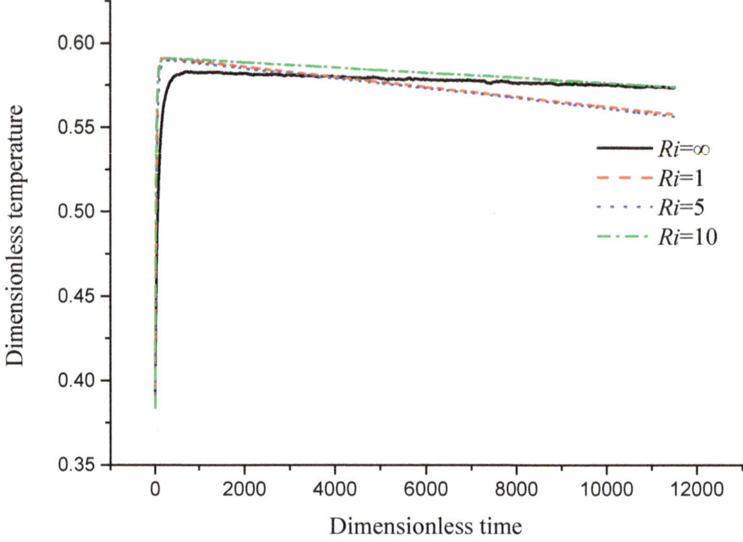

Figure 10. Comparison of temperature drop among cases 1 to 4.

For further analysis of the effect of rolling on the thermal process, the cycle-averaged Nusselt numbers on some typical boundaries shown in Figure 11 at five different time instants were compared among cases 1 to 4 in Figure 12. It was found that the largest Nusselt number corresponded to the lowest Ri, and the smallest Nusselt number corresponded to the largest Ri. However, for moderate Ri, such an obvious characteristic was not found, since the influence of angle effect on natural convection plays an inevitable role.

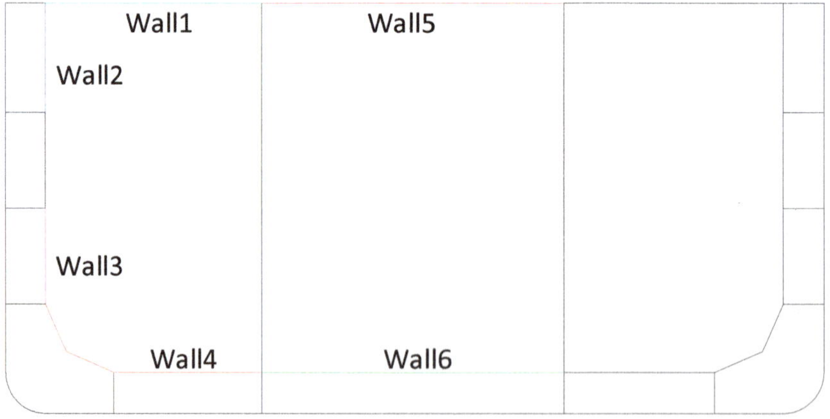

Figure 11. Walls under observation for Nusselt numbers.

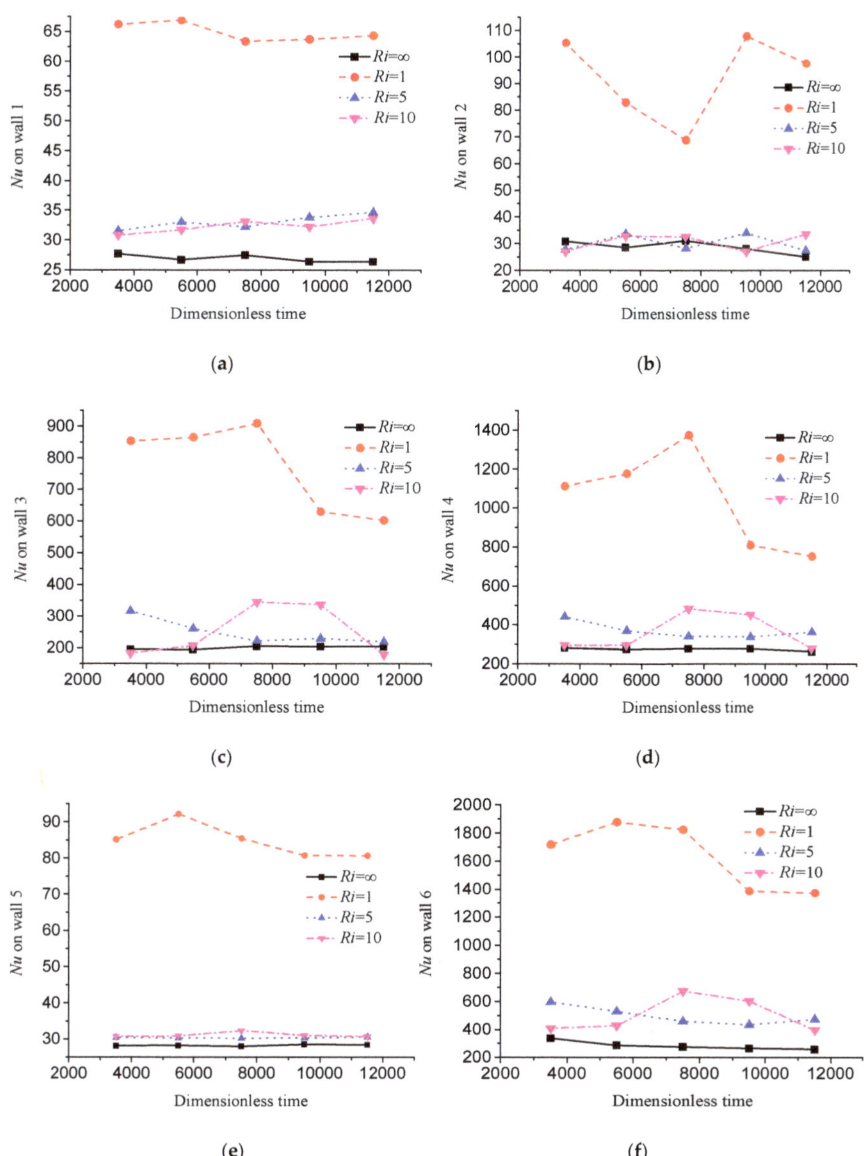

Figure 12. The *Nu* number on different walls in cases 1, 2, 3, and 4: (**a**) *Nu* on wall 1, (**b**) *Nu* on wall 2, (**c**) *Nu* on wall 3, (**d**) *Nu* on wall 4, (**e**) *Nu* on wall 5, (**f**) *Nu* on wall 6.

4.2. The Effect of ω^* with Fixed Ri

For the rolling motion encountered by the oil tanker, the rotational angular velocity is small and usually changes in a narrow range. Therefore, large variation in ω^* is usually caused by the change of geometry size. Therefore, ω^* is dominated by the geometry size, which is different from a small-size model. In this part, three different ω^* were investigated with fixed *Ri* in cases 3, 5, and 6. The results were compared with each other, and some instantaneous temperature and velocity fields are shown in Figure 13. It was shown that the temperature drops faster at lower ω^*, since lower ω^* corresponded to smaller geometry size within a small angular velocity range, and thus led to a faster temperature drop. Figure 14 further indicates that lower ω^* corresponded to a faster temperature drop.

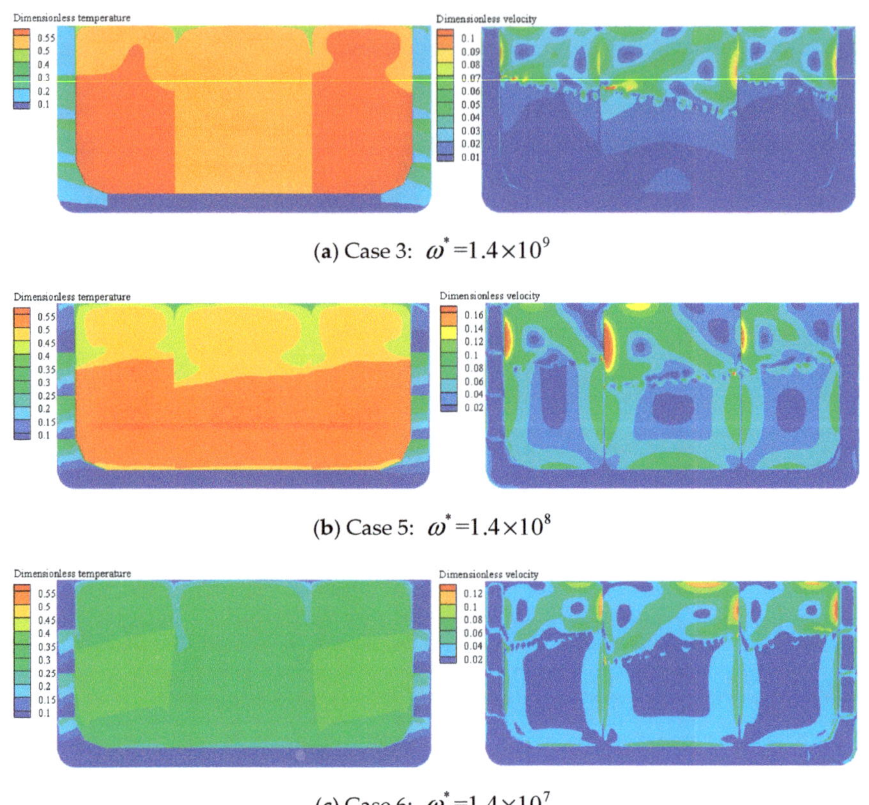

(**a**) Case 3: $\omega^* = 1.4 \times 10^9$

(**b**) Case 5: $\omega^* = 1.4 \times 10^8$

(**c**) Case 6: $\omega^* = 1.4 \times 10^7$

Figure 13. The temperature field and the velocity field for cases 3, 5, and 6 at $\tau = 11{,}500$: (**a**) Case 1, (**b**) Case 2, (**c**) Case 3.

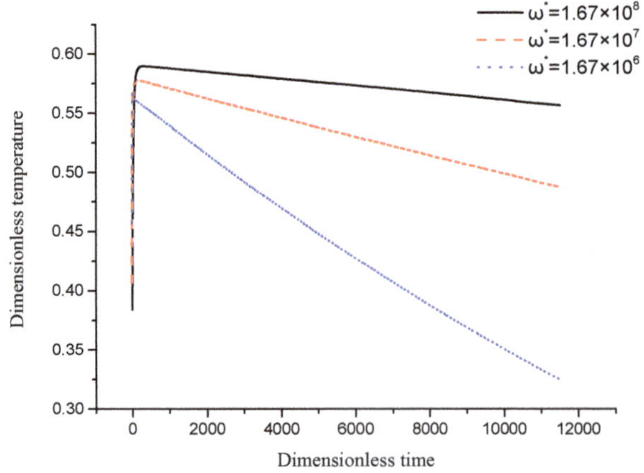

Figure 14. Comparison of temperature drop among cases 3, 5, and 6.

The comparisons of boundary Nusselt numbers are shown in Figure 15. It was also found that lower ω^* corresponded to a larger Nusselt number.

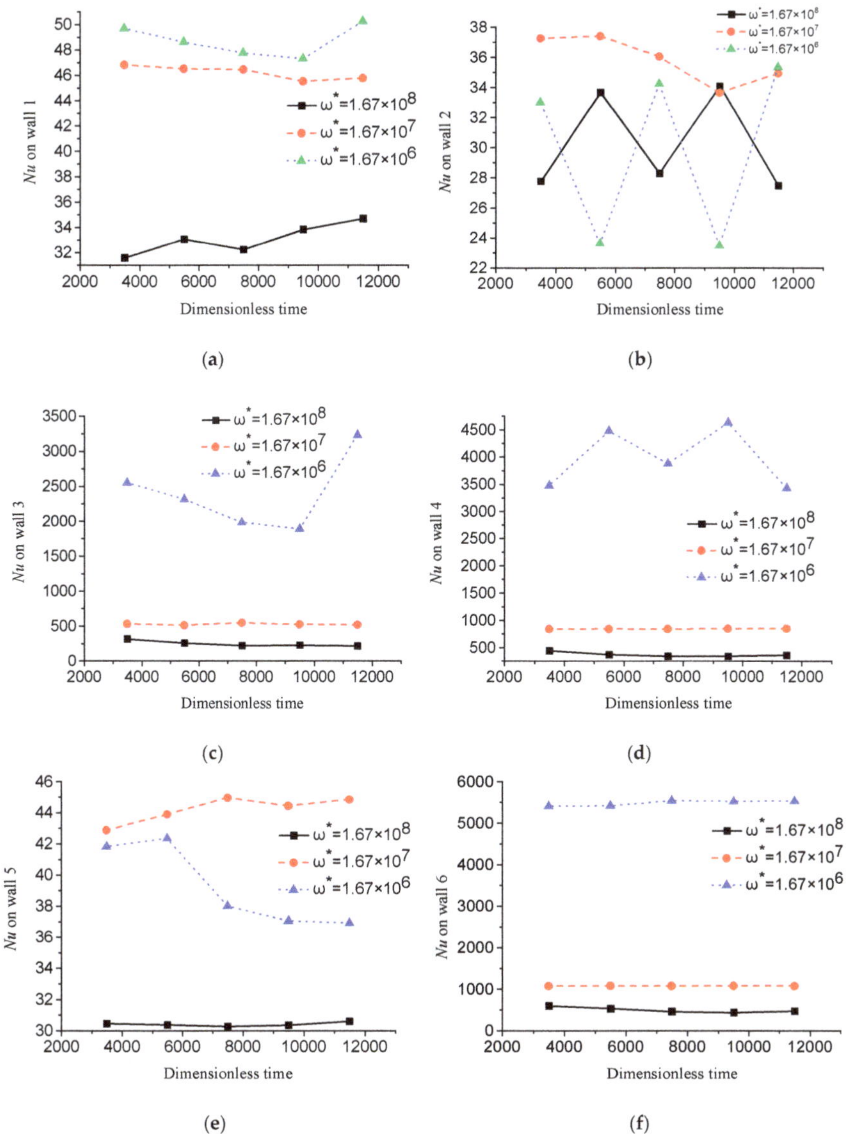

Figure 15. Comparison of boundary Nusselt numbers between cases 3, 5, and 6: (**a**) *Nu* on wall 1, (**b**) *Nu* on wall 2, (**c**) *Nu* on wall 3, (**d**) *Nu* on wall 4, (**e**) *Nu* on wall 5, (**f**) *Nu* on wall 6.

4.3. The Effect of Non-Newtonian Behavior with Fixed ω^* and Ri

When the fluid was subjected to a rolling motion, the flow velocity was larger, which can be found in Figures 9 and 13. Thus, the shear rate in the rolling case should be large, which influences the viscosity of non-Newtonian fluids. In this part, the effect of different non-Newtonian behavior, represented by the flow-behavior index *n*, was studied for the static case and the rolling case, and the influence of non-Newtonian behavior on thermal-hydraulic process was clarified. The viscosity of non-Newtonian behavior is described by a power law shown in Equation (20).

$$\mu = \begin{cases} 0.07, & \gamma \leq 1 \\ 0.07\gamma^{n-1}, & \gamma > 1 \end{cases} \tag{20}$$

Figure 16 compares the temperature and the velocity fields at $\tau = 11,500$ among cases 7 to 9 which correspond to the rolling case, and Figure 17 compares those among cases 10 to 12 which correspond to the static case. It was found that with these rolling cases that the stronger the non-Newtonian behavior (lower n), the faster the temperature drop.

Obviously, as n increases the average temperature becomes higher. In case 7, the crude oil flow velocity field is slightly stronger. Comparison of average temperature drop also validates this conclusion. However, it can be found in both Figures 16 and 18 that different non-Newtonian behavior does not induce obvious differences in the thermal process for the static case.

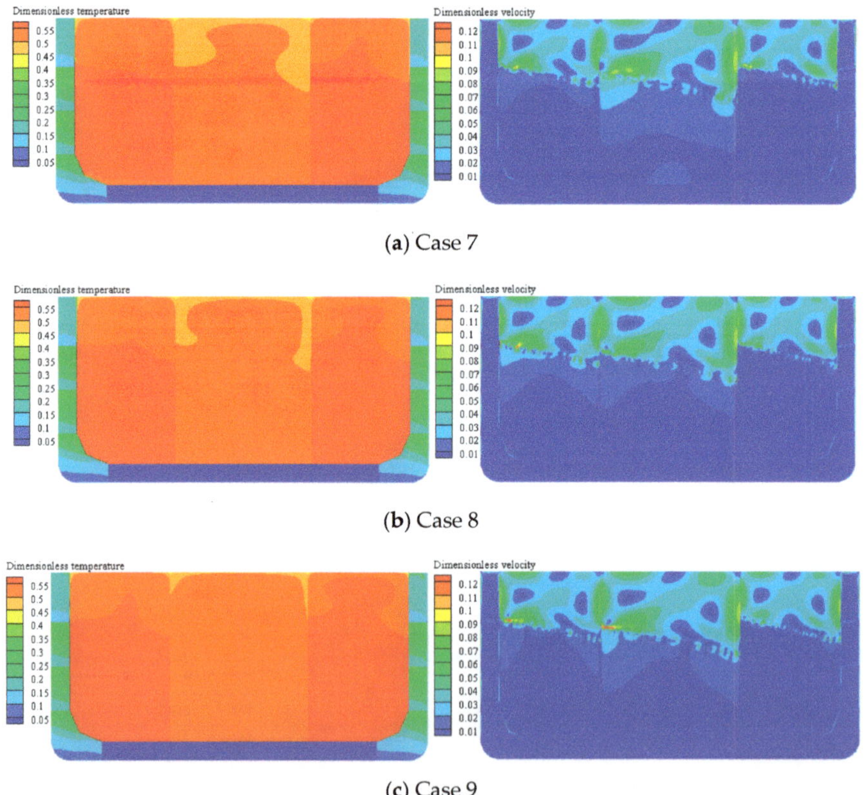

(a) Case 7

(b) Case 8

(c) Case 9

Figure 16. The temperature field and the velocity field (from top to bottom, respectively, are cases 7, 8, and 9): (**a**) Case 7, (**b**) Case 8, (**c**) Case 9.

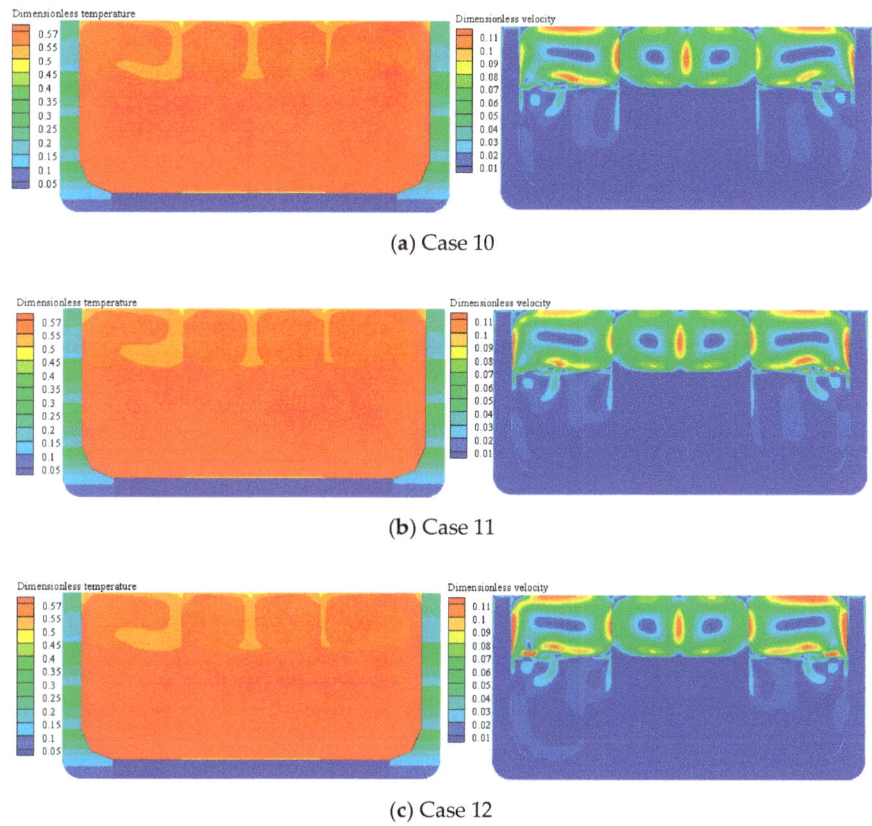

(**a**) Case 10

(**b**) Case 11

(**c**) Case 12

Figure 17. The temperature field and the velocity field (from top to bottom, respectively, are cases 10, 11, and 12): (**a**) Case 10, (**b**) Case 11, (**c**) Case 12.

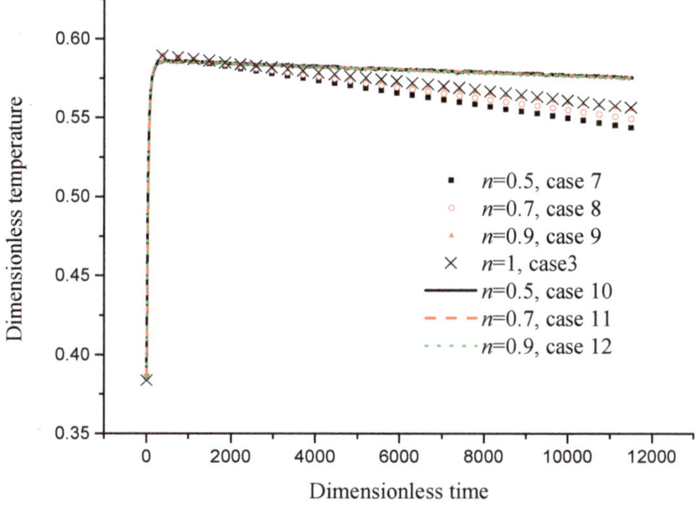

Figure 18. Comparison of temperature drop among case 2 and cases 7–12.

5. Conclusions

In this paper, the effects of Ri, ω^*, and non-Newtonian behavior on the thermal-hydraulic process of crude oil subjected to a rolling motion were investigated numerically. The following conclusions were obtained with the designed cases:

(1) A rolling motion induces mixed convection, which intensifies the heat transfer depending on the Richardson number Ri; the lower the Ri, the faster the heat transfer.

(2) For a large-size geometry model, such as an oil tanker, the rotation-strength number ω^* has a negative effect on heat transfer, i.e., the larger the ω^*, the weaker the heat transfer, since ω^* is dominated by geometry size.

(3) The total effect of the rolling motion on thermal-hydraulic behavior is determined by the combined effect of Ri and ω^*, the effect of their combination may be different for small-size cases and large-size cases.

(4) Non-Newtonian behavior plays an important role in thermal-hydraulic process of crude oil in rolling cases, and a smaller flow-behavior index corresponds to stronger heat transfer; however its influence in the static case is weak.

Author Contributions: G.Y. directed the research and revised the paper writing. S.J. analyzed the data and wrote the original manuscript. Y.G. carried out the numerical simulation.

Funding: This research was funded by National Science Foundation of China (No. 51606117), and China Postdoctoral Science Foundation funded project (Nos. 2017M621473 and 2018T110396).

References

1.   Wawrzyński, W. Area of the unstable solution of rolling equation-jumps of the oscillations amplitude. *J. KONES* **2018**, *3*, 489–496.

2.   Hiroharu, K. Effects of Rolling on the Heat Transfer from Cango Oils of Tankers. *J. Jpn. Soc. Naval Archit. Ocean Eng.* **2009**, *1969*, 421–430.

3.   Yu, G.J.; Yang, Q.L.; Dai, B.; Fu, Z.G.; Lin, D.L. Numerical Study on the Characteristic of Temperature Drop of Crude Oil in a Model Oil Tanker Subjected to Oscillating Motion. *Energies* **2018**, *11*, 1229. [CrossRef]

4.   Doerffer, S.; Mikielewicz, J. The influence of oscillations on natural convection in ship tanks. *Int. J. Heat Fluid Flow* **1986**, *7*, 49–60. [CrossRef]

5.   Akagi, S.; Kato, H. Numerical analysis of mixed convection heat transfer of a high viscosity fluid in a rectangular tank with rolling motion. *Int. J. Heat Mass Transf.* **1987**, *30*, 2423–2432. [CrossRef]

6.   Tan, S.C.; Su, G.H.; Gao, P.Z. Experimental and theoretical study on single-phase natural circulation flow and heat transfer under rolling motion condition. *Appl. Therm. Eng.* **2009**, *29*, 3160–3168. [CrossRef]

7.   Tan, S.C.; Su, G.H.; Gao, P.Z. Heat transfer model of single-phase natural circulation flow under a rolling motion condition. *Nucl. Eng. Des.* **2009**, *239*, 2212–2216.

8.   Yu, S.; Yan, C.; Wang, J.; Cao, X.; Zhu, Z.; Wang, C. Experimental and numerical study on fluctuations and distributions of fluid temperature under rolling motion conditions. *Ann. Nucl. Energy* **2017**, *110*, 384–399. [CrossRef]

9.   Liu, D.; Tian, W.; Xi, M.; Chen, R.; Qiu, S.; Su, G.H. Study on safety boundary of flow instability and CHF for parallel channels in motion. *Nucl. Eng. Des.* **2018**, *335*, 219–230. [CrossRef]

10. Wu, P.; Shan, J.; Xiang, X.; Zhang, B.; Gou, J.; Zhang, B. The development and application of a sub-channel code in ocean environment. *Ann. Nucl. Energy* **2016**, *95*, 12–22. [CrossRef]

11. Xing, D.; Yan, C.; Sun, L. Flow fluctuation behaviors of single-phase forced circulation under rolling conditions. *Ocean Eng.* **2014**, *82*, 115–122. [CrossRef]

12. Hu, Z.Q.; Wang, S.Y.; Chen, G.; Chai, S.H.; Jin, Y.T. The effects of LNG-tank sloshing on the global motions of FLNG system. *Int. J. Nav. Archit. Ocean* **2017**, *9*, 114–125. [CrossRef]

13. Grotle, E.L.; Æsøy, V. Dynamic modelling of the thermal response enhanced by sloshing in marine LNG fuel tanks. *Appl. Therm. Eng.* **2018**, *135*, 512–520. [CrossRef]

14. Jiang, S.C.; Teng, B.; Bai, W.; Gou, Y. Numerical simulation of coupling effect between ship motion and liquid sloshing under wave action. *Ocean Eng.* **2015**, *108*, 140–154. [CrossRef]
15. Cercos-Pita, J.L.; Bulian, G.; Pérez-Rojas, L.; Francescutto, A. Coupled simulation of nonlinear ship motions and a free surface tank. *Ocean Eng.* **2016**, *120*, 281–288. [CrossRef]
16. Yang, S. *Heat Transfer*, 4th ed.; Higher Education Press: Beijing, China, 2006.
17. Olinde, R. Des lois géometriques qui regissent les déplacements d'un systéme solide dans l'espace, et de la variation des coordonnées provenant de ces déplacement considérées indépendant des causes qui peuvent les produire. *J. Math. Pures Appl.* **1840**, *5*, 380–440.
18. Brackbill, J.U.; Kothe, D.B.; Zemach, C. A continuum method for modeling surface tension. *J. Comput. Phys.* **1992**, *100*, 335–354. [CrossRef]
19. Ker, Y.T.; Lin, T.F. A combined numerical and experimental study of air convection in a differentially heated rotating cubic cavity. *Int. J. Heat. Mass. Trans.* **1996**, *39*, 3193–3210. [CrossRef]
20. Patankar, S.V. *Numerical Heat Transfer and Fluid Flow*; Hemisphere Publishing: Washington, DC, USA, 1980.
21. Tso, C.P.; Jin, L.F.; Tou, S.K.W. Numerical Segregation of the Effects of Body Forces in a Rotating, Differentially Heated Enclosure. *Numer. Heat. Trans. A-Appl.* **2007**, *51*, 85–107. [CrossRef]
22. Saleh, H.; Hashim, I. Numerical Analysis of Nanofluids in Differentially Heated Enclosure Undergoing Orthogonal Rotation. *Adv. Math. Phys.* **2014**, *2014*, 1–11. [CrossRef]

Journal of
*Marine Science
and Engineering*

Article

Numerical Study on the Natural Convection of Air in a Cubic Cavity Subjected to a Yawing Motion

Guojun Yu [1,2], Lizhi Zhang [2], Sheng Jia [2], Yanting Geng [2] and Jie Liu [1,*]

[1] School of Design, Shanghai Jiao Tong University, Shanghai 200240, China
[2] Merchant Marine College, Shanghai Maritime University, Shanghai 201306, China
* Correspondence: jackliu@sjtu.edu.cn

Received: 27 May 2019; Accepted: 27 June 2019; Published: 1 July 2019

Abstract: Natural convections subjected to multi-Degree of Freedom (DoF) motion are much more complex than those in static case, and those subjected to yawing motion are the simplest and ideal case for investigating their characteristics due to less interactive parameters. In this paper, the characteristic of natural convection under yawing motion was studied systematically to clarify the interaction between yawing motion and thermal-dynamic behavior. First of all, the mathematical model was established in a non-inertial coordinate system, and the dimensionless governing equations were derived. Subsequently, the governing equations were discretized in the framework of the finite volume method, and a computer code was developed and validated. After that, the natural convection under yawing motion was calculated with different combinations of dimensionless parameters, and the influence of rotation frequency and amplitude on heat and mass transfer was investigated. It was found that the yawing motion plays a notable role in flow and heat transfer, depending on the relative magnitudes of rotation-induced velocity and buoyancy-induced velocity: At a lower Rayleigh number of 10^4, the Nusselt number on hot boundary is enhanced by approximately 25% when the rotation period is changed from 12 s to 2 s; while the changing in rotation period from 12 s to 2 s did not induce obvious difference in hot-boundary Nusselt number for a higher Rayleigh number of 10^5. It is concluded that the vertical-axis harmonic rotation enhances heat transfer if the rotation-induced velocity dominates the flow. The clarification of natural convection characteristic in yawing motion provides convenience for analyzing that in other multi-DoF systems.

Keywords: natural convection; yawing motion; vertical-axis harmonic rotation; non-inertial coordinate system

---

1. Introduction

Multi-DoF motion is frequently encountered for different types of systems during sea transportation. Natural convection in multi-DoF motion system is frequently encountered with the heat exchangers mounted in marine or vehicles, either in heat exchangers of a ship, or the heating system in an oil tanker, or the circulation pipe in the nuclear reactor core. The six-DoF motions include rolling, pitching, yawing, swaying, heaving, and surging, as shown in Figure 1. Due to the influence of additional forces (Coriolis, tangential, centrifugal forces, etc.) caused by oscillating motion, the natural convection in such a system is more complex than that in a static system. The effects of ocean conditions on the thermal-hydraulic characteristics have attracted interest in recent years, some experiments and numerical simulations have been carried out, however, the simulations were only for some simple cases. Therefore, the general heat transfer characteristic of the system under multi-DoF motion is not clear and needs to be investigated. Natural convection in yawing motion is much simpler than that in other DoF-motion, since there is no angle effect. So in the paper, the characteristic of natural convection in yawing motion will be investigated systematically so as to investigate more complex heat transfer in other DoF motion.

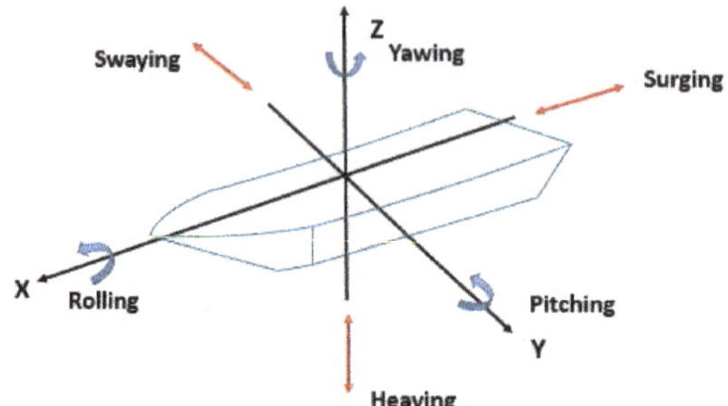

Figure 1. Six-Dof movements of a ship during sea transportation.

1.1. Experimental and Numerical Simulations for Specific Applications

Most of the researches regarding the effect of harmonic rotation motion on heat and mass transfer were by experiments and confined to some specialized system (i.e., there were not fundamental researches). Although a few numerical researches were found in the open literature, they were limited to one-dimensional or two-dimensional cases, which were not sufficient to analyze the detailed flow and heat transfer behavior involved in the multi-DoF system. A large proportion of these researches concerned the flow and heat transfer in pipe flow or channel flow of a certain system under rolling or pitching motion. The experimental studies of Murata et al. [1] and Tan et al. [2] showed that the heat transfer coefficient of rolling conditions is greater than that of the non-rolling case, and the heat transfer coefficient of the rolling case increases with the increase in rolling amplitude and frequency. Yan [3,4] studied the heat transfer with laminar pulsating flow in a channel or tube in rolling motion, showing that the oscillating amplitude of Nusselt number increases with the increase of Prandtl number, and there is an approximate linear relation between the oscillating amplitude of Nusselt number and rolling frequency. Wang et al. [5] experimentally studied the single-phase heat transfer characteristics of pulsating flows induced by rolling motion in a circular pipe, indicating that the relative pulsation amplitude of Nusselt number increases with the increase of the maximum rolling angle, and decreases with the increase of rolling period and cycle-averaged flow rate, while the pulsating flow induced by rolling motion does not lead to significant variations in cycle-averaged heat transfer characteristics. Yuan et al. [6] found from their experiment that the average Nusselt number increases with the rolling amplitude and frequency in transitional flow regime, however rolling motion does not affect the Nusselt number in turbulent regime obviously. Yu [7] also found that rolling motion has influence on heat and fluid flow, but this influence is also dependent of the Reynolds number or flow regime. Yan's numerical investigation [8] on the flowing and heat transfer characteristics of turbulent flow in tubes and rectangular channels in rolling motion also indicate that the Nusselt number is affected by rolling frequency, but the degree of influence also depend on the flow velocity.

1.2. Fundamental Researches

Fundamental study on the natural convection in rotation systems mainly concerned the constant-velocity rotation system, which is a special case for the harmonic rotation in this research. In these researches, the cavities rotate about either a horizontal axis [9–15] (orthogonal to the gravity axis) or a vertical axis [16–19]. These vertical rotations are the special cases (constant angular velocity) of yawing motion, which will be investigated in this study. It was reported in [15] that the effect of Coriolis force and centrifugal force are small but are differentiated from those of other forces. Baig [9] studied the natural convection in the horizontal rotating and differentially heated square enclosure, indicating that significant enhancement in heat transfer is achieved due to rotational effects. Jin [11] also showed that rotation enhances the heat transfer. Hamidy's experimental and numerical research showed that that the Coriolis and centrifugal buoyancy forces arising from rotation have a remarkable influence on the local heat transfer when compared with the nonrotating results [10].

However, Tso [15] found that the effects of Coriolis force and centrifugal force on the natural convection are small by their numerical investigation on the natural convection in a horizontal-axis rotation cavity. Kumar [13] numerically studied the effect of Rayleigh number with rotation on the flow and heat transfer characteristics in a differentially heated enclosure rotating about the horizontal axis. For the vertical-axis rotation cases, it is also found that rotation has effect influence on the natural convection. Ker's results [18] clearly revealed that significant reduction of heat transfer only occurs at low rotation rate, while high-speed rotation results in increase of heat transfer. Lee [19] numerically studied the transient three-dimensional mixed convection of air in a differentially heated vertical cubic cavity rotating about a vertical axis through the cavity center. The centrifugal and Coriolis forces were found to exhibit significant effects on the flow and heat transfer in the cavity when they are high enough. It was also found that the thermal buoyancy driven flow is strengthened at increasing centrifugal force but weakened in the near-wall regions by the increasing Coriolis force. The space average heat transfer from the isothermal plates is substantially reduced by the Coriolis force but is insensitive to the centrifugal force. Chori [16] investigated the laminar natural convection heat transfer in three-dimensional molten Lithium filled differentially heated enclosure, rotating about the vertical ridge is studied numerically. It was found that increase of the Taylor number significantly reduce the heat transfer, however, high rotational Rayleigh number values are found to significantly enhance the heat transfer. These constant-angular-velocity vertical rotations are the special cases of the yawing motion, which will be investigated in this study.

It is concluded from the above analysis that there are some experiment researches focusing on the natural convection undergoing oscillations, but these researches concerned some specialized cases which are not convenient for extrapolation and reference for application in different research area, and that numerical studies were one-dimensional or two-dimensional, and yawing motion cannot be described. Although the investigation for constant-velocity rotation is very mature and sufficient, large difference exists with the harmonic rotation. The harmonic rotation has some influencing factors such as Eulerian force and time-vary Coriolis force, so the characteristic should be much different. In this paper, a three-dimensional mathematical model for the natural convection under yawing motion will be established first of all, and the computer code will be developed and validated thereafter. With the validated code, the thermal and hydraulic characteristic of natural convection in a vertical-axis rotation system will be investigated; the influencing factors and sensitivity of natural convection to which will be clarified. This fundamental research should provide an effective reference for researchers in different research areas, since it is easy to extend to other fields, and the results are easy for reference in a different area.

2. Mathematical Modeling and Solution Method

2.1. Mathematical Model

The schematic representation of the physical system under investigation is shown in Figure 2. A cubic cavity which contains air as the working fluid is considered. Initially at time $t = 0$, the enclosure is rolling about the z axis at $\varphi = 0$ and is isothermal at temperature T_0. At time $t > 0$, the wall temperatures of the enclosure are suddenly raised and lowered to some uniform constant temperature conditions in the following manner. The surface at $y = 0.5L$ is subjected to a uniform temperature T_h and the surface at $y = -0.5L$, is subjected to a different uniform temperature T_c. The rest surfaces are thermally well insulated from the surroundings.

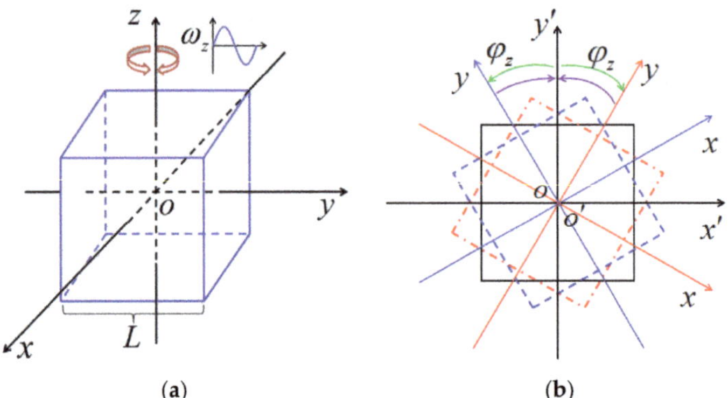

Figure 2. Schematic of the vertical-axis harmonic rotation: (**a**) 3d model, (**b**) top view.

It is usually assumed that yawing motion is set as a sine function, i.e., $\varphi = \varphi_m \sin(2\pi/t_c t)$, where φ is the angular position of the enclosure, φ_m is the amplitude of the rolling, and t_c is the period of the harmonic rotation. It should be noted that, the angular velocity is defined positive when it rotates counterclockwise, and the φ is defined positive in the first half of the cycle, and negative in the second half of the cycle.

Thus the air flow inside the enclosure is simultaneously driven by the harmonic rotation and the thermal buoyancy. By using the Boussinesq approximation [20] in which linear density variation with temperature (i.e., $\rho = \rho_0[1 - \beta(T - T_0)]$) is considered in both the body force and centrifugal force terms, the governing equations and boundary conditions become

$$\frac{\partial(\rho_0 u)}{\partial x} + \frac{\partial(\rho_0 v)}{\partial y} + \frac{\partial(\rho_0 w)}{\partial z} = 0 \tag{1}$$

$$\frac{\partial(\rho_0 u)}{\partial t} + \frac{\partial(\rho_0 uu)}{\partial x} + \frac{\partial(\rho_0 uv)}{\partial y} + \frac{\partial(\rho_0 uw)}{\partial z} = -\frac{\partial p_{eff}}{\partial x} + \frac{\partial}{\partial x}\left(\mu\frac{\partial u}{\partial x}\right) + \frac{\partial}{\partial y}\left(\mu\frac{\partial u}{\partial y}\right) + \frac{\partial}{\partial z}\left(\mu\frac{\partial u}{\partial z}\right)$$
$$- \rho_0\Omega_z^2 x\beta(T - T_0) + \rho_0\frac{d\Omega_z}{dt}y + 2\rho_0\Omega_z v - \rho_0\frac{d\Omega_z}{dt}y\beta(T - T_0) - \rho_0\Omega_z v\beta(T - T_0) \tag{2}$$

$$\frac{\partial(\rho_0 v)}{\partial t} + \frac{\partial(\rho_0 vu)}{\partial x} + \frac{\partial(\rho_0 vv)}{\partial y} + \frac{\partial(\rho_0 vw)}{\partial z} = -\frac{\partial p_{eff}}{\partial y} + \frac{\partial}{\partial x}\left(\mu\frac{\partial v}{\partial x}\right) + \frac{\partial}{\partial y}\left(\mu\frac{\partial v}{\partial y}\right) + \frac{\partial}{\partial z}\left(\mu\frac{\partial v}{\partial z}\right)$$
$$- \rho_0\Omega_z^2 y\beta(T - T_0) - \rho_0\frac{d\Omega_z}{dt}x - 2\rho_0\Omega_z u + \rho_0\frac{d\Omega_z}{dt}x\beta(T - T_0) + 2\rho_0\Omega_z u\beta(T - T_0) \tag{3}$$

$$\frac{\partial(\rho_0 w)}{\partial t} + \frac{\partial(\rho_0 wu)}{\partial x} + \frac{\partial(\rho_0 wv)}{\partial y} + \frac{\partial(\rho_0 ww)}{\partial z} = -\frac{\partial p_{eff}}{\partial z} + \frac{\partial}{\partial x}\left(\mu\frac{\partial w}{\partial x}\right) + \frac{\partial}{\partial y}\left(\mu\frac{\partial w}{\partial y}\right) + \frac{\partial}{\partial z}\left(\mu\frac{\partial w}{\partial z}\right)$$
$$+ \rho_0 g\beta(T - T_0) \tag{4}$$

$$\frac{\partial p_{eff}}{\partial x} = \frac{\partial p}{\partial x} - \rho_0\Omega_z^2 x, \ \frac{\partial p_{eff}}{\partial y} = \frac{\partial p}{\partial y} - \rho_0\Omega_z^2 y, \ \frac{\partial p_{eff}}{\partial z} = \frac{\partial p}{\partial z} - \rho_0 g \tag{5}$$

$$p_{eff} = p - \rho_0\left[xy\frac{d\Omega_z}{dt} + 0.5\Omega_z^2(x^2 + y^2) + gz\right] \tag{6}$$

Energy equation:

$$\frac{\partial(\rho_0 T)}{\partial t} + \frac{\partial(\rho_0 uT)}{\partial x} + \frac{\partial(\rho_0 vT)}{\partial y} + \frac{\partial(\rho_0 wT)}{\partial z} = \frac{\partial}{\partial x}\left(\frac{\lambda}{c_p}\frac{\partial T}{\partial x}\right) + \frac{\partial}{\partial y}\left(\frac{\lambda}{c_p}\frac{\partial T}{\partial y}\right) + \frac{\partial}{\partial z}\left(\frac{\lambda}{c_p}\frac{\partial T}{\partial z}\right) \tag{7}$$

In Equations (1)~(7), u, v, and w are three velocity components defined on the coordinates x, y, and z rotating with the cavity, as shown in Figure 2, Ω denotes the time-varying rotation angular velocity, T represents the temperature of the fluid, λ and c_p are thermal conductivity and heat capacity respectively, p_{eff} is the effective pressure, ρ_0 and β are reference density and thermal expansion factor of the air respectively. The forces on the right-hand side of Equations (2) and (3) will be explained in Equations (10) and (11).

The corresponding initial and boundary conditions are

$$
\begin{aligned}
t < 0, \quad & u = 0, v = 0, w = 0 \ and \ T = T_0 = T_h \ for \ all \ x, y, z \\
t \geq 0, \quad & x = L/2 & u = v = w = 0, \tfrac{\partial T}{\partial x} = 0 \\
& x = -L/2 & u = v = w = 0, \tfrac{\partial T}{\partial x} = 0 \\
& y = L/2 & u = v = w = 0, T = T_h \\
& y = -L/2 & u = v = w = 0, T = T_c \\
& z = L/2 & u = v = w = 0, \tfrac{\partial T}{\partial z} = 0 \\
& z = -L/2 & u = v = w = 0, \tfrac{\partial T}{\partial z} = 0
\end{aligned}
\tag{8}
$$

To derive the non-dimensionless governing equations, the following non-dimensional variables are defined.

$$
Y = \tfrac{x}{L}, \ Y = \tfrac{y}{L}, \ Z = \tfrac{z}{L}, \ U = \tfrac{uL}{\alpha}, \ V = \tfrac{vL}{\alpha}, \ W = \tfrac{wL}{\alpha}, \ \Theta = \tfrac{T-T_c}{T_h-T_c}, \ \tau = \tfrac{t\alpha}{L^2},
$$
$$
P = p_{eff} L^2 / \rho_0 \alpha^2, \ \widetilde{Ta} = \tfrac{4(2\pi/t_c)^2 L^4}{\nu^2}, \ Ra = \tfrac{g\beta(T_h-T_c)L^3}{\nu\alpha}, \ \widetilde{Ra_\omega} = \tfrac{(2\pi/t_c)^2 H\beta(T_h-T_c)L^3}{\nu\alpha}.
$$

Here, X, Y and Z indicate the dimensionless coordinates; U, V and W are dimensionless velocity profiles; α is the thermal diffusivity ($\alpha = \lambda/(\rho c_p)$), Θ denotes the dimensionless temperature; $\widetilde{Ta}$ is defined as the nominal Taylor number describing the general strength of the harmonic rotation motion, in a similar way with the definition of Taylor number for constant-velocity rotation; Ra_ω is defined as the nominal rotation Rayleigh number in the harmonic rotation system.

With the dimensionless parameters defined above, the governing equations, initial and boundary conditions then become

$$
\frac{\partial U}{\partial X} + \frac{\partial V}{\partial Y} + \frac{\partial W}{\partial Z} = 0
\tag{9}
$$

$$
\frac{\partial U}{\partial \tau} + \frac{\partial (UU)}{\partial X} + \frac{\partial (UV)}{\partial Y} + \frac{\partial (UW)}{\partial Z} = -\frac{\partial P}{\partial X} + Pr\left(\frac{\partial^2 U}{\partial X^2} + \frac{\partial^2 U}{\partial Y^2} + \frac{\partial^2 U}{\partial Z^2}\right)
$$
$$
\underbrace{-PrRa_\omega \Theta X \varphi_m^2 \cos^2\!\left(0.5\widetilde{Ta}^{0.5} Pr\tau\right)}_{centrifugal\ buoyancy\ force} \underbrace{-0.25\widetilde{Ta}Pr^2 Y \varphi_m \sin\!\left(0.5\widetilde{Ta}^{0.5} Pr\tau\right)}_{tangential\ force}
$$
$$
\underbrace{+\widetilde{Ta}^{0.5} Pr V \varphi_m \cos\!\left(0.5\widetilde{Ta}^{0.5} Pr\tau\right)}_{Coriolis\ force} + \underbrace{PrRa_\omega \Theta Y \varphi_m \sin\!\left(0.5\widetilde{Ta}^{0.5} Pr\tau\right)}_{tangential\ buoyancy\ force}
$$
$$
\underbrace{-4\widetilde{Ta}^{-0.5} Ra_\omega \Theta V \varphi_m \cos\!\left(0.5\widetilde{Ta}^{0.5} Pr\tau\right)}_{Coriolis\ buoyancy\ force}
\tag{10}
$$

$$
\frac{\partial V}{\partial \tau} + \frac{\partial (VU)}{\partial X} + \frac{\partial (VV)}{\partial Y} + \frac{\partial (VW)}{\partial Z} = -\frac{\partial P}{\partial Y} + Pr\left(\frac{\partial^2 V}{\partial X^2} + \frac{\partial^2 V}{\partial Y^2} + \frac{\partial^2 V}{\partial Z^2}\right)
$$
$$
\underbrace{-PrRa_\omega \Theta Y \varphi_m^2 \cos^2\!\left(0.5\widetilde{Ta}^{0.5} Pr\tau\right)}_{centrifugal\ boyancy\ force} \underbrace{+0.25\widetilde{Ta}Pr^2 X \varphi_m \sin\!\left(0.5\widetilde{Ta}^{0.5} Pr\tau\right)}_{tangential\ force}
$$
$$
\underbrace{-\widetilde{Ta}^{0.5} Pr U \varphi_m \cos\!\left(0.5\widetilde{Ta}^{0.5} Pr\tau\right)}_{Coriolis\ force} \underbrace{-PrRa_\omega \Theta X \varphi_m \sin\!\left(0.5\widetilde{Ta}^{0.5} Pr\tau\right)}_{tangential\ buoyancy\ force}
$$
$$
\underbrace{+4\widetilde{Ta}^{-0.5} Ra_\omega \Theta U \varphi_m \cos\!\left(0.5\widetilde{Ta}^{0.5} Pr\tau\right)}_{Coriolis\ buoyancy\ force}
\tag{11}
$$

$$
\frac{\partial W}{\partial \tau} + \frac{\partial (WU)}{\partial X} + \frac{\partial (WV)}{\partial Y} + \frac{\partial (WW)}{\partial Z} = -\frac{\partial P}{\partial Z} + Pr\left(\frac{\partial^2 W}{\partial X^2} + \frac{\partial^2 W}{\partial Y^2} + \frac{\partial^2 W}{\partial Z^2}\right)
$$
$$
+PrRa\Theta
\tag{12}
$$

$$\frac{\partial(\Theta)}{\partial\tau} + \frac{\partial(U\Theta)}{\partial X} + \frac{\partial(V\Theta)}{\partial Y} + \frac{\partial(W\Theta)}{\partial Z} = \frac{\partial^2\Theta}{\partial X^2} + \frac{\partial^2\Theta}{\partial Y^2} + \frac{\partial^2\Theta}{\partial Z^2} \tag{13}$$

Accordingly, the dimensionless boundary conditions become

$$
\begin{aligned}
t < 0, &\quad U = 0, V = 0, W = 0 \ and \ \Theta = 0 \ for \ all \ X, Y, Z \\
t \geq 0, &\quad X = 0.5 &\quad U = V = W = 0, \ \tfrac{\partial\Theta}{\partial X} = 0 \\
&\quad X = -0.5 &\quad U = V = W = 0, \ \tfrac{\partial\Theta}{\partial X} = 0 \\
&\quad Y = 0.5 &\quad U = V = W = 0, \ \Theta = 1 \\
&\quad Y = -0.5 &\quad U = V = W = 0, \ \Theta = 0 \\
&\quad Z = 0.5 &\quad U = V = W = 0, \ \tfrac{\partial\Theta}{\partial Z} = 0 \\
&\quad Z = -0.5 &\quad U = V = W = 0, \ \tfrac{\partial\Theta}{\partial Z} = 0
\end{aligned}
\tag{14}
$$

The above formulations indicate that the flow and heat transfer to be examined is governed by the Prandtl number Pr, nominal Taylor number $\widetilde{Ta}$, Rayleigh number Ra and nominal rotational Rayleigh number $\widetilde{Ra_\omega}$, here $\widetilde{Ra_\omega} = Ra(2\pi/t_c)^2 L/g$. Hence the rotational buoyancy becomes important when the rotating speed is high or when cavity size is large, that is when $(2\pi/t_c)^2 L$ is much larger than g. In addition to examining the time evolution of the velocity and temperature fields, results for the local and space-average Nusselt numbers Nu and $\overline{Nu}$ on the heated or cooled plates are important in thermal design and can be evaluated from Equations (15) and (16), respectively.

$$\mathrm{Nu} = -\left.\frac{\partial\Theta}{\partial Z}\right|_{Z=Z_{\min}, Z=Z_{\min}} \tag{15}$$

$$\overline{\mathrm{Nu}} = \int_{Y_{\min}}^{Y_{\max}} \int_{X_{\min}}^{X_{\max}} \frac{\partial\Theta}{\partial Z} dX dY. \tag{16}$$

2.2. Solution Method

The highly nonlinear and coupled Equations (9)–(13), with the corresponding boundary conditions, Equation (12), are discretized in the framework of the finite volume method on a 3D non-staggered grid shown in Figure 3. To resolve the steep velocity and temperature gradients in the near-wall region, non-uniform grid was used. Specifically, the grid lines passing through volume size in each direction satisfies a sine function, XCV(I) = 1.0×SIN(FLOAT(I)/L1×3.14159), and then XCV(I) = XCV(I) ×∑XCV(I)/, here L1is the number of the nodes in each dimension (x, y, z) and I is the sequence number of the nodes. The convection terms are discretized by a stability-guaranteed second-order difference (SGSD) scheme [21], and the diffusion terms are discretized by the central differencing scheme. The resulting discretized equations were solved by a computer code which was developed by the author based on the IDEAL algorithm [22], which is found to show good convergence.

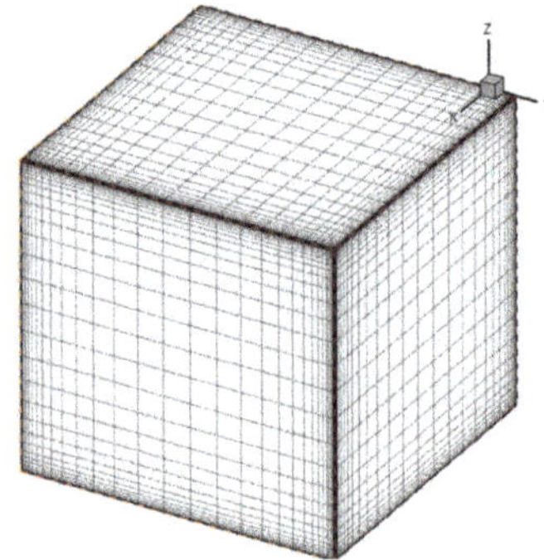

Figure 3. Grid system employed in the calculation.

2.3. Code Validation

In view of the complex flow and heat transfer to be investigated, stringent validation tests should be performed to verify the computer code developed. For this purpose, our numerical code is validated against well-known Ansys Fluent results in yawing motion and against those reported in the literature regarding the natural convection flow within cubic enclosure filled with air undergoing vertical-axis rotation with constant-velocity. The vertical-axis rotation with constant-velocity is a special case of the harmonic rotation in this research.

First, the code results were compared with Ansys Fluent results calculated based on dynamic mesh, with $Ra = 1 \times 10^6$ and $t_c = 1s$. Since dynamic mesh method and inertial coordinate system method are two independent methods, this validation is meaningful. Figure 4 shows the comparison between code results and Ansys Fluent results, in which the left column corresponds to the Ansys Fluent results and right column corresponds to the code results. In this figure, (+) and (-) indicates the clockwise rotation and counterclockwise rotation respectively. Good agreement can be found from this comparison. In addition, similar numerical simulations are performed with the same parameters in Lee's research [19]. The Nusselt number on both hot boundary and cold boundary are compare in Table 1. It is found that both results are in good agreement.

Table 1. Comparison of computed Nusselt number at hot wall between the present work and those from Lee and Lin [19].

	$\overline{Nu}$ at $X = 0.5$		$\overline{Nu}$ at $X = -0.5$	
τ	**Present**	**Lee and Lin [19]**	**Present**	**Lee and Lin [19]**
0.05	4.712	4.727	3.698	3.625
0.10	4.172	4.152	3.872	3.834
0.15	4.007	3.987	3.925	3.894
0.20	3.956	3.939	3.942	3.912
Steady-state	3.933	3.926	3.949	3.916

With the validation above, the results of the numerical procedure proposed in this research is believable. With the validated code, the test cases in Table 2 will be studied one by one. The results and findings are discussed in the rest part of this paper.

Table 2. Parameters in calculation.

Cases	Ra	t_c	$\widetilde{Ra_w}$	$\widetilde{Ta}$	φ_m
1	1×10^4	2 s	512	1,109,892	$\pi/4$
2	1×10^4	6 s	57	123,321	$\pi/4$
3	1×10^4	12 s	14	30,830	$\pi/4$
4	1×10^6	2 s	51161	1,109,892	$\pi/4$
5	1×10^6	6 s	5985	123,321	$\pi/4$
6	1×10^6	12 s	1421	30,830	$\pi/4$
7	1×10^4	2 s	512	1,109,892	$\pi/8$
8	1×10^4	6 s	57	123,321	$\pi/8$
9	1×10^4	12 s	14	30,830	$\pi/8$
10	1×10^6	2 s	51161	1,109,892	$\pi/8$
11	1×10^6	6 s	5985	123,321	$\pi/8$
12	1×10^6	12 s	1421	30,830	$\pi/8$

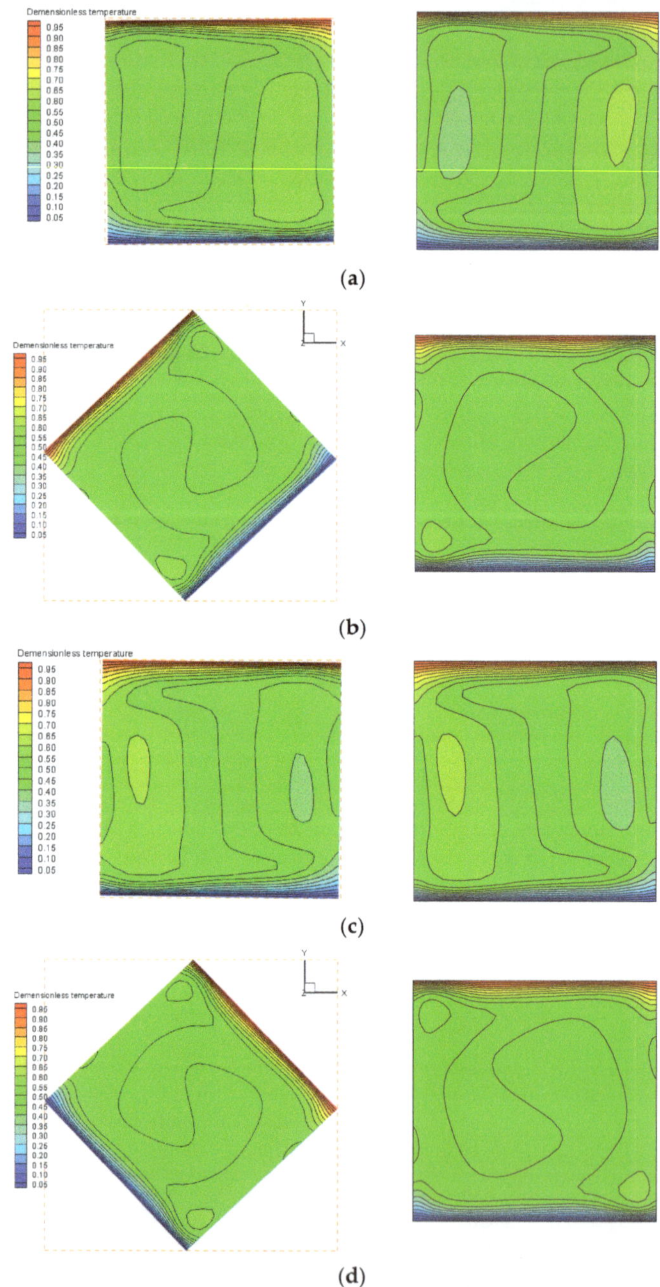

Figure 4. Comparison between code results (**right**) and Ansys-Fluent results with Dynamic mesh (**left**) with Ra = 10^6 and t_c = 1s: (**a**) $\varphi = 0\ (+)$; (**b**) $\varphi = \varphi_m$; (**c**) $\varphi = 0\ (-)$; (**d**) $\varphi = -\varphi_m$.

3. Results and Discussions

In contrast to the stationary case, the harmonic rotation introduces Coriolis forces, centrifugal forces, the Eulerian forces, and the buoyancy caused by these forces. So the characteristic of the natural convection under the interactive effect of these forces is much more complex than that in static case. Therefore, in this part, the characteristic of the natural convection under the coupling effect of these forces will be studied and the effect of the harmonic rotation on the flow and heat transfer will be clarified. According to Equations (9)–(13), the dimensionless parameters governing this kind of flows are Rayleigh number (Ra), Prandtl number (Pr), nominal rotational Rayleigh number (Raw), and nominal Taylor number (Ta). The Pr is chosen to be 0.71, which corresponds to the air. An appropriate choice of other parameters is required to obtain stable and converged solutions. Hence, for this purpose we have chosen these parameters as listed in Table 2, in which 12 cases are designed with different combinations of influencing factors. The 12 cases are calculated and the results are analyzed.

Figure 5 shows the velocity vector distributions in the x-o-y plane at $Z = 0.5$ for Case 1. In the figure, φ is the rotation angle as shown in Figure 2b, and (+) and (-) indicates the clockwise rotation and counterclockwise rotation respectively. It can be seen from the figure that due to the periodic change in the magnitude and direction of the Eulerian force, the magnitude and direction of the velocity also change correspondingly. The direction of the Eulerian force is opposite to the direction of the tangential acceleration, so in the first half of the cycle, the counterclockwise angular velocity gradually increases, and when it reaches 1/2 cycle, it increases to the maximum (Figure 5e). In the second half of the cycle, due to the change of the direction of the Eulerian force, the clockwise velocity began to accumulate, and when it reaches the whole cycle, the clockwise velocity reaches the maximum (Figure 5a). Form counterclockwise circulation to clockwise circulation, there is a transition state at which the counterclockwise velocity was almost weakened by the opposite-direction velocity, as shown in Figure 5c,g. After that, the clockwise velocity dominates the whole flow gradually. Since the Eulerian forces are larger at the position farther away from the rotation axis, so the transition appear firstly near the wall and then in the core region. It can also be found from the figure that the thickness of the flow boundary layer increases or decreases periodically, being thinnest at the amplitude position and the largest at the equilibrium position. This is because the Coriolis force, pointed to the rotation axis, is the largest and smallest at the equilibrium position and amplitude position, respectively.

Due to the angular velocity in x-o-y plane, the temperature fields in this plane also show the rotation effect, as shown in Figure 6. Figure 6 shows the temperature fields in x-o-y plane at $Z = 0.5$. It can be seen that the hot and cold boundaries present a larger influencing region due to the rotation effect. It can also be found that the thickness of the thermal boundary layer changes periodically with the rotation.

The periodical clockwise and counterclockwise flows in the x-o-y plane will be subjected to a Coriolis force pointing to the rotation axis. The counterclockwise velocity induced by gravitational buoyancy in the y-o-z (higher temperature on the right) will be pulled towards the rotation axis (z axis) periodically due to rotation, presenting an elliptical motion trajectory as shown in Figure 7. This motion trajectory is different with those in static case. As can be seen from Figure 7, when the Coriolis force is large, the flows near the left and right boundaries move closer to the rotation axis, resulting in a thicker flow boundary layer. When the Coriolis force is zero, the flow boundary layer becomes very thin, and the thickness of the boundary layer changes periodically. Similarly the temperature field also presents a thermal boundary layer changing periodically, as shown in Figure 8.

Figure 9 shows the velocity field in the x-o-z plane at $y = 0.5$. It can be seen from the figure that the velocity vector is mainly dominated by the natural convection in the square cavity, which is still counterclockwise. Such a velocity field results in the flow field as shown in Figure 10.

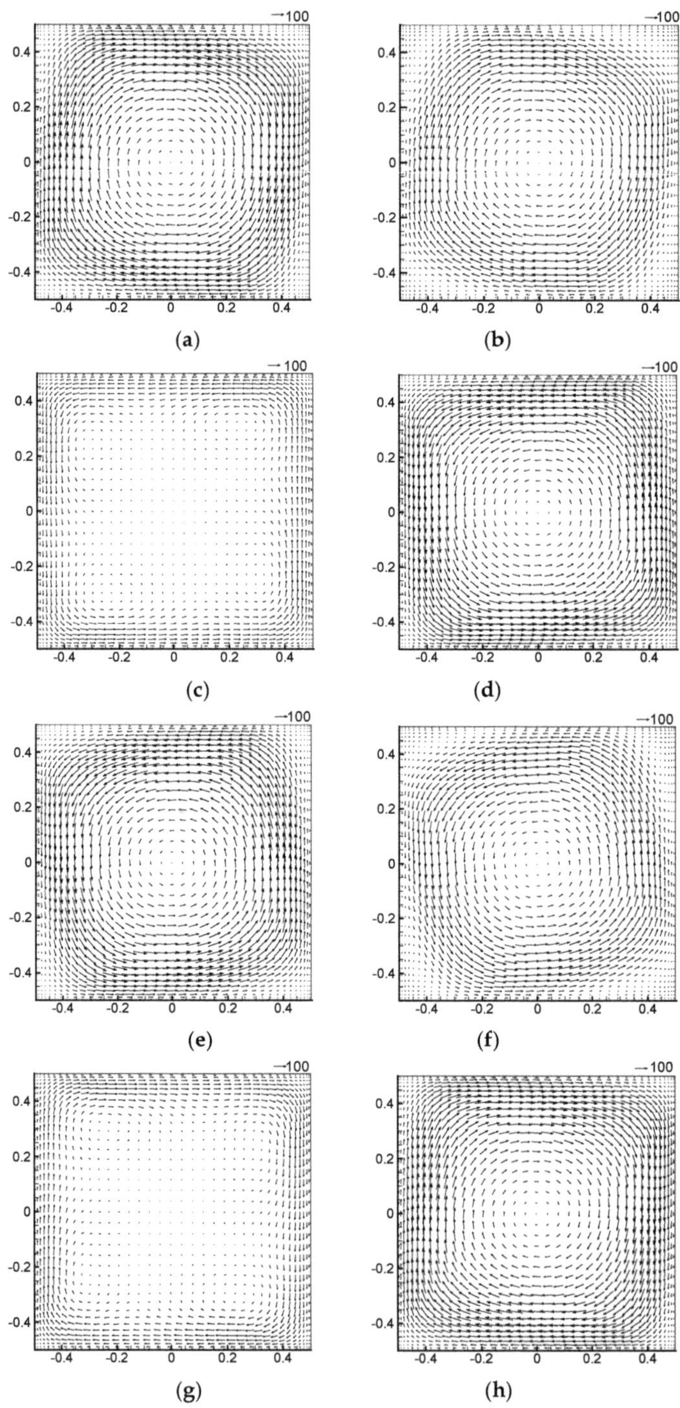

Figure 5. Velocity vector at x-o-y plane for Case 1: (**a**) $\varphi = 0$ (+); (**b**) $\varphi = \varphi_m/2$ (+); (**c**) $\varphi = \varphi_m$; (**d**) $\varphi = \varphi_m/2$ (−); (**e**) $\varphi = 0$ (−); (**f**) $\varphi = -\varphi_m/2$ (−); (**g**) $\varphi = -\varphi_m$; (**h**) $\varphi = -\varphi_m/2$ (+).

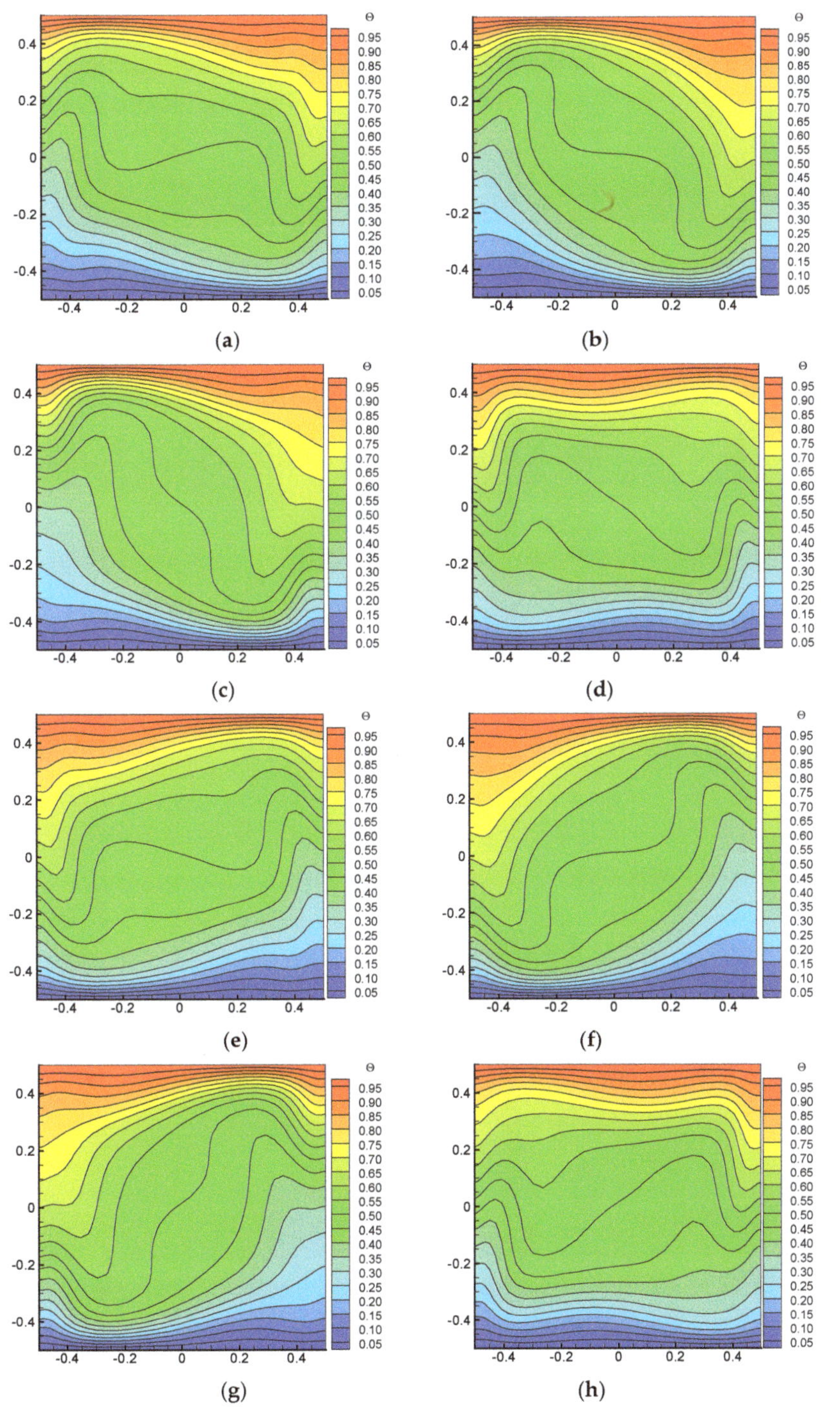

Figure 6. Temperature field evolution at x-o-y plane for Case 1: (**a**) $\varphi = 0$ (+); (**b**) $\varphi = \varphi_m/2$ (+); (**c**) $\varphi = \varphi_m$; (**d**) $\varphi = \varphi_m/2$ (−); (**e**) $\varphi = 0$ (−); (**f**) $\varphi = -\varphi_m/2$ (−); (**g**) $\varphi = -\varphi_m$; (**h**) $\varphi = -\varphi_m/2$ (+).

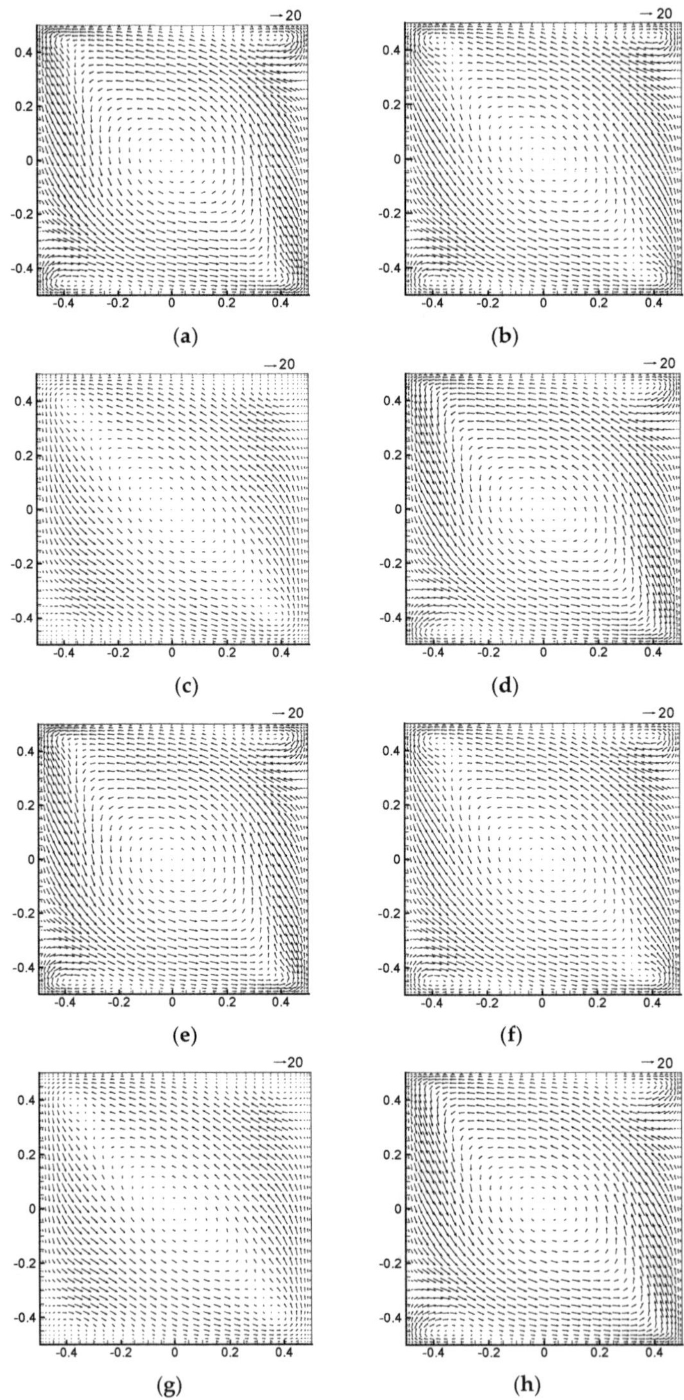

Figure 7. Velocity vector at y-o-z plane for Case 1: (**a**) $\varphi = 0$ (+); (**b**) $\varphi = \varphi_m/2$ (+); (**c**) $\varphi = \varphi_m$; (**d**) $\varphi = \varphi_m/2$ (−); (**e**) $\varphi = 0$ (−); (**f**) $\varphi = -\varphi_m/2$ (−); (**g**) $\varphi = -\varphi_m$; (**h**) $\varphi = -\varphi_m/2$ (+).

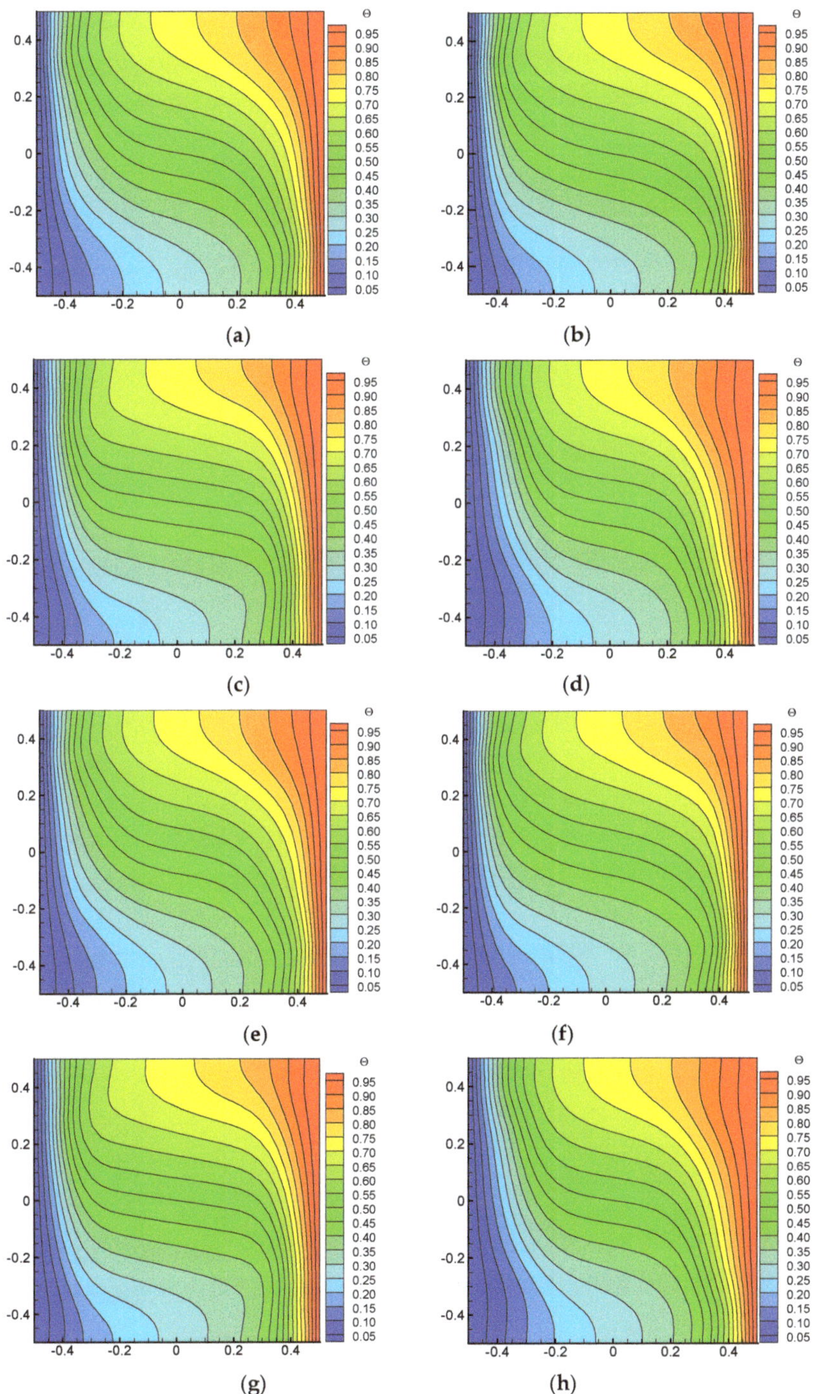

Figure 8. Temperature field evolution at y-o-z plane for Case 1: (**a**) $\varphi = 0 \; (+)$; (**b**) $\varphi = \varphi_m/2 \; (+)$; (**c**) $\varphi = \varphi_m$; (**d**) $\varphi = \varphi_m/2 \; (-)$; (**e**) $\varphi = 0 \; (-)$; (**f**) $\varphi = -\varphi_m/2 \; (-)$; (**g**) $\varphi = -\varphi_m$; (**h**) $\varphi = -\varphi_m/2 \; (+)$.

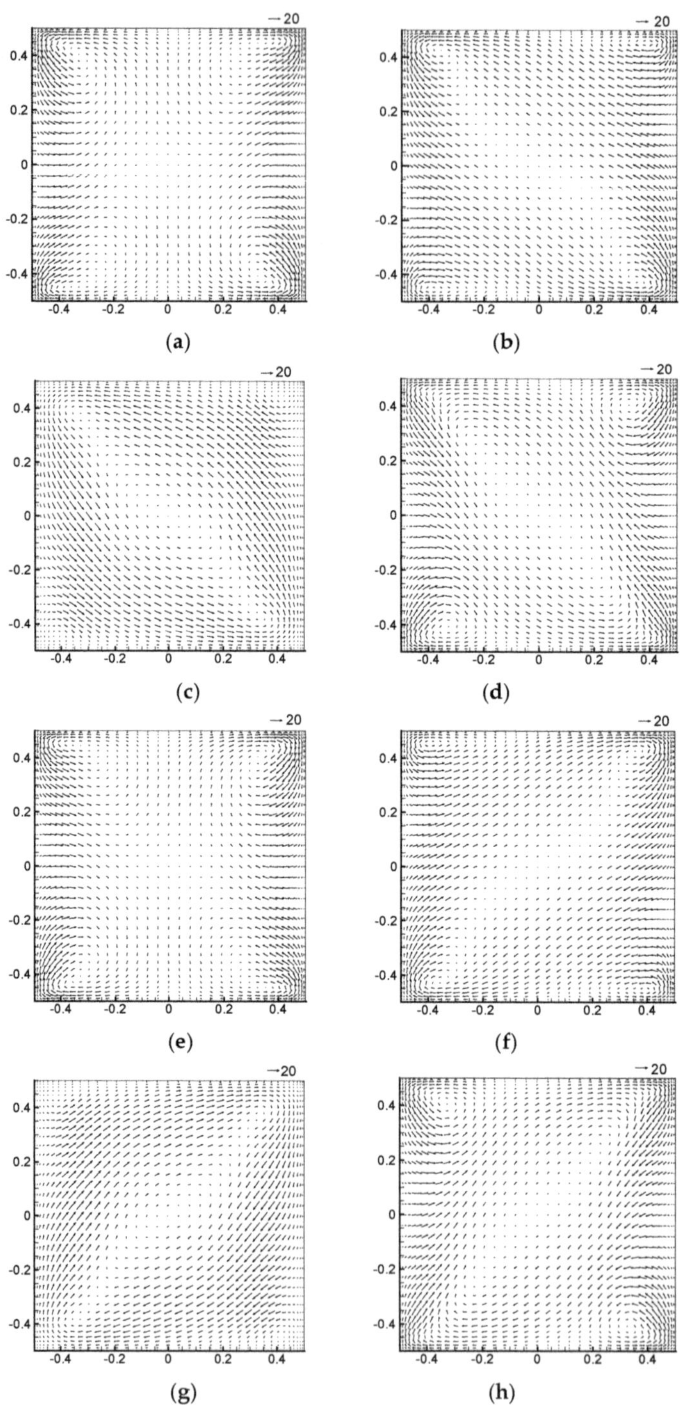

Figure 9. Velocity vector evolution at x-o-z plane for Case 1: (**a**) $\varphi = 0$ (+); (**b**) $\varphi = \varphi_m/2$ (+); (**c**) $\varphi = \varphi_m$; (**d**) $\varphi = \varphi_m/2$ (−); (**e**) $\varphi = 0$ (−); (**f**) $\varphi = -\varphi_m/2$ (−); (**g**) $\varphi = -\varphi_m$; (**h**) $\varphi = -\varphi_m/2$ (+).

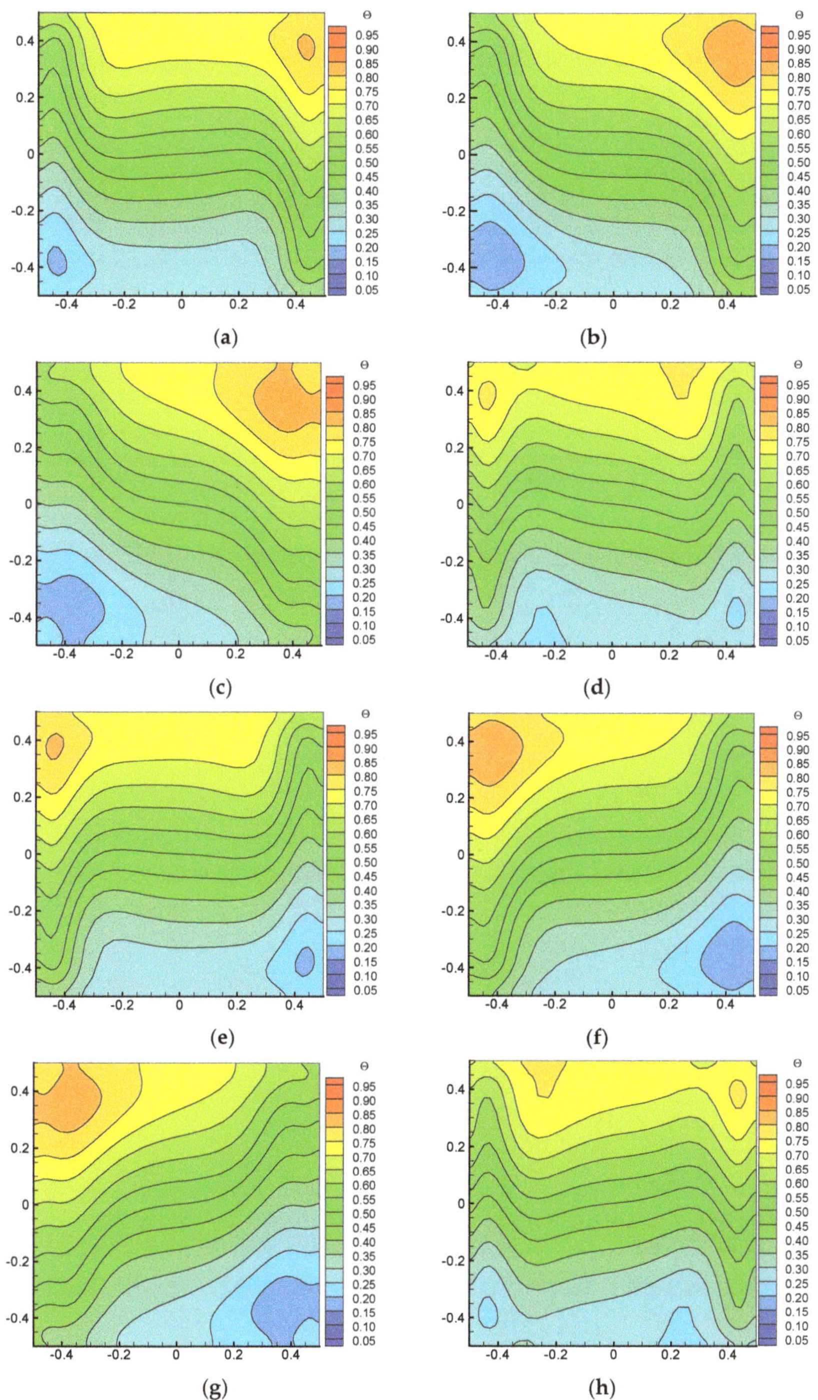

Figure 10. Temperature field evolution at x-o-z plane for Case 1: (**a**) $\varphi = 0$ (+); (**b**) $\varphi = \varphi_m/2$ (+); (**c**) $\varphi = \varphi_m$; (**d**) $\varphi = \varphi_m/2$ (−); (**e**) $\varphi = 0$ (−); (**f**) $\varphi = -\varphi_m/2$ (−); (**g**) $\varphi = -\varphi_m$; (**h**) $\varphi = -\varphi_m/2$ (+).

As can be seen from the vector distributions in the three planes, the Eulerian force-induced velocity is far greater than gravitational buoyancy-induced velocity. Therefore, the change of Eulerian force will seriously affect the change of flow field.

The above analysis is regarding the flow and heat transfer at low Rayleigh number, and in the following part the situation of larger Ra will be discussed. Figures 11–13 show the velocity vector distributions and temperature fields in plane x-o-y, plane y-o-z and plane x-o-z, respectively, when the Ra increases to 10^6 (Case 4). Compared with Figure 5, it can be seen from the Figures 11–13 that when the Rayleigh number is larger, the flow boundary layer and the thermal boundary layer will become thinner due to the enhancement of natural convection, as shown in Figures 11 and 12. Moreover, the temperature field at y-o-z plane is stratified, resulting in a uniform temperature distribution in the core region in x-o-y plane. Due to the strong flow generated by natural convection in the square cavity, the flow field formed in the x-o-z plane is also relatively complex. Moreover, with the influence of periodical Coriolis, the velocity is periodically pulled towards the rotation axis. It can also be seen from the figure that the flow intensity of natural convection is slightly higher than that generated by the Eulerian force, so with this Rayleigh number, the flow is dominated by natural convection.

From the perspective of velocity fields and temperature fields, the flow and heat transfer characteristics in a vertical-axis rotation system are comparatively studied between a higher Ra and a lower Ra. It can be found from the figure that the flow and heat transfer characteristics are more complicated than those of static case due to the Eulerian force, Coriolis force, centrifugal force generated by rotation and the buoyancy force introduced by these forces. When the rotation-induced velocity is stronger than that induced by gravitational buoyancy, the flow and heat transfer characteristics are seriously affected by rotation. However, when natural convection is dominated, the effect of rotation is not as significant as that of lower Ra. In order to compare the velocity dominance under different Ra conditions, velocity profiles on three center lines of the cavity shown in Figure 14 are compared. Figure 15 shows the velocity profiles on the three center lines when Ra is 1.0×10^4. It can be seen from the figure that the velocities U and V in x-o-y plane caused by rotation dominate the flow, while the flow caused by natural convection and other forces are obviously weaker than that generated by rotation. Figure 16 shows the velocity distribution on three center lines when Ra is 1.0×10^6. As can be seen from the figure, the z-direction velocity caused by natural convection dominates the flow, followed by the velocity caused by rotation. Therefore, when the Rayleigh number is 1.0×10^4, rotation-induced velocity dominates the flow, while at 1.0×10^6, natural convection-induced velocity dominates the flow.

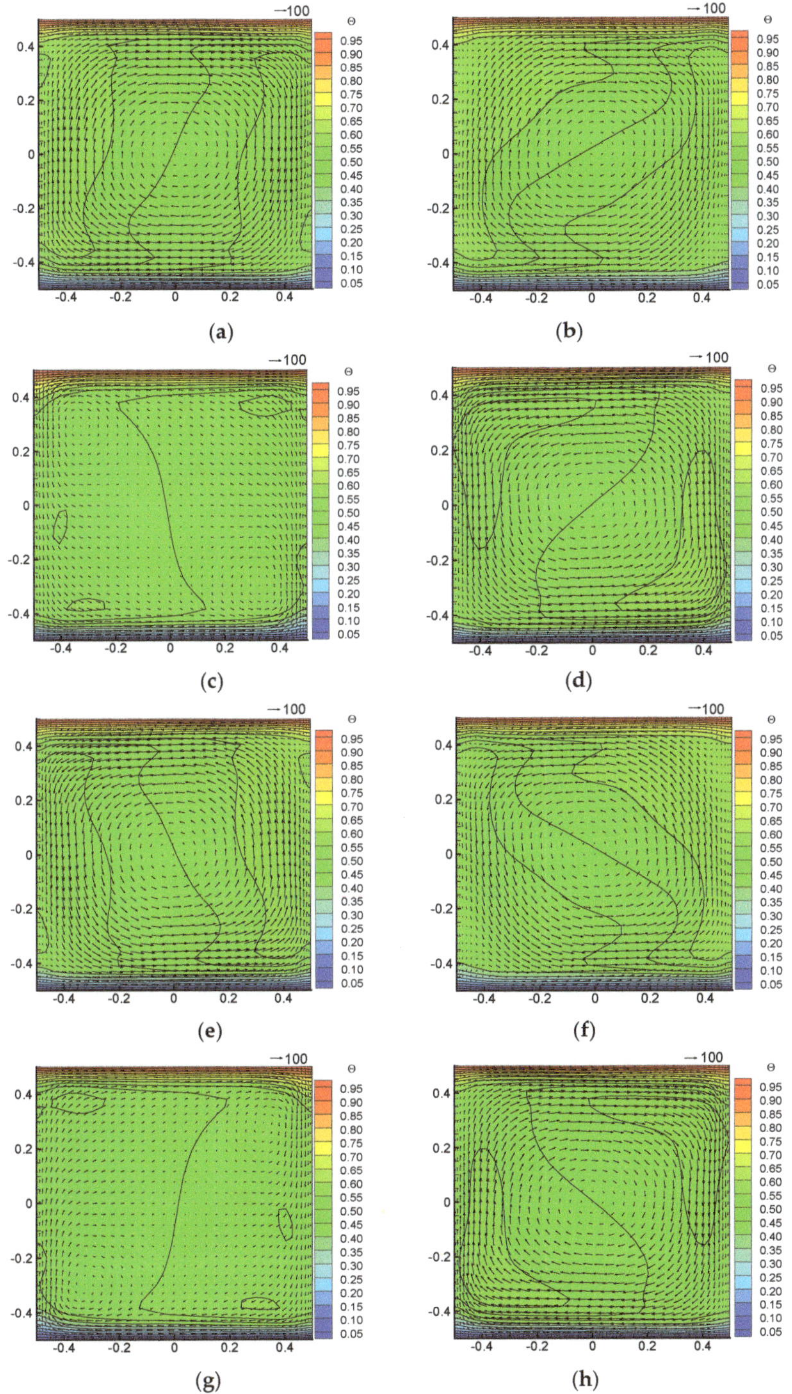

Figure 11. Temperature field evolution at x-o-z plane for Case 4: (**a**) $\varphi = 0$ (+); (**b**) $\varphi = \varphi_m/2$ (+); (**c**) $\varphi = \varphi_m$; (**d**) $\varphi = \varphi_m/2$ (−); (**e**) $\varphi = 0$ (−); (**f**) $\varphi = -\varphi_m/2$ (−); (**g**) $\varphi = -\varphi_m$; (**h**) $\varphi = -\varphi_m/2$ (+).

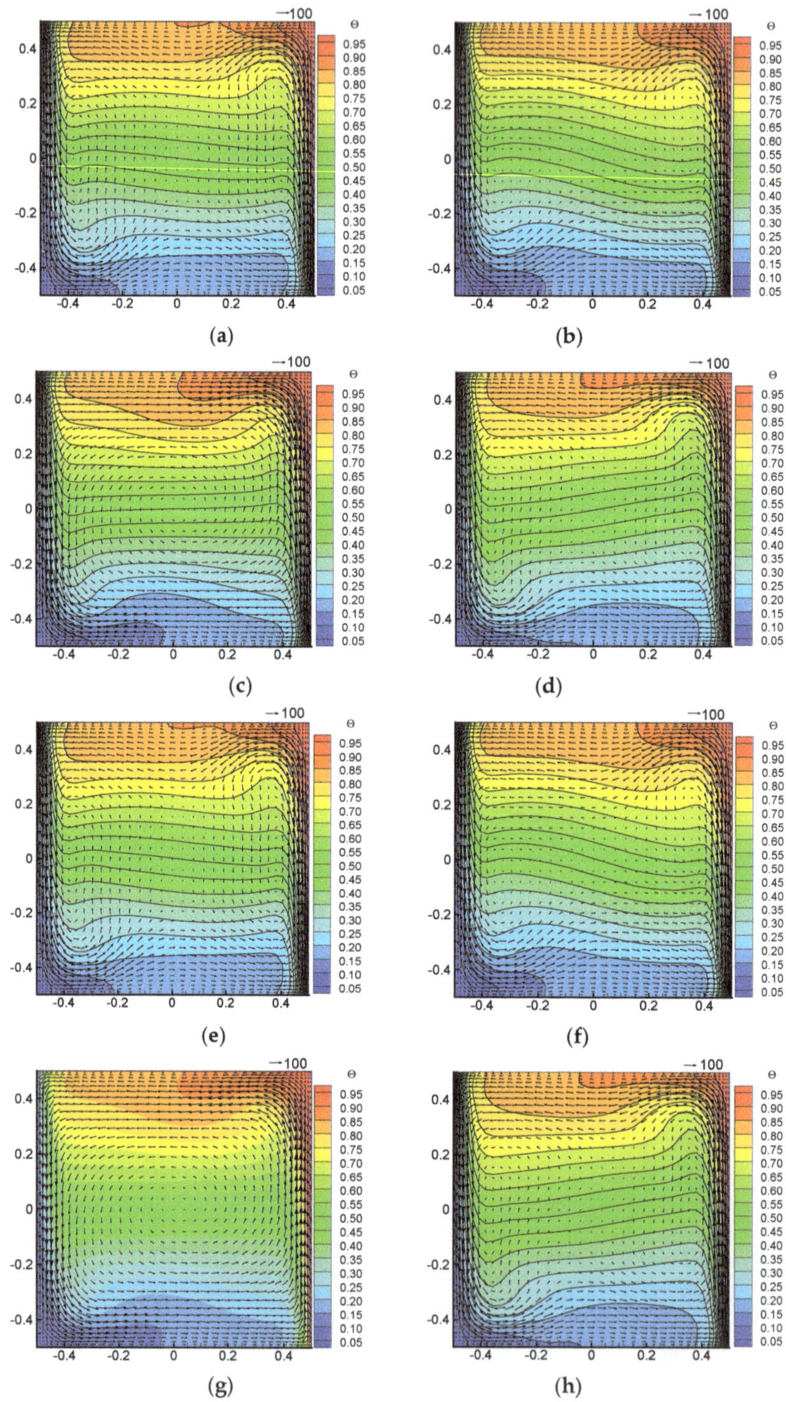

Figure 12. Temperature field evolution at x-o-z plane for Case 4: (**a**) $\varphi = 0$ (+); (**b**) $\varphi = \varphi_m/2$ (+); (**c**) $\varphi = \varphi_m$; (**d**) $\varphi = \varphi_m/2$ (−); (**e**) $\varphi = 0$ (−); (**f**) $\varphi = -\varphi_m/2$ (−); (**g**) $\varphi = -\varphi_m$; (**h**) $\varphi = -\varphi_m/2$ (+).

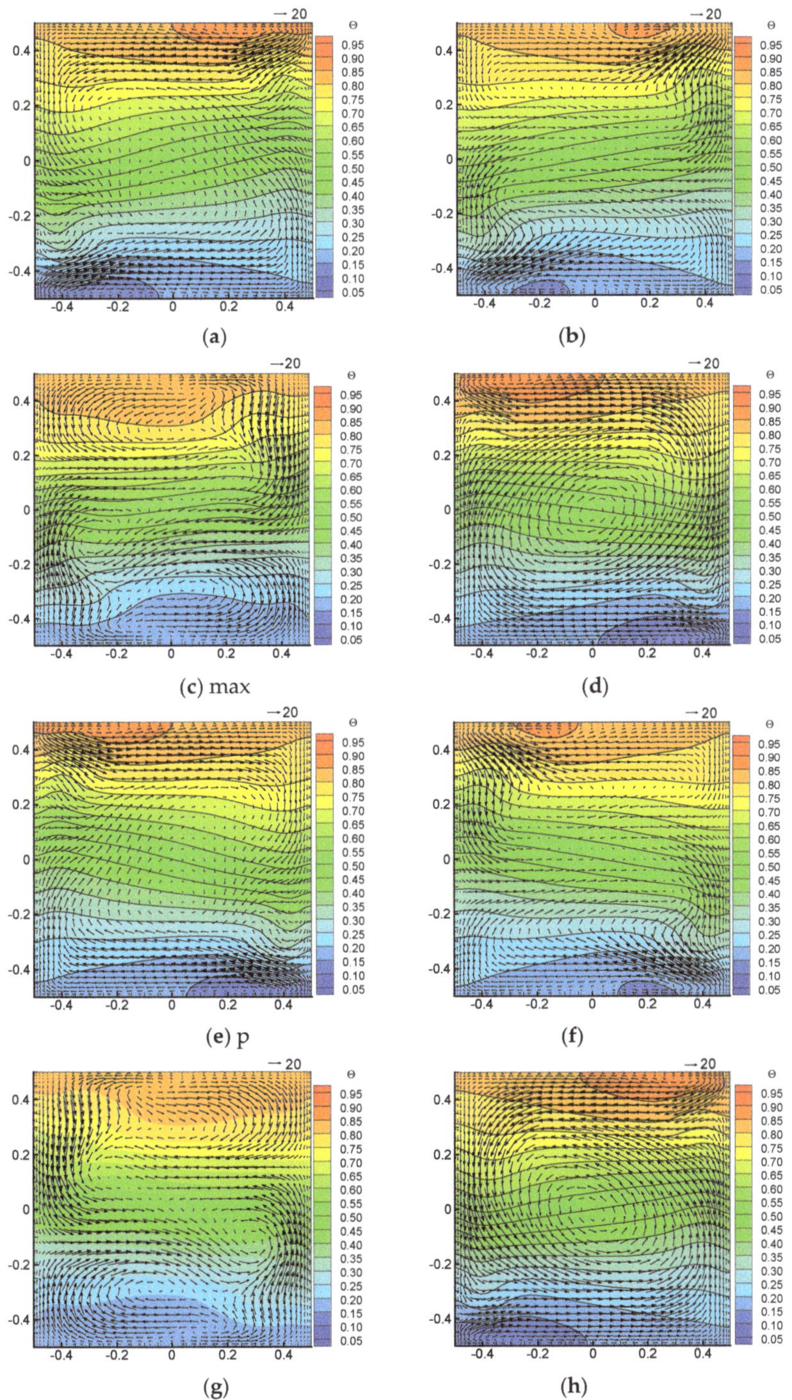

Figure 13. Temperature field evolution at x-o-z plane for Case 4: (**a**) $\varphi = 0$ (+); (**b**) $\varphi = \varphi_m/2$ (+); (**c**) $\varphi = \varphi_m$; (**d**) $\varphi = \varphi_m/2$ (−); (**e**) $\varphi = 0$ (−); (**f**) $\varphi = -\varphi_m/2$ (−); (**g**) $\varphi = -\varphi_m$; (**h**) $\varphi = -\varphi_m/2$ (+).

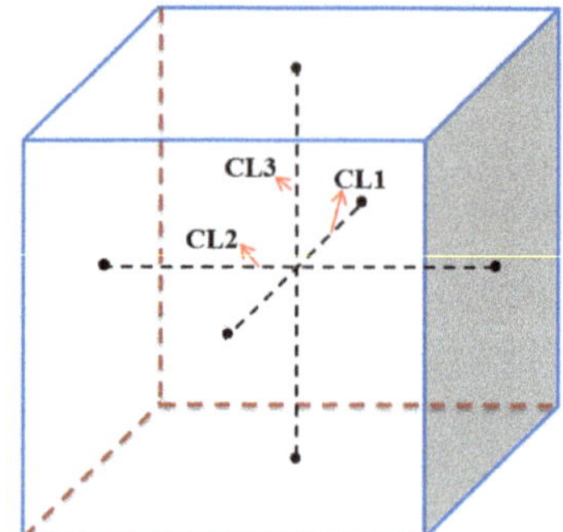

Figure 14. Observed centerlines for the cavity.

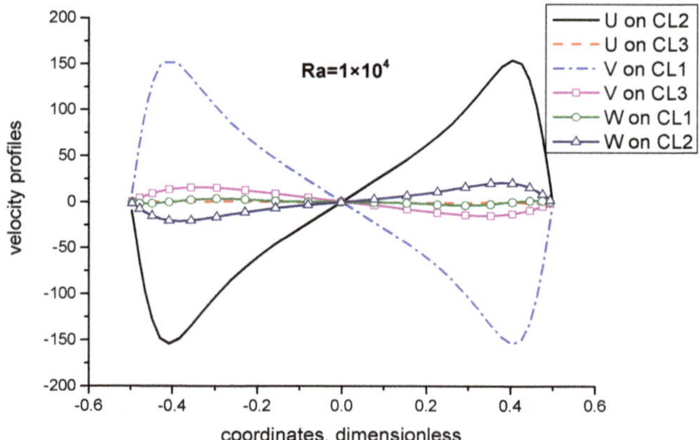

Figure 15. Velocity profiles on the center lines shown in Figure 14 for Case 1.

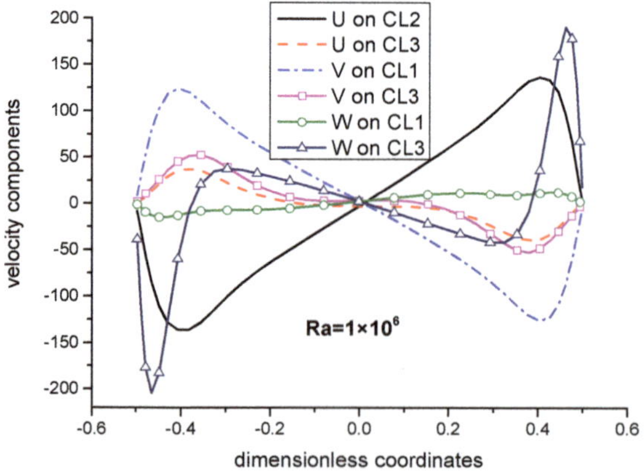

Figure 16. Velocity profiles on the center lines shown in Figure 14 for Case 4.

In the following part, the heat transfer characteristics of natural convection under different Ra, rotation period t_c and amplitude φ_m will be comparatively investigated from the perspective of boundary heat transfer rate. Figure 17 compares the hot-boundary Nu under different Ra and rotation periods with amplitude of $\pi/4$. The left and right columns of the figure correspond to Ra of 1.0×10^4 and 1.0×10^6, respectively. As can be seen from the figure, when the Rayleigh number is low, the hot-boundary Nu decreases with the increase of the period, that is, the hot-boundary Nu is proportional to the rotation frequency; however, when Ra is higher, changing the period of the rotation within a certain range has no obvious influence on Nu on the hot boundary. This is because the flow rate is dominated by rotation at low Ra, while the flow is dominated by natural convection at high Ra. In addition, under the same Rayleigh number, the smaller the period, the greater the range of the hot-boundary Nu. Figure 18 shows the results for a smaller amplitudes, $\pi/8$. Compared with Figure 17, it can be found that the Nu is larger if the amplitude is larger, with other parameters remaining same.

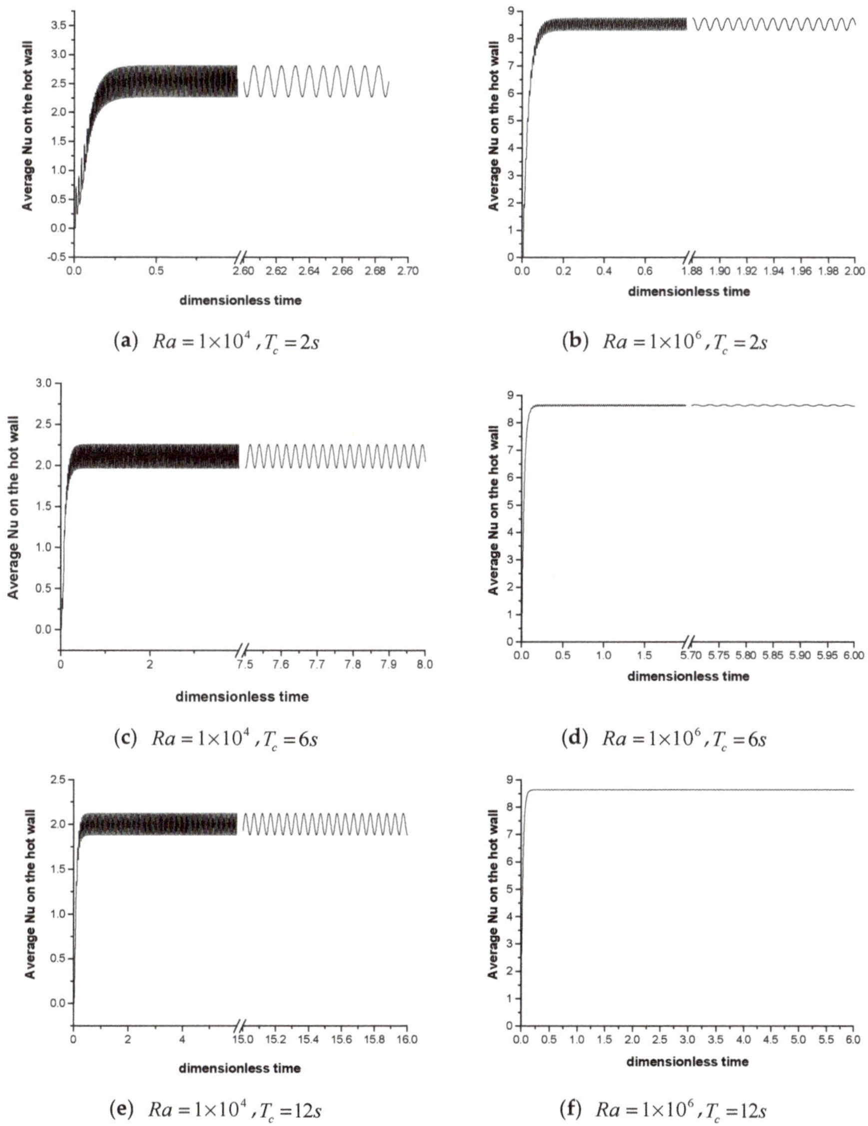

Figure 17. Average Nu on the hot face under different cases at $\varphi_m = \pi/4$: (**a**) $Ra = 1 \times 10^4$, $T_c = 2\,s$; (**b**) $Ra = 1 \times 10^6$, $T_c = 2\,s$; (**c**) $Ra = 1 \times 10^4$, $T_c = 6\,s$; (**d**) $Ra = 1 \times 10^6$, $T_c = 6\,s$; (**e**) $Ra = 1 \times 10^4$, $T_c = 12\,s$; (**f**) $Ra = 1 \times 10^6$, $T_c = 12\,s$.

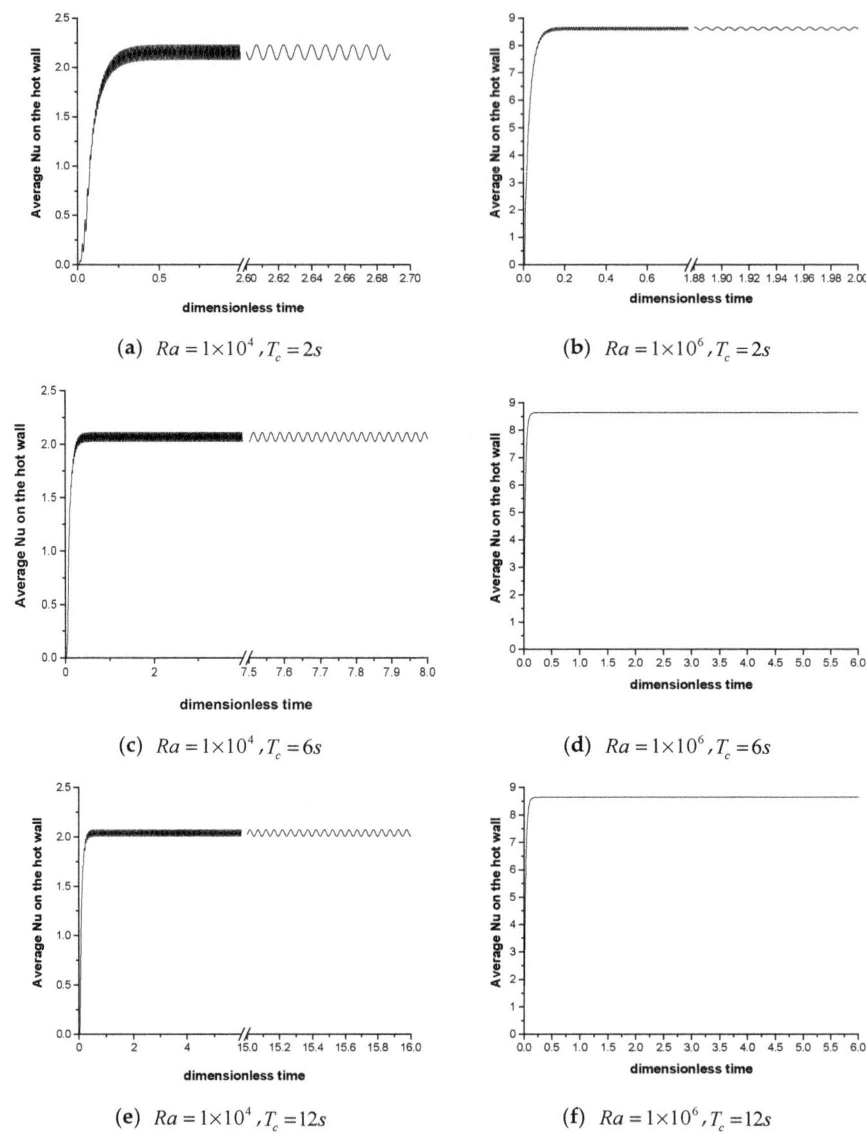

(a) $Ra = 1 \times 10^4, T_c = 2s$

(b) $Ra = 1 \times 10^6, T_c = 2s$

(c) $Ra = 1 \times 10^4, T_c = 6s$

(d) $Ra = 1 \times 10^6, T_c = 6s$

(e) $Ra = 1 \times 10^4, T_c = 12s$

(f) $Ra = 1 \times 10^6, T_c = 12s$

Figure 18. Average Nu on the hot face under different cases at $\varphi_m = \pi/8$: (**a**) $Ra = 1 \times 10^4$, $T_c = 2s$; (**b**) $Ra = 1 \times 10^6$, $T_c = 2s$; (**c**) $Ra = 1 \times 10^4$, $T_c = 6s$; (**d**) $Ra = 1 \times 10^6$, $T_c = 6s$; (**e**) $Ra = 1 \times 10^4$, $T_c = 12s$; (**f**) $Ra = 1 \times 10^6$, $T_c = 12s$.

4. Conclusions

The tangential force (Euler force) and Coriolis force induced by harmonic rotation have important effect on the natural convection in a vertical-axis harmonic rotation system. The periodical-changing tangential force results in circle-like flow circulation in x-o-y plane, and this circulation is either clockwise or clockwise or a transition one. This rotation-induced circulation will be subjected to Coriolis force which is pointed to rotation axis. As a result, the velocities induced by gravitational buoyancy near the hot and cold wall will be pulled periodically towards the rotation axis due to the

Coriolis force, and the boundary layer becomes thicker or thinner periodically due to the change in Coriolis force. Consequently, the Nusselt number on the hot and cold boundary varies periodically: At $\varphi = 0$, Coriolis force is the largest and the boundary layer is the thickest, and at $\varphi = \varphi_m$, the Coriolis is the smallest, so the boundary layer is thinnest and thus the Nusselt number is highest. On the other hand, the Nusselt number is also positively influenced by the angular velocity near the hot boundary. With the same Ra, the larger the rotation frequency, the higher the Nusselt number. However, the effect of rotation on natural convection also depends on the relative magnitudes of rotation-induced velocity and buoyancy-induced velocity: when the rotation-induced velocity dominates the flow, the harmonic rotation has obvious effect on flow and heat transfer; otherwise the rotation has limited effect on natural convection.

Author Contributions: G.Y. carried out the numerical simulation and wrote the draft paper. L.Z. drew the figures and analyzed the data. S.J. performed validations of the code results. Y.G. made the literature review and revised the manuscript. J.L. designed the cases and directed the whole research.

Funding: This research was funded by National Science Foundation of China (No. 51606117), and China Postdoctoral Science Foundation funded project (Nos. 2017M621473 and 2018T110396).

Nomenclature

c_p	heat capacity
g	magnitude of gravity acceleration
L	the length of the cavity
Nu	local Nusselt number
P	dimensionless pressure
p	dimensional pressure
p_{eff}	dimensional effective dynamic pressure
Pr	Prandtl number
Ra	Rayleigh number
$\widetilde{\mathrm{Ra}}_\omega$	nominal rotational Rayleigh number
T	temperature
t_c	period of the harmonic rotation
$\widetilde{\mathrm{Ta}}$	nominal Taylor number
T_h	temperature on the hot wall
T_c	temperature on the cold wall
T_0	initial fluid temperature
t, τ	dimensional and dimensionless time
u, v, w	dimensional velocities in x, y, z directions
U, V, W	dimensionless velocities in X, Y, Z directions
X, Y, Z	dimensionless coordinate variables
x, y, z	dimensional coordinate variables
Greek symbols	
α	thermal diffusivity, $\alpha = \lambda / (\rho c_p)$
β	thermal expansion coefficient
λ	thermal conductivity
ν	kinematic viscosity
ρ	air density
ρ_0	reference air density
φ	rolling angle, $\varphi = \varphi_m \cos(\omega t)$
φ_m	amplitude of rolling
Ω	magnitude of rotation angular velocity, $\Omega = \frac{d\varphi}{dt}$
$\boldsymbol{\Omega}$	vector of rotation angular velocity
Θ	dimensionless temperature

References

1. Murata, H.; Sawada, K.-I.; Kobayashi, M. Experimental Investigation of Natural Convection in a Core of a Marine Reactor in Rolling Motion. *J. Nucl. Sci. Technol.* **2000**, *37*, 509–517. [CrossRef]
2. Tan, S.-C.; Su, G.H.; Gao, P.-Z. Heat transfer model of single-phase natural circulation flow under a rolling motion condition. *Nucl. Eng. Des.* **2009**, *239*, 2212–2216.
3. Yan, B.H.; Yu, L.; Yang, Y.H. Heat transfer with laminar pulsating flow in a channel or tube in rolling motion. *Int. J. Therm. Sci.* **2010**, *49*, 1003–1009. [CrossRef]
4. Yan, B.H.; Yu, L.; Yang, Y.H. Heat transfer of laminar flow in a channel in rolling motion. *Prog. Nucl. Energy* **2010**, *52*, 596–600. [CrossRef]
5. Wang, C.; Gao, P.; Wang, S.; Li, X.; Fang, C. Experimental study of single-phase forced circulation heat transfer in circular pipe under rolling motion. *Nucl. Eng. Des.* **2013**, *265*, 348–355. [CrossRef]
6. Yuan, H.; Tan, S.; Zhuang, N.; Lan, S. Flow and heat transfer in laminar–turbulent transitional flow regime under rolling motion. *Ann. Nucl. Energy* **2016**, *87*, 527–536. [CrossRef]
7. Yu, Z.; Lan, S.; Yuan, H.; Tan, S. Temperature fluctuation characteristics in a mini-rectangular channel under rolling motion. *Prog. Nucl. Energy* **2015**, *81*, 203–216. [CrossRef]
8. Yan, B.-H.; Gu, H.-Y.; Yang, Y.-H.; Yu, L. Theoretical models of turbulent flow in rolling motion. *Prog. Nucl. Energy* **2010**, *52*, 563–568. [CrossRef]
9. Baig, M.F.; Masood, A. Natural Convection in a Two-Dimensional Differentially Heated Square Enclosure Undergoing Rotation. *Numer. Heat Transf. A Appl.* **2001**, *40*, 181–202.
10. Hamady, F.J.; Lloyd, J.R.; Yang, K.T.; Yang, H.Q. A Study of Natural Convection in a Rotating Enclosure. *J. Heat Transf.* **1994**, *116*, 136–143. [CrossRef]
11. Jin, L.F.; Tou, K.W.; Tso, C.P. Experimental and Numerical Studies an a Rotating Cavity with Discrete Heat Sources with Conjugate Effects. *Exp. Heat Transf.* **2005**, *18*, 259–277. [CrossRef]
12. Mandal, J.C.; Li, X.; Sonawane, C.R. Simulation of flow inside differentially heated rotating cavity. *Int. J. Numer. Methods Heat Fluid Flow* **2013**, *23*, 23–54. [CrossRef]
13. Kumar, M.N.; Pundrika, G.; Narasimha, K.R.; Seetharamu, K.N. Effect of Rayleigh Number with Rotation on Natural Convection in Differentially Heated Rotating Enclosure. *J. Appl. Fluid Mech.* **2017**, *10*, 1125–1138.
14. Saleh, H.; Hashim, I. Numerical Analysis of Nanofluids in Differentially Heated Enclosure Undergoing Orthogonal Rotation. *Adv. Math. Phys.* **2014**, *2014*, 874132. [CrossRef]
15. Tso, C.P.; Jin, L.F.; Tou, S.K.W. Numerical Segregation of the Effects of Body Forces in a Rotating, Differentially Heated Enclosure. *Numer. Heat Transf. A Appl.* **2007**, *51*, 85–107. [CrossRef]
16. Chokri, R.; Brahim, B.-B. Three-dimensional natural convection of molten Lithium in a differentially heated rotating cubic cavity about a vertical ridge. *Powder Technol.* **2016**, *291*, 97–109. [CrossRef]
17. Govender, S. Thermal Instability of Convection in a Rotating Nanofluid Saturated Porous Layer Placed at a Finite Distance From the Axis of Rotation. *J. Heat Transf.* **2016**, *138*, 102402. [CrossRef]
18. Ker, Y.T.; Lin, T.F. A combined numerical and experimental study of air convection in a differentially heated rotating cubic cavity. *Int. J. Heat Mass Transf.* **1996**, *39*, 3193–3210. [CrossRef]
19. Lee, T.L.; Lin, T.F. Transient three-dimensional convection of air in a differentially heated rotating cubic cavity. *Int. J. Heat Mass Transf.* **1996**, *39*, 1243–1255. [CrossRef]
20. Kleinstreuer, C. *Engineering Fluid Dynamics: An Interdisciplinary Systems Approach*; Cambridge University Press: Cambridge, UK, 1997.
21. Li, Z.Y.; Tao, W.Q. A new stability-guaranteed second-order difference scheme. *Numer. Heat Transf. B Fundam.* **2002**, *42*, 349–365. [CrossRef]
22. Sun, D.L.; Qu, Z.G.; He, Y.L.; Tao, W.Q. An efficient segregated algorithm for incompressible fluid flow and heat transfer problems—IDEAL (inner doubly iterative efficient algorithm for linked equations) Part I: Mathematical formulation and solution procedure. *Numer. Heat Transf. B Fundam.* **2008**, *53*, 1–17. [CrossRef]

Journal of
*Marine Science
and Engineering*

Article

Vibration Control of Marine Top Tensioned Riser with a Single Tuned Mass Damper

Jixiang Song [1,2], Tao Wang [3], Weimin Chen [1,2,*], Shuangxi Guo [1,2] and Dingbang Yan [1,2]

[1]   Key Laboratory for Mechanics in Fluid-Solid Coupling Systems, Institute of Mechanics, Chinese Academy of Science, Beijing 100190, China; songjixiang18@mails.ucas.ac.cn (J.S.); guoshuangxi2012@126.com (S.G.); yandingbang18@mails.ucas.ac.cn (D.Y.)

[2]   School of Engineering Science, University of Chinese Academy of Sciences, Beijing 100049, China

[3]   Aker Solutions, Houston, TX 77042, USA; tao.wang@akersolutions.com

*   Correspondence: wmchen@imech.ac.cn

Received: 17 August 2020; Accepted: 28 September 2020; Published: 9 October 2020

Abstract: The study of the Tuned Mass Damper (TMD) on Top Tensioned Risers (TTRs) through the application of numerical analysis is of great significance for marine engineering. However, to the best knowledge of the author, neither the in-field riser data nor the ocean current data used in published papers were from engineering design, so the research results provide limited guidance to the actual engineering project. In view of this problem, this study designed a single TMD to suppress the vibration of the engineering TTR under the action of the actual ocean current. First, the dynamic model of a riser-TMD system was established, and the modal superposition method was used to calculate the model. The non-resonant modal method of the flexible structure was used to design the TMD parameters for the engineering riser. Ocean current loading in the South China Sea was then applied to the riser. The vibration of the riser without and with TMD was compared. The result showed that TMD could effectively reduce the vibration response of the riser. When compared without TMD, the maximum value of displacement envelope and the RMS displacement were reduced by 26.70% and 17.83% in the in-line direction, respectively. Moreover, compared to without TMD, the maximum value of displacement envelope and RMS displacement were decreased by 17.01% and 22.05% in the cross-flow direction, respectively. In the in-line direction, the installation position of TMD on the riser was not sensitive to the effect of the displacement response; meanwhile, in the cross-flow direction the installation position of TMD on the riser was more sensitive to the effect of the displacement response.

Keywords: tuned mass damper (TMD); top tensioned riser; numerical analysis; marine dynamics

---

1. Introduction

Suppressing the vibration of marine risers is a popular research direction. Spiral strakes, fairings, or other forms of passive vibration suppression devices are usually installed on the riser to suppress vibration [1]. The most commonly installed type is the helical strake. Strakes can destroy vortex shedding in the flow direction and reduce the effect of the vortex on the riser, which is a method to reduce the energy input. In contrast to passive vibration suppression devices, another device such as the tuned mass damper (TMD) can increase the energy consumption. TMDs are widely used in the field of civil engineering, such as in truss bridges [2], high-rise buildings [3], high towers [4], and structures.

Although some scholars have conducted TMD research on marine risers, they are still in the research state of numerical analysis and preliminary laboratory tests, and there is still a long way to go before applications [1,5,6]. Jaiswal [5] carried out an experimental study in a towed pool at MIT and numerically analyzed the effect of the stiffness of the TMD on the vibration displacement of the flexible riser. The results showed that the TMD could reduce the vibration displacement of the flexible riser. However, the author did not show the detailed experimental results so far. The slenderness ratio of the experimental model was 7.62 = 0.381 m/0.05 m and the Cauchy number was $C_Y = \rho L^3 U^2 / (16B) = 4.06e - 6$ [7], similar to that of a rigid cylinder. Nikoo [6] used a pipe-in-pipe (PIP) instead of a traditional TMD to conduct numerical research on vortex-induced vibration suppression, and the results showed that 84% of the vortex-induced vibration could be suppressed. The slenderness ratio of its numerical model was in the range of 5–13, which is different from the slenderness ratio of engineering marine flexible risers, 10^2–10^3. Obviously, in the above article, the riser was a rigid structure, but the actual marine riser was a flexible structure, which caused the TMD to suppress the vibration effect of the riser to be exaggerated. As the TMD is usually designed for a certain resonant mode (target resonant mode), when the structure is flexible the tuned absorber not only experiences the support motion of the resonant mode, but also the support motion of other usually higher-frequency modes, and only a part of the force transmitted by the absorber enters the target resonant mode. When the structure is rigid, the tuned absorber only experiences the support motion of the resonant mode [8]. Therefore, the TMD suppresses the vibration effect of rigid risers better than the vibration effect of flexible risers.

Chaojun [1] studied the suppression of experimental riser vibration in the Gulf of Mexico using TMD and a semi-active tuned mass damper (STMD) through numerical analysis, and analyzed the dynamic response displacements under uniform flow and the Gulf stream. The results showed that the STMD's vibration suppression effect was better than the TMD's vibration suppression effect. However, the TMDs and STMDs were evenly distributed along the entire length of the flexible riser, which would lead to the installation of many damper components. The increase in the cross-sectional area of the damper cannot be ignored. This results in a larger load on the riser, a greater installation difficulty, difficulty in maintenance, and a higher cost.

In summary, due to the lack of engineering data in the numerical calculation, the actual effect of TMDs in suppressing riser vibration cannot be accurately quantified, so the practical guidance of the calculation results for the project is also limited. Accurate numerical calculation is the basis of engineering application. Therefore, this study uses a TMD to suppress marine flexible riser vibration in the actual ocean current load, and quantitatively analyzes the vibration suppression effect of the TMD.

The structure of the paper is as follows. Section 2 establishes the dynamic model of the TMD acting on flexible marine risers. Section 3 uses the modal superposition method to carry out numerical analysis on a flexible marine riser such as a Top Tensioned Risers (TTR). In Section 4, the numerical results are discussed. The conclusion is in Section 5.

2. Dynamic Model of TTR Marine Riser with TMD

The tuned mass damper (TMD) consists of a ring, a viscous damper, and a spring. It is installed on the outer surface of the riser, as shown below in Figure 1.

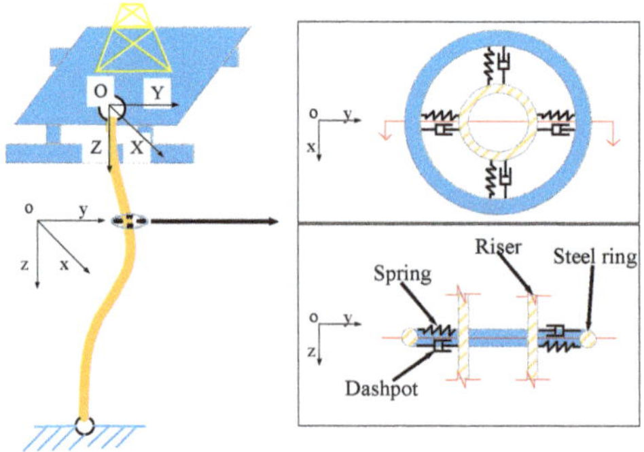

Figure 1. Schematic of the TMD and TTR model.

Vibration control equation in the XY plane:

$$EI\frac{\partial^4 y(z,t)}{\partial z^4} - T\frac{\partial^2 y(z,t)}{\partial z^2} + m_z\frac{\partial^2 y(z,t)}{\partial t^2} + c\frac{\partial y(z,t)}{\partial t} - f(z,t) - \delta(z-s)f_{TMD}(z,t) = 0, \tag{1}$$

$$f_{TMD}(s,t) = -k_{TMD}(y_{TMD}(s,t) - y(s,t)) - c_{TMD}\left(\frac{\partial y_{TMD}(s,t)}{\partial t} - \frac{\partial y(s,t)}{\partial t}\right) = m_{TMD}\frac{\partial^2 y_{TMD}(s,t)}{\partial t^2}, \tag{2}$$

$$\delta(z-s) = \left\{\begin{array}{cc} \infty & z=s \\ 0 & z \neq s \end{array}\right\}, \int_0^L \delta(z-s)f(z)dz = f(s), \tag{3}$$

where y represents the transverse displacement of the riser. The vibration control equation is also valid in the x-direction. z represents the length position. t represents the time. EI represents the bending stiffness, and T is the tension. m_z represents the uniform mass per unit length of the riser. c represents the structural damping coefficient. $f(z,t)$ represents the transverse force per unit length. $f_{TMD}(z,t)$ represents the transverse force of TMD. δ is the Dirac function. k_{TMD} represents the spring stiffness of the TMD, and m_{TMD} represents the mass of TMD. $y_{TMD}(s,t)$ and $y(s,t)$ represent the TMD transverse displacement and the transverse displacement of the riser at time t and riser position s, respectively. c_{TMD} represents the damping coefficient of TMD.

The transverse force per unit length $f(z,t)$ of Equation (1) can be decomposed into the drag force in the in-line direction and the lift force in the cross-flow direction [9,10], as follows:

$$\begin{array}{l} f_D(z,t) = \frac{1}{2}\rho_f C_D(z,t)U(z,t)^2 D + A_D\cos(4\pi f_v t + \beta) \\ f_L(z,t) = \frac{1}{2}\rho_f C_L(z,t)U(z,t)^2 D\cos(2\pi f_v t + \alpha) \end{array}, \tag{4}$$

where ρ_f is the sea water density. $C_D(z,t)$, $C_L(z,t)$ is the drag force and lift force coefficient. $U(z,t)$ is the velocity of the current. D is the diameter of the riser. The non-dimensional vortex shedding frequency is $f_v = \frac{S_t U}{D}$. S_t is the St Number, usually taken as 0.2. A_D is 20% of the first term of f_D (z,t). α and β are the phase angles.

The boundary conditions are simply supported; the displacement is 0, and the bending moment is 0, as shown below:

$$\begin{array}{l} y(0,t) = 0,\ EI\frac{\partial^2 y(0,t)}{\partial z^2} = 0 \\ y(L,t) = 0,\ EI\frac{\partial^2 y(L,t)}{\partial z^2} = 0 \end{array}. \tag{5}$$

Since the marine riser's vibration is a small amplitude, the modal superposition method is used to solve the above partial differential equation. First, the natural frequency and mode are calculated, and the non-conservative force in Equation (1) is set as 0, then one obtains:

$$EI\frac{\partial^4 y(z,t)}{\partial z^4} - T\frac{\partial^2 y(z,t)}{\partial z^2} + m_z\frac{\partial^2 y(z,t)}{\partial t^2} = 0. \tag{6}$$

Assume that the solution of formula (6) has the form:

$$y(z,t) = \phi(z)q(t), \tag{7}$$

where $\phi(z)$ represents the mode shape. $q(t)$ represents the mode displacement.

Inserting Equation (7) into Equation (6),

$$EI\phi''''(z)q(t) - T\phi''(z)q(t) + m_z\phi(z)\ddot{q}(t) = 0$$

$$\frac{EI\phi''''(z) - T\phi''(z)}{\phi(z)} = \frac{-m_z\ddot{q}(t)}{q(t)} = m_z w^2$$

where w is unknown. "'" denotes the derivative of position z, and "." denotes the derivative of time. t.

$$EI\phi''''(z) - T\phi''(z) - m_z w^2 \phi(z) = 0. \tag{8}$$

The boundary condition Equation (5) becomes:

$$\begin{aligned} \phi(0) &= 0, \phi''(0) = 0 \\ \phi(L) &= 0, \phi''(L) = 0 \end{aligned}. \tag{9}$$

Introducing hypothesis $\phi(z) = Ce^{\lambda z}$ into Equation (8), we obtain:

$$EI\lambda^4 - T\lambda^2 - m_z w^2 = 0$$

The above Equation is a quartic equation of one element, and the four roots can be solved.

$$\lambda_1 = -i\sqrt{\frac{\sqrt{T^2+4EIm_z w^2}-T}{2EI}}, \lambda_2 = i\sqrt{\frac{\sqrt{T^2+4EIm_z w^2}-T}{2EI}}$$

$$\lambda_3 = -\sqrt{\frac{T+\sqrt{T^2+4EIm_z w^2}}{2EI}}, \lambda_4 = \sqrt{\frac{T+\sqrt{T^2+4EIm_z w^2}}{2EI}}$$

Thus, $\phi(z) = C_1 e^{\lambda_1 z} + C_2 e^{\lambda_2 z} + C_3 e^{\lambda_3 z} + C_4 e^{\lambda_4 z}$. After further simplification, using trigonometric and hyperbolic functions to replace the exponential form, we get:

$$\phi(z) = A\sin\delta z + B\cos\delta z + C\sinh\varepsilon z + D\cosh\varepsilon z$$

where $\delta = \sqrt{\frac{\sqrt{T^2+4EIm_z w^2}-T}{2EI}}$, $\varepsilon = \sqrt{\frac{T+\sqrt{T^2+4EIm_z w^2}}{2EI}}$.

Bringing the boundary condition Equation (9) into the above formula, we get $\sin\delta L = 0$. Because $\sin n\pi = 0$, $n = 1,2,\cdots,\infty$, so $\delta_n L = n\pi$. Further, we get the natural frequency w_n,

$w_n = \frac{n\pi}{L}\sqrt{\left(\frac{n\pi}{L}\right)^2 \frac{EI}{m_z} + \frac{T}{m_z}}$ and the mode shape $\phi_n(z) = \sin\frac{n\pi z}{L}, n = 1,2,\cdots,\infty$.

Since the displacement $y(z,t)$ is a combination of any number of modals, the expression of $y(z,t)$ is:

$$y(z,t) = \sum_{n=1}^{\infty} q_n(t)\phi_n(z). \tag{10}$$

Bringing the mode shape $\phi_n(z)$ into Equation (10), one obtains:

$$y(z,t) = \sum_{i=n}^{\infty} q_n(t)\sin\frac{n\pi z}{L}. \tag{11}$$

Taking Equation (11) into Equation (1) and simplifying it, we obtain:

$$m_z\sum_{n=1}^{\infty} q_n(t)\phi_n(z)w_n^2 + m_z\sum_{n=1}^{\infty}\ddot{q}_n(t)\phi_n(z) + c\sum_{n=1}^{\infty}\dot{q}_n(t)\phi_n(z) - f(z,t) - \delta(z-s)f_{TMD}(z,t) = 0. \tag{12}$$

To simplify the calculation complexity, we set the structural damping c in the modal damping ratio:

$$c = 2m_z w_n\xi. \tag{13}$$

According to the characteristics of trigonometric functions,

$$\int_0^L \phi_n(z)\phi_m(z)dz = \begin{cases} 0 & n \neq m \\ L/2 & n = m \end{cases}. \tag{14}$$

Multiply each term of Equation (12) by $\phi_m(z)$, integrate from 0 to L, and take Equations (13) and (14) into Equation (12), and we get:

$$m_z q_n(t) \int_0^L \phi_n^2(z) w_n^2 dz + m_z \ddot{q}_n(t) \int_0^L \phi_n^2(z) dz + 2m_z w_n \xi \dot{q}_n(t) \int_0^L \phi_n^2(z) dz \\ - \int_0^L f(z,t) \phi_n(z) dz - f_{TMD}(z,t) \phi_n(s) = 0 \tag{15}$$

After further simplification, we can obtain:

$$\ddot{q}_n(t) + 2w_n \xi \dot{q}_n(t) + w_n^2 q_n(t) = \frac{1}{M_n} P_n(t), \tag{16}$$

where the Nth order mode mass is $M_n = \int_0^L \phi_n^2(z) m_z dz$, and the Nth order mode force is $P_n(t) = \int_0^L f(z,t) \phi_n(z) dz + f_{TMD}(z,t) \phi_n(s)$.

The displacement of the riser connected to the TMD is $y(s,t) = \sum\limits_{i=n}^{\infty} q_n(t) \sin \frac{n\pi s}{L}$. Bring the displacement into Equation (2), where $c_{TMD} = 2m_{TMD} w_{nTMD} \xi_{TMD}$, and we obtain:

$$f_{TMD}(s,t) = -k_{TMD}\left(y_{TMD}(s,t) - \sum\limits_{n=1}^{\infty} q_n(t) \sin \frac{n\pi s}{L}\right) \\ -2m_{TMD} w_{nTMD} \xi_{TMD}\left(\frac{\partial y_{TMD}(s,t)}{\partial t} - \sum\limits_{n=1}^{\infty} \dot{q}_n(t) \sin \frac{n\pi s}{L}\right) = m_{TMD} \frac{\partial^2 y_{TMD}(s,t)}{\partial t^2} \tag{17}$$

Simultaneously in (16) and (17), the dynamic equations of the system can be obtained with infinite degrees of freedom:

$$\begin{cases} \ddot{q}_n(t) + 2w_n \xi \dot{q}_n(t) + w_n^2 q_n(t) = \frac{1}{M_n} P_n(t) \\ f_{TMD}(s,t) = -k_{TMD}\left(y_{TMD}(s,t) - \sum\limits_{n=1}^{\infty} q_n(t) \sin \frac{n\pi s}{L}\right) \\ -2m_{TMD} w_{nTMD} \xi_{TMD}\left(\frac{\partial y_{TMD}(s,t)}{\partial t} - \sum\limits_{n=1}^{\infty} \dot{q}_n(t) \sin \frac{n\pi s}{L}\right) = m_{TMD} \frac{\partial^2 y_{TMD}(s,t)}{\partial t^2} \end{cases} \tag{18}$$

3. Numerical Analysis

3.1. Establishing Numerical Equations

When the mode superposition method is used in the project, the mode mass participation coefficient is required to be above 90% [11], so N is taken as the first five orders, and the mode mass participation coefficient is 92.31%.

$$\ddot{q}_1(t) + 2w_1 \xi \dot{q}_1(t) + w_1^2 q_1(t) = \frac{1}{M_1} P_1(t) \\ \ddot{q}_2(t) + 2w_2 \xi \dot{q}_2(t) + w_2^2 q_2(t) = \frac{1}{M_2} P_2(t) \\ \ddot{q}_3(t) + 2w_3 \xi \dot{q}_3(t) + w_3^2 q_3(t) = \frac{1}{M_3} P_3(t) \\ \ddot{q}_4(t) + 2w_4 \xi \dot{q}_4(t) + w_4^2 q_4(t) = \frac{1}{M_4} P_4(t) \\ \ddot{q}_5(t) + 2w_5 \xi \dot{q}_5(t) + w_5^2 q_5(t) = \frac{1}{M_5} P_5(t) \\ m_{TMD} \frac{\partial^2 y_{TMD}(s,t)}{\partial t^2} + k_{TMD}\left(y_{TMD}(s,t) - \sum\limits_{n=1}^{5} q_n(t) \sin \frac{n\pi s}{L}\right) \\ +2m_{TMD} w_{nTMD} \xi_{TMD}\left(\frac{\partial y_{TMD}(s,t)}{\partial t} - \sum\limits_{n=1}^{5} \dot{q}_n(t) \sin \frac{n\pi s}{L}\right) = 0 \tag{19}$$

Sorting out the above formula, we get the following:

$$M\ddot{U} + C\dot{U} + KU = P, \tag{20}$$

where $U = \left\{ \begin{array}{cccccc} q_1(t) & q_2(t) & q_3(t) & q_4(t) & q_5(t) & y_{TMD}(s,t) \end{array} \right\}^T$.

$$M = \begin{bmatrix} 1 & 0 & 0 & 0 & 0 & 0 \\ 0 & 1 & 0 & 0 & 0 & 0 \\ 0 & 0 & 1 & 0 & 0 & 0 \\ 0 & 0 & 0 & 1 & 0 & 0 \\ 0 & 0 & 0 & 0 & 1 & 0 \\ 0 & 0 & 0 & 0 & 0 & m_{TMD} \end{bmatrix}$$

$$C = \begin{bmatrix} 2w_1\xi & 0 & 0 & 0 & 0 & 0 \\ 0 & 2w_2\xi & 0 & 0 & 0 & 0 \\ 0 & 0 & 2w_2\xi & 0 & 0 & 0 \\ 0 & 0 & 0 & 2w_2\xi & 0 & 0 \\ 0 & 0 & 0 & 0 & 2w_2\xi & 0 \\ -A\sin\frac{1\pi s}{L} & -A\sin\frac{2\pi s}{L} & -A\sin\frac{3\pi s}{L} & -A\sin\frac{4\pi s}{L} & -A\sin\frac{5\pi s}{L} & A \end{bmatrix},$$

$$A = 2m_{TMD}w_{nTMD}\xi_{TMD}$$

$$K = \begin{bmatrix} w_1^2 & 0 & 0 & 0 & 0 & 0 \\ 0 & w_2^2 & 0 & 0 & 0 & 0 \\ 0 & 0 & w_3^2 & 0 & 0 & 0 \\ 0 & 0 & 0 & w_4^2 & 0 & 0 \\ 0 & 0 & 0 & 0 & w_5^2 & 0 \\ -k_{TMD}\sin\frac{1\pi s}{L} & -k_{TMD}\sin\frac{2\pi s}{L} & -k_{TMD}\sin\frac{3\pi s}{L} & -k_{TMD}\sin\frac{4\pi s}{L} & -k_{TMD}\sin\frac{5\pi s}{L} & k_{TMD} \end{bmatrix}$$

$$P = \left\{ \begin{array}{ccccccc} \frac{1}{M_1}P_1(t) & \frac{1}{M_2}P_2(t) & \frac{1}{M_3}P_3(t) & \frac{1}{M_4}P_4(t) & \frac{1}{M_5}P_5(t) & 0 \end{array} \right\}^T$$

3.2. Select Model Parameters

The design of TMD has a large degree of freedom. In this paper, TMD is designed according to the design method proposed in [8], which considers the non-resonant modal contribution of flexible structures. The first-order modal damping ratio of a system consisting of a riser and a TMD is set to $\zeta_d = 0.10$, and the remaining parameters are shown in Tables 1 and 2 below.

Table 1. Risers and TMD parameter table.

Parameters	Value	Parameters	Value
Riser Length L (m)	1000	Outside Diameter of Riser (m)	0.2
Mass Per Unit Length (kg/m)	15	Sea Water Density (kg/m^3)	1024
Structural damping ratio %	5	Flexural rigidity EI (N/m^2)	4×10^9
Pretension F (N)	1.2×10^6	Drag coefficient C_d	1
Drag force amplitude A_D	0.2	phase angle α	0
Lift force coefficient C_L	1	phase angle β	0
TMD Mass (kg)	306	TMD damping ratio %	10.1
TMD spring stiffness (N/m)	245	TMD mass/Riser full length mass	0.02

Table 2. Natural vibration frequency of the riser (Hz).

Modal Order	1	2	3	4	5
Natural vibration frequency	0.1437	0.3009	0.4830	0.6989	0.9546

The data is based on actual measurement data in the South China Sea for one year, and the depth is close to 1000 m, as shown in Figure 2. The data is a mixture of waves and currents. Since it cannot be separated and the current is dominantly away from the water surface, it is considered to be all ocean currents. The direction of the current is not exactly the same in the whole water depth, and the direction of the current is opposite in most months. Most ocean currents are negative at a depth of

(−300,0) m and positive at a range of (−1000,−300) m. The negative velocity is greater than the positive velocity, the (−100,0) m velocity near the sea surface has the highest velocity, and the (−800, −1000) m velocity near the seafloor is almost zero. The max speed of the current in December is −35 cm/s. Compared with the uniform flow and shear flow often used previously in riser calculation, the actual operating current velocity in the field is not large, but the flow direction in full depth is more than a single direction and can be the opposite. It can be seen that the actual current is complicated.

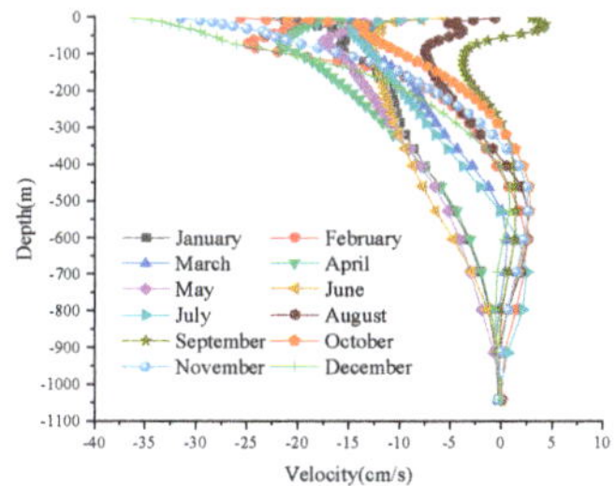

Figure 2. Typical current velocity map from January to December in the South China Sea.

The riser calculation is based on the most unfavorable December current data. The two directions' dynamic response are calculated, respectively, for the in-line (y) direction and the cross-flow (x) direction. We take the drag force f_D (z,t) of Equation (4) into Equation (20) to calculate the dynamic response in the in-line direction, and take the lift force f_L (z,t) of Equation (4) into Equation (20) to calculate the dynamic response in the cross-flow direction. The initial state is static, the displacement is 0, the velocity is 0, and the acceleration is 0. Using the numerical integration Newmark-β method, $\alpha = 0.5$, $\beta = 0.25$, and the average constant acceleration method is unconditionally stable.

4. Numerical Results

4.1. In-Line Direction

The influence of the TMD installation position on the vibration of the riser is analyzed. Figure 3 shows the displacement envelope without and with the TMD installed in different positions (100–900 m, every 100 m). It can be seen that, without TMD, the maximum value of displacement envelope is 0.0876 m at 550 m, and the minimum displacement envelope value is −0.0003 m at 750 m.

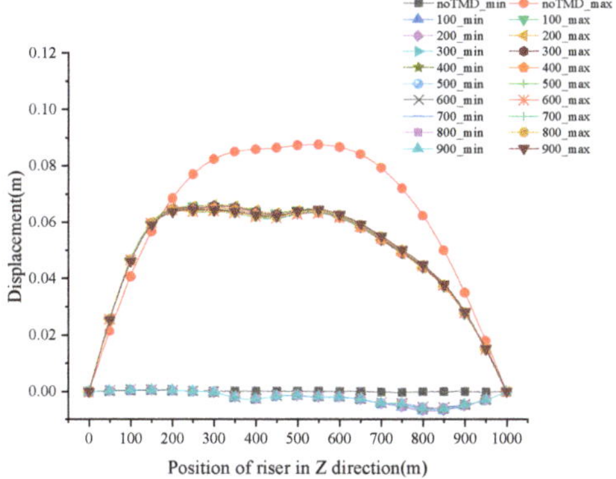

Figure 3. Envelope diagram of the vibration displacement of the riser in the in-line direction when the TMD is installed in different positions.

The target mode of TMD design is mode 1. The maximum displacement of the mode shape of mode 1 is at the center of the riser (500 m). Therefore, in theory, the TMD installed in the center of the riser is the best place to control the riser vibration. However, the calculation shows that the maximum value of displacement envelope of TMD installed at 800 m is the smallest. Moreover, the displacement value at 550 m is 0.0642 m, which is 26.70% lower than that of the riser without TMD. The best installation of TMD at 800 m instead of at 500 m is due to the unevenly distributed current.

The riser displacement envelope value of TMD installed at any position of (0,1000) m is smaller than the riser displacement envelope value without TMD. Besides this, the difference in the displacement envelope value of the riser installed at any position is small and almost coincident. Moreover, the maximum difference is only 3.49% when the TMD is installed at 200 m and 700 m.

Figure 4 is the root mean square (RMS) diagram of the vibration displacement of the riser without the TMD and the riser with the TMD installed in different positions (100–900 m, every 100 m). The maximum displacement response without the TMD is 0.0559 m at 350 m, while the maximum displacement response with the TMD is 0.0459 m at 200 m, and the maximum reduction in the displacement is 17.83%. Thus, the position of the maximum displacement of the two is different. The displacement of the TMD installed at (0,150) m is slightly larger than that without the TMD, while the displacement of the TMD installed at (150,1000) m is significantly smaller than that without the TMD. It can be seen that the TMD installation can effectively reduce the RMS displacement of the riser.

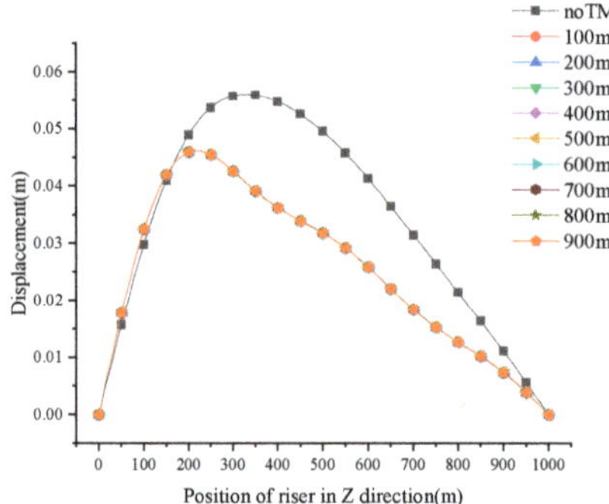

Figure 4. RMS of the vibration displacement of the riser in the in-line direction when the TMD is installed at different positions.

4.2. Cross-Flow Direction

Figure 5 shows the envelope diagram of vibration displacement of riser when the TMD is installed at different positions. Because there is only current and no wave, the envelope diagram shows a symmetrical figure with the displacement of 0 axes—that is, the positive and negative envelope values are symmetrical. The displacement envelope value without TMD completely envelops the displacement envelope value with TMD. Therefore, the TMD reduces the displacement response. Specifically, when the TMD is not installed, the maximum value of displacement envelope is 0.0911 m at 700 m, and the minimum value is −0.0911 m at 700 m. While the TMD is installed at 400 m, the maximum value of displacement envelope is 0.0756 m, which is 17.01% lower than the displacement without TMD, and the minimum displacement envelope value is −0.0762 m in the same position, which is 16.36% lower than the displacement without the TMD.

Figure 6 shows the RMS diagram of the vibration displacement of the riser in the cross-flow direction when the TMD is installed at different positions. When the TMD is not installed, the vibration displacement is the largest, which is 0.0390 m at 700 m. While the TMD is installed, the vibration displacement becomes smaller. In other words, with the TMD installed at 400 m, it is 0.0304 m at 700 m in the z-direction, which is 22.05% lower than that without the TMD. Besides this, the displacement without the TMD at 500 m in the z-direction is 0.0370 m, and the displacement with the TMD installed at 500 m in the z-direction is 0.0234 m, which is 38.38% lower than the displacement without the TMD.

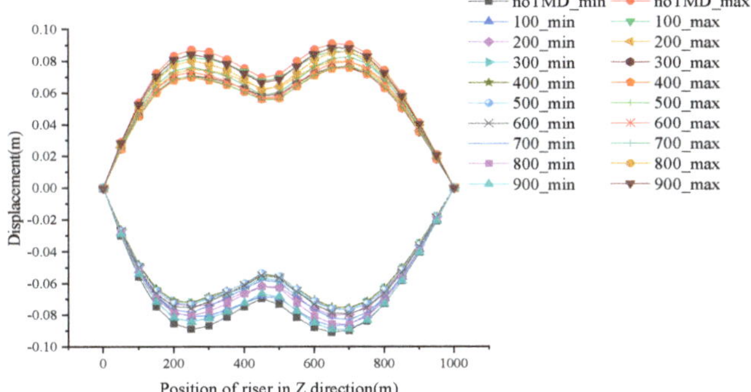

Figure 5. Envelope diagram of vibration displacement of the riser in the cross-flow direction when the TMD is installed at different positions.

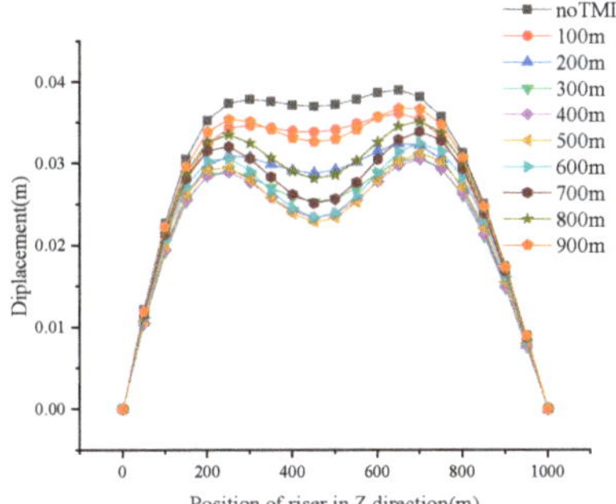

Figure 6. RMS of the vibration displacement of the riser in the cross-flow direction when the TMD is installed at different positions.

The TMD installation position has different effects on the displacement suppression in the in-line and cross-flow directions. Specifically, in the in-line direction, the installation position of the TMD is not sensitive to the vibration displacement control of the standpipe. However, in the cross-flow direction, the installation position of the TMD is more sensitive to the vibration displacement control of the riser. For example, in Figure 5, when the TMD is installed at 100 and 400 m in the z-direction, its maximum displacement envelope is 0.0865 and 0.0756 m, respectively. Compared with the displacement without the TMD, the decrease is 5.05% and 17.01%. The displacement reduction effect of installing TMD at 400m is 3.37 times as much as that of installing TMD at 100m.

5. Conclusions

In this study, a single TMD was used to suppress the vibration of the engineering TTR under the action of the actual ocean current. The dynamic model of a riser-TMD system was established, and the model was calculated using the modal superposition method. The influence of TMD on the vibration displacement of the riser under the South China Sea current was numerically and quantitatively analyzed, which is very helpful for engineering applications. We drew the following conclusions:

1. The TMD could effectively restrain the vibration displacement of the riser. When compared to the condition without the TMD, the maximum value of displacement envelope and the RMS

displacement were reduced by 26.70% and 17.83% in the in-line direction, respectively. Besides, compared to without the TMD, the maximum value of displacement envelope and the RMS displacement were decreased by 17.01% and 22.05% in the cross-flow direction, respectively.

2. In the cross-flow direction, the vibration displacement control was sensitive to the TMD installation position, and the TMD installation should be near the best place (at 400 m in the z-direction of a 1000 m riser); but in the in-line flow direction, the vibration displacement control was not sensitive to the TMD installation position.

The next step is to prepare for experimental verification.

Author Contributions: Conceptualization, J.S., T.W., and W.C.; data curation, J.S.; formal analysis, J.S. and T.W.; funding acquisition, W.C.; investigation, S.G.; methodology, J.S. and D.Y.; project administration, W.C.; resources, W.C.; software, J.S. and D.Y.; supervision, W.C.; validation, T.W. and S.G.; visualization, J.S. and D.Y.; writing—original draft, J.S. and T.W.; writing—review and editing, W.C., J.S., S.G., and D.Y. All authors have read and agreed to the published version of the manuscript.

Funding: This research was funded by the Strategic Priority Research Program of the Chinese Academy of Sciences, grant number XDA22000000.

Acknowledgments: The authors thank the editors and anonymous reviewers for their assistance.

Conflicts of Interest: The authors declare no conflict of interest.

References

1. Chaojun, H. Structural Health Monitoring System for Deepwater Risers with Vortex-Induced Vibration: Nonlinear Modeling, Blind Identification, Fatigue/Damage Estimation and Vibration Control. Ph.D. Thesis, Rice University, Houston, TX, USA, 2013.
2. Hoang, N.; Fujino, Y.; Warnitchai, P. Optimal tuned mass damper for seismic applications and practical design formulas. *Eng. Struct.* **2008**, *30*, 707–715. [CrossRef]
3. Elias, S.; Matsagar, V. Wind response control of tall buildings with a tuned mass damper. *J. Build. Eng.* **2018**, *15*, 51–60. [CrossRef]
4. Kwok, K.C.S.; Samali, B. Performance of tuned mass dampers under wind loads. *Eng. Struct.* **1995**, *17*, 655–667. [CrossRef]
5. Jaiswal, V. Effect of Traveling Waves on Vortex-Induced Vibration of long Flexible Cylinders. Ph.D. Thesis, Massachusetts Institute of Technology, Cambridge, MA, USA, 2009.
6. Nikoo, H.M.; Bi, K.; Hao, H. Passive vibration control of Pipe-In-Pipe (PIP) systems subjected to Vortex Induced Vibration (VIV). In Proceedings of the 27th International Ocean and Polar Engineering Conference, San Francisco, CA, USA, 25–30 June 2017; International Society of Offshore and Polar Engineers: Mountain View, CA, USA, 2017; p. 6.
7. Gosselin, F.; de Langre, E.; Machado-Almeida, B.A. Drag reduction of flexible plates by reconfiguration. *J. Fluid Mech.* **2010**, *650*, 319–341. [CrossRef]
8. Krenk, S.; Høgsberg, J. Tuned mass absorber on a flexible structure. *J. Sound Vib.* **2014**, *333*, 1577–1595. [CrossRef]
9. Faltinsen, O.M. *Sea Loads on Ships and Offshore Structures*; Faltinsen, O.M., Ed.; Cambridge University Press: Cambridge, UK; New York, NY, USA, 1990.
10. Blevins, R.D. *Flow-Induced Vibration*; Van Nostrand Reinhold: New York, NY, USA, 1977.
11. Wilson, E.L. *Three-Dimensional Static and Dynamic Analysis of Structures*; Computers and Structures: Berkeley, CA, USA, 2002.

Journal of
Marine Science and Engineering

Article

Experimental and Numerical Model Investigations of the Underwater Towing of a Subsea Module

Yingfei Zan [1], Ruinan Guo [1], Lihao Yuan [1,*] and Zhaohui Wu [2]

[1] College of Shipbuilding Engineering, Harbin Engineering University, Harbin 150001, China; zanyingfei@hrbeu.edu.cn (Y.Z.); guoruinan@hrbeu.edu.cn (R.G.)
[2] Offshore Oil Engineering Co., Ltd., Tianjin 300451, China; zhaohuiwu112@163.com
* Correspondence: yuanlihao@hrbeu.edu.cn

Received: 26 September 2019; Accepted: 23 October 2019; Published: 29 October 2019

Abstract: In underwater towing operations, the drag force and vertical offset angle of towropes are important considerations when choosing and setting up towing equipment. The aim of this paper is to study the variation in drag force, vertical offset angle, resistance, and attitude for towing operations with a view to optimizing these operations. An underwater experiment was conducted using a 1:8 scale physical model of a subsea module. A comprehensive series of viscous Computational Fluid Dynamics (CFD) simulations were carried out based on Reynolds-averaged Navier–Stokes equations for uniform velocity towing. The results of the simulation were compared with experimental data and showed good agreement. Numerical results of the vorticity field and streamlines at the towing speeds were presented to analyze the distribution of vortexes and flow patterns. The resistance components were analyzed based on the numerical result. It was found that the lateral direction was a better direction for towing operations because of the smaller drag force, resistance, and offset angle. Similar patterns and locations of streamlines and vortexes were present in both the longitudinal and lateral directions, the total resistance coefficient decreases at a Reynolds number greater than that of a cylinder.

Keywords: drag model test; vertical offset angle; drag force; resistance coefficient; CFD simulation; flow field

---

1. Introduction

The oil and gas industry are engaging in offshore operations in deeper and more distant areas. As a result, subsea modules are becoming larger and more complex [1–3]. A key issue is how to transport subsea modules with the lowest risk and cost. Most approaches to addressing these issues can be divided into two categories: (1) the subsea module can be transported on the deck of a vessel and lowered through the surface near the subsea module site, and (2) in wet tow methods, the module can be lowered through the splash zone at inshore sheltered areas and towed on the surface of the water or underwater [4]. When the body is towed on the surface, the floating state, stability, and drag force of the towed body should meet the requirements of the towing operation and take into account the effects of wind and waves. The platform is one of the most commonly towed bodies, and excessive roll and pitch motions in stormy seas can lead to damage to the structure and to the overturning of the platform. For bodies with a large height, the wind load is also an important factor. The towed body can be linked to the vessels either directly by towropes or indirectly by pencil buoys when it is towed underwater [5]. The underwater towing method is used in marine exploration to obtain high-resolution seismic images of the subsurface, including shallow sediments in a deep-sea environment [6]. Unlike the surface towing method, the body is totally underwater and away from the surface, hence the wave and wind loads are no longer significant parameters. Subsea towing does not require vessels with large decks and high lifting capacity and avoids the effects of working in extreme sea states [7].

In recent years, several studies on towing operations have been carried out using numerical or experimental methods. Kang et al. [8] analyzed a specific jack-up model during surface towing by considering the roll, pitch, and heave motions in a stochastic wave input process. They created a reliability-based stochastic analysis method and considered the probability of deck overtopping and instability for roll and pitch. Zhang et al. [9] studied the surface towing dynamic behaviors of an offshore integrated meteorological mast (OIMM) with different towing conditions in various wind and wave conditions using MOSES software developed by Ultramarine. The OIMM with a low draft had the most significant pitch motion fluctuations because of the small righting force, but heave motion showed increases in middle draft states. Ding et al. [10] found that higher mooring positions and towing velocities can achieve moderate dynamic response amplitudes. Their results also showed that the wind load is dominant when the drag force of the towed body is small and that the drag resistance fluctuates greatly because of the influence of the free surface. In the underwater towing method, the drag force is an important parameter in towing operations. Rattanasiri et al. [11] investigated the viscous interaction between autonomous underwater vehicles (AUVs) and studied the influence of their shape using ANSYS CFX 12.1 software, and found that the spacing between hulls determines the drag force of AUVs, and that increasing the spacing results in a lower interaction. The configuration's shape had no positive effect on the drag force for the fleet. Wu et al. [12] designed a controllable underwater towing system and examined the hydrodynamic and control behaviors using towing experiments and the Charge Coupled Device (CCD) underwater photogrammetric technique. Go and Ahn [13] proposed a method for determining hydrodynamic coefficients by using the Computational Fluid Dynamics (CFD) method to simulate the working conditions of a towed-fish. The hydrodynamic model obtained can be used to simulate the motion of a towed-fish. Due to the effect on the stability of the towed body, the hydrodynamic performance and shape variation in the towrope should be considered in the towing system. Sun et al. [14] carried out a new nodal position finite element method to avoid the accumulated errors from time steps over a long-time simulation, and used the new method for towing simulations. To improve the stability of the towed vehicle, a two-part underwater towing method has been developed experimentally and numerically [15,16]. It is preferable to select a sufficiently long secondary cable to improve the hydrodynamic behavior of the towed vehicle for heave and pitch motions [17]. The vertical offset angle of the towrope is limited by the size of the moon-pool, or by heave compensation equipment in subsea towing, and thus the vertical offset angle of the towrope is an important factor to be considered, in addition to the drag force. Jacobsen and Leira [18] investigated the variation in dynamic drag force and offset angle for towropes in different wave and heave periods using both towing experiments and software simulation of marine operations (SIMO) developed by Det Norske Veritas. They also studied the influence of the bottom proximity effects of the added mass.

The aforementioned research provides a description of towing methods, including surface towing and subsea towing. There are two important parameters in the subsea towing method: the drag force and the vertical offset angle of towropes. To study the hydrodynamic performance and characteristics of a subsea module, a subsea towing analysis should be conducted. Different towing directions matched with different towing speeds can form different towing cases. Studies should be carried out to determine which case is better for underwater towing of this subsea module, and to explain why this is so. This involves studying how the towing conditions affect the drag force and offset angle. In the present study, numerical results of the vorticity field and streamlines at different towing speeds were presented to analyze the distribution of vortexes and flow patterns and to explain the effects of drag force, offset angle, and resistance coefficients. The main objective of the present study is to analyze the drag force, vertical offset angle, attitude of the module, and flow characteristics for uniform towing motions using physical towing experiments and numerical simulations (STAR-CCM+ 13.04).

The paper is organized as follows: first, module configuration and towing tests in a calm water are described, and the experimental results of drag force coefficients, vertical offset angle, and the attitude of the module are presented. A brief description of the numerical method used is then presented, followed by a description of the numerical model setup, mesh generation, and validation of the numerical results. Subsequently, the relationships between the drag force coefficients, vertical offset angle, and towing velocities are analyzed. Results of the flow characteristic studies based on the numerical analysis are then presented. In the last section, the components of resistance were analyzed, and the frictional and pressure resistance coefficients were obtained.

2. Tests Using a Physical Model

2.1. Geometry and General Parameters

The model of the subsea module is shown in Figure 1. The subsea module was designed by Offshore Oil Engineering Co., Ltd. for the subsea pipeline from the Wenchang gas fields to the Yacheng pipeline. The data presented in this paper have been rescaled according to Froude's similarity law with a scale ratio $\lambda_m = 8$. The test model is made of Q235 carbon steel. The model consisted of a set of intersecting steel pipes that were welded together between two trusses, with plates that were divided into three parts and welded below the trusses. The trusses were made using a universal beam, using H300A steel for the upper trusses and H300B for the lower ones. Three types of steel pipes were used: $\varphi273 \times 13$, $\varphi168 \times 9$, and $\varphi140 \times 8$. The steel pipes were welded at the inner panel point. The lower truss and perforated plate create a large amount of rectangular space. In accordance with actual construction requirements, it was necessary to punch in each rectangular space. As shown in Figure 2, the diameter of each drain hole on the lowest surface plate was $\varphi = 6.250$ mm. Three perforated plates were enclosed by skirt plates and form a large rectangular cavity under the perforated plate. The thickness of the perforated plate and skirt is 94 mm. The model was 2.210 m long (L), 1.500m wide (B), and 0.440 m deep (H), and its total weight was 104 kg. The submerged weight is 852 N. In this experiment, the Reynolds number Re = $3.78 \times 10^5 > 3.5 \times 10^5$.

Figure 1. Photograph of the physical model.

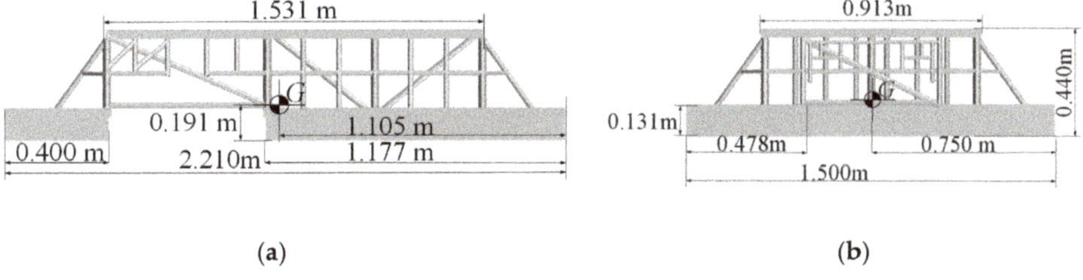

Figure 2. *Cont.*

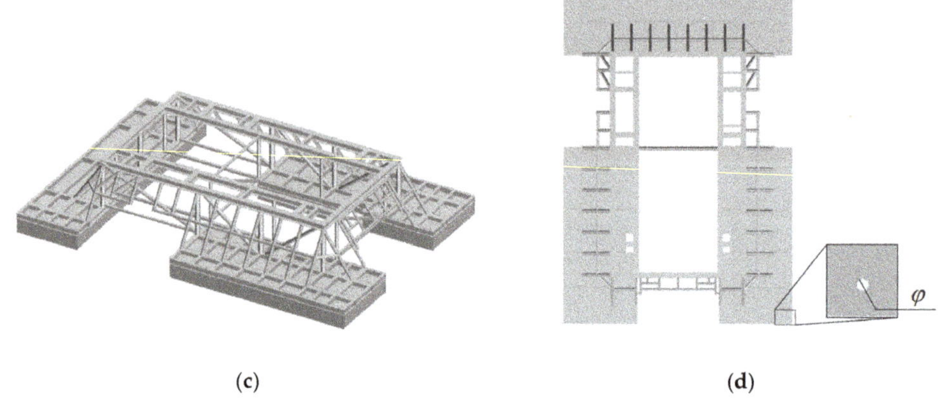

(c)   (d)

Figure 2. Principal layout of the test module: (**a**) side view, (**b**) front view, (**c**) isometric drawing, (**d**) top view.

2.2. Experimental Setup

The model tests were conducted in the towing tank at Harbin Engineering University. The tank is 110 m long, 7 m wide, and 3.5 m deep, and the water temperature was 14 °C. The vertical position of the center of gravity from the bottom was 0.191 m and the horizontal position from the head was 1.105 m. The arrangement of the experimental equipment, north-east-down coordinate system (O-xyz), and the coordinate system of gravity (G-xyz) are shown in Figure 3. The Ox axis is along the longitudinal of the water tank, O is the location of the spin center, and G is the center of gravity. An offset angle transducer was fixed on the carriage and was connected to the main towrope with a force ring. The offset angle transducer can only measure the angle of the main towrope about the Oy axis, the range of measurement is ±20°, and the accuracy is 0.01°. The model was suspended from the force ring through four branches, each 10 mm thick. The branches of the towropes were linked to four points equidistant from the center of gravity on the frame. The coordinates of the four points were (0.760, 0.440, −0.249) m, (0.760, −0.440, −0.249) m, (−0.760, −0.440, −0.249) m, and (−0.760, 0.440, −0.249) m. The model was underwater at all times. The force ring used in this experiment was an elastic force sensor. When the force sensor was stretched or compressed axially, the force signal was converted into a voltage signal, and the numerical value was output by the data acquisition system. The force sensor has a measuring voltage of 5 V, a measuring range of 10,000 N, and an accuracy of 1 N. A camera was used to record the attitude of the model. The camera is installed on the carriage in the third octant of the initial position of G-xyz.

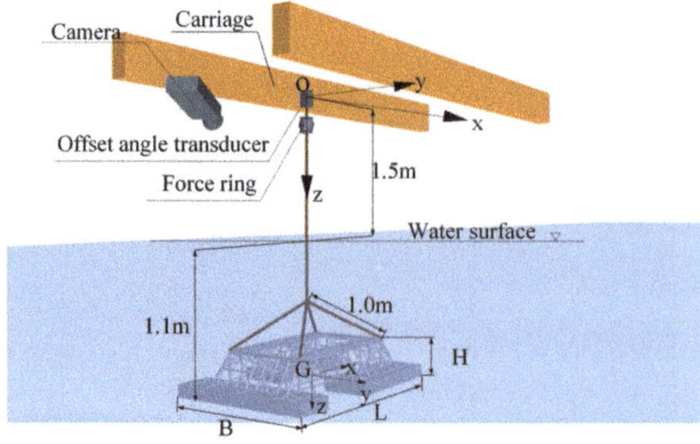

Figure 3. Arrangement of experimental equipment.

To precisely investigate the variation of drag force and offset angle with velocities, each case was applied twice. The test conditions are shown in Table 1. The longitudinal (Gx) direction of the model was along the lengthwise direction of the water tank (Ox) and stable before the X-direction cases. The lateral (Gy) of the model direction was along the lengthwise direction of the water tank (Ox) and stable before the Y-direction cases. The data for the force ring were cleared before each case, and thus the wet weight was excluded.

Table 1. Towed test cases.

Towing Direction: X	Velocity/m·s^{-1}	Towing Direction: Y	Velocity/m·s^{-1}
Case 1	0.2	Case 6	−0.2
Case 2	0.3	Case 7	−0.3
Case 3	0.4	Case 8	−0.4
Case 4	0.5	Case 9	−0.5
Case 5	0.6	Case 10	−0.6

2.3. Physical Modeling of the Module

The force analysis of an object with a submerged weight W suspended by a wire is shown in Figure 4. At any vertical position z of the wire, the system satisfied the following equilibrium function in the Oz direction, with a towing speed of U:

$$F(z) \cos \alpha = W - F_L + \rho g a z + mgl(z) - \int_0^{l(z)} q \sin \alpha \, ds \tag{1}$$

where $F(z)$ is the drag force at the end of the wire, q and mg are the drag force and submerged weight per unit length, respectively, α is the offset angle, a is the cross-sectional area of wire with a length of $l(z)$, and F_L indicates lift force. In the Ox direction, the equilibrium function is

$$F(z) \sin \alpha = F_t + \int_0^{l(z)} q \cos \alpha \, ds - p_0(z) \sin \alpha \tag{2}$$

The first term is total resistance (frictional resistance F_f and pressure resistance F_{pv}). The second term is the horizontal component of the drag force acting on the wire and $p_0(z)$ is the pressure at level z.

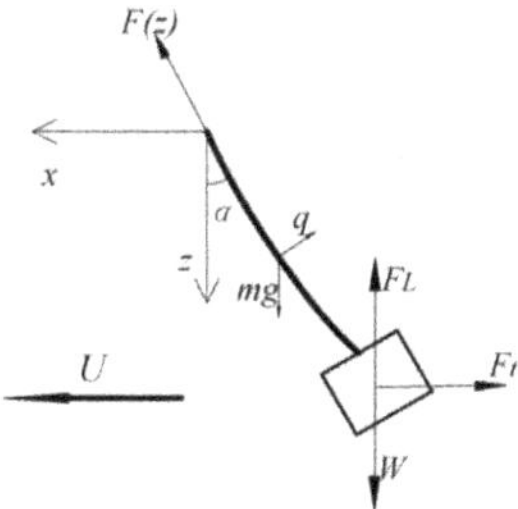

Figure 4. The force analysis of an object suspended by a wire.

Equations (1) and (2) can be approximated and simplified to (3) and (4). The length of wire is small and its hydrodynamic force has little effect when compared with the drag force and weight of the module. Thus, the weight and drag force of the wire can be ignored.

$$F(z) \cos \alpha = W - F_L \tag{3}$$

$$(F(z) + p_0(z)a) \sin \alpha = F_t \tag{4}$$

2.4. Experimental Results

In this section, the drag force and offset angle are analyzed considering the submerged weight of the module. The data obtained from the two cases in the same state were in good agreement with each other. As shown in Figure 5a,b, the drag force ($F_{X,EXP}$) and the offset angle ($\alpha_{X,EXP}$) in the X-direction tow increased rapidly as the model accelerated and reached a maximum, $F_{X,EXP}$ and $\alpha_{X,EXP}$ then decreased slowly. The maximum drag force increased as the towing velocities increased, but the maximum in Case 1 was not obvious. There was a peak after the maximum in Cases 2, 3, 4, and 5, and the peak value also increased as the velocities increased. Finally, $F_{X,EXP}$ and $\alpha_{X,EXP}$ reduced with the model acceleration until stable. After considering the submerged weight of the module, the maximum drag forces in the X and Y directions increased nonlinearly with the towing speed, and the values were very similar. The maximum relative difference was 8.6%. This occurred when the drag speed was 0.4 m/s. The maximum drag force of 1119.037 N occurred when the speed was 0.6 m/s in the X-direction. The maximum drag force was 131.3% of the submerged weight, while the minimum drag force of 917.615 N occurred at a speed of 0.2 m/s, also in the X-direction, and was 107.7% of the submerged weight. The average drag force for a uniform state in the X-direction increased noticeably at two-speed ranges: 0.2 m/s–0.3 m/s and 0.5 m/s–0.6 m/s, and the maximum rate of increase was 66.1%. The minimum rate of increase was 7.4% and appeared in the speed range of 0.3 m/s–0.5 m/s. The average drag force in the Y-direction showed a different trend: a peak occurred at a speed of 0.3 m/s with a value of 908.026 N, while the drag force increased nonlinearly after 0.4 m/s. The model was affected by vortex-induced oscillation, and hence the measurements, especially for drag force, are oscillating when the model is towed in a uniform motion. The amplitude of oscillation increased with increased velocities. The maximum oscillation rate was 6.45% and appeared in Case 3. The maximum change in the drag force was 3.662 N and appeared in Case 5. Vortex-induced vibration had little effect on the offset angle. The maximum relative fluctuation rate was 1.28% and appeared in Case 1. The maximum vertical deflection angle was 0.0977° and occurred in Case 5.

Figure 5c,d shows time series results from the experiments for the drag force ($F_{Y,EXP}$) and offset angle ($\alpha_{Y,EXP}$) in the Y-direction tow. Compared with the X-direction tow, the maximum $F_{Y,EXP}$ value is close to maximum $F_{X,EXP}$, and their maximum difference is 8.6%, which occurred when the velocity was 0.4 m/s. $F_{Y,EXP}$ and $\alpha_{Y,EXP}$ still have oscillations under stable towing, but $F_{Y,EXP}$ and $\alpha_{Y,EXP}$ are closed under different velocities. The values of $\alpha_{Y,EXP}$ show different trends to those of as $\alpha_{X,EXP}$: the values of $\alpha_{Y,EXP}$ stop reducing for a period of time in Cases 8, 9, and 10. In this paper, stable towing means that the drag force and offset angle of the module vary within a certain range. The maximum oscillation rate was 59.8% and appeared in Case 3, and the maximum change in drag force was 29.908 N. The maximum relative fluctuation rate of 1.77% appeared in Case 1, and the maximum vertical deflection angle was 0.0488°. The maximum offset angle for all towing tests appeared at the speed of 0.6 m/s for $\alpha_{X,max,EXP} = 15.879°$. The minimum offset angle appeared at the same speed for $\alpha_{Y,max,EXP} = 2.863°$. The maximum difference between maximum offset angles ($\alpha_{X,max,EXP}$ and $\alpha_{Y,max,EXP}$) was 14.5%, and the minimum was 1.0%.

When the carriage stopped, the offset angle decreased at first, and then increased to the peak value in the opposite direction. The drag force appeared as another peak and finally decreased rapidly. These phenomena occurred because the carriage stopped with a large acceleration in a short time. In the X-direction towing cases, these peak values were greater than the average offset angle in Cases 1 and 2, and the maximum rate was 142.6%. These peak values were less than the average offset angle in Cases 3, 4, and 5, and the maximum rate was 88.3%. Except for Cases 1 and 2, the drag forces do not show peak value as the offset angles, they only decrease rapidly when the carriage stops because the oscillation is greater. The peak values were 100.4% and 101.1% of the average drag forces in Cases 1 and 2, respectively. In the Y-direction cases, the peak values of the offset angle in the stopping state were greater than the peaks in the X-direction and the average values in the Y-direction. The maximum difference of the peaks in the X and Y-directions was 100.4% at a speed of 0.6 m/s, and the minimum difference was 38.4% at a speed of 0.2 m/s. As for the drag force, peak values occurred in Cases 6, 7, and 8. The peak values were 100.9%, 102.4%, and 101.3% of the average values, respectively. It can, therefore, be concluded that the drag force and offset angle in the Y-direction were more easily affected by centripetal forces in the stopping state.

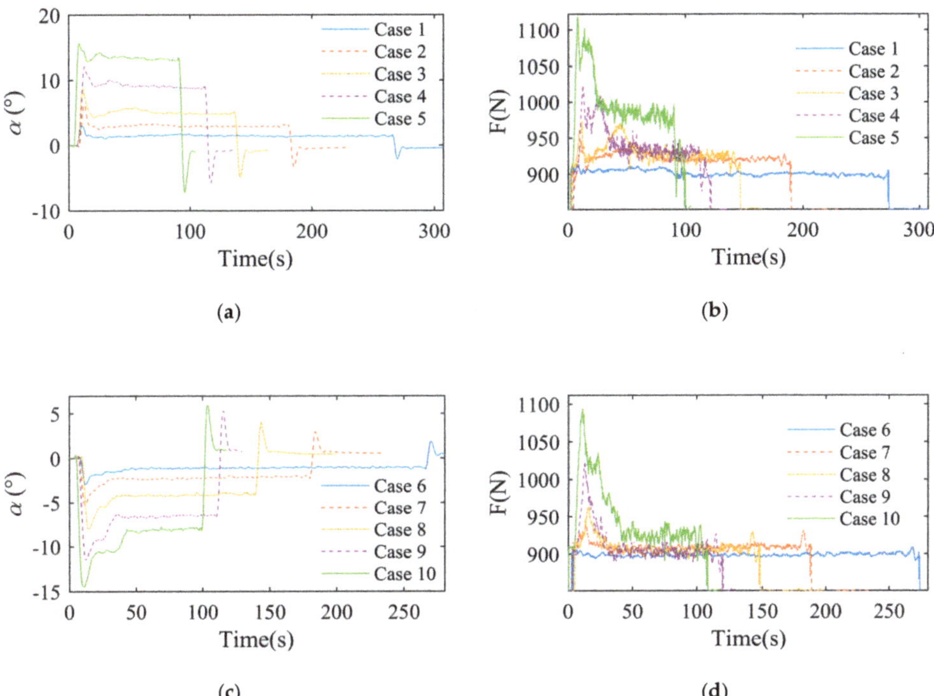

Figure 5. Time histories of drag force and vertical offset angle with different speeds: (**a**) Offset angle variation for X-direction towing, (**b**) Drag force variation for X-direction towing, (**c**) Offset angle variation for Y-direction towing, (**d**) Drag force variation for Y-direction towing.

The relationship between the average drag coefficient (C_d), the average drag force (F), the wetted area (A), and drag velocity (U) is given by [19]:

$$C_d = \frac{2F}{\rho A U^2} \tag{5}$$

where ρ is the density of water. The module is too irregular to be treated as a cylinder and has a large scale like that of a ship. Consequently, the wetted area was selected as a non-dimensional parameter where $A = 17.290 \text{ m}^2$.

The comparisons of variation in the average drag coefficient and offset angle with velocity in different towing directions are shown in Table 2 and Figure 6. The average drag force was calculated from the average of $F_{Y,EXP}$ and $F_{X,EXP}$ in a stable towing state. As illustrated in Table 2, the drag coefficients decrease with velocity, and they change little in high-velocity states. The maximum average drag coefficients occurred at a speed of 0.2 m/s ($C_{dX,EXP} = 2.607$ and $C_{dY,EXP} = 2.603$) and the minimum coefficients occurred at a speed of 0.6 m/s (0.316 and 0.299). The average drag force coefficients $C_{dX,EXP}$ and $C_{dY,EXP}$ are very similar but $C_{dX,EXP}$ is larger than $C_{dY,EXP}$ for the same towing speed. The maximum difference between $C_{dX,EXP}$ and $C_{dY,EXP}$ was only 5.4% at the speed of 0.6 m/s. The reason for this difference is that the resistance in the X-direction is greater than in the Y-direction, especially for pressure resistance. A more detailed discussion is provided in Section 4.4.

Table 2. Variation in drag coefficient and offset angle with velocity.

Velocity/m·s^{-1}	$C_{dX,EXP}$	$C_{dY,EXP}$	$\Delta C_d\%$	$\alpha_{X,EXP}/°$	$\alpha_{Y,EXP}/°$	$\Delta\alpha\%$
0.2	2.607	2.603	0.15	1.490	1.063	28.7
0.3	1.180	1.167	1.1	2.794	2.168	22.4
0.4	0.667	0.652	2.2	5.301	4.072	23.2
0.5	0.431	0.419	2.8	9.053	6.535	27.8
0.6	0.316	0.299	5.4	13.430	9.073	32.4

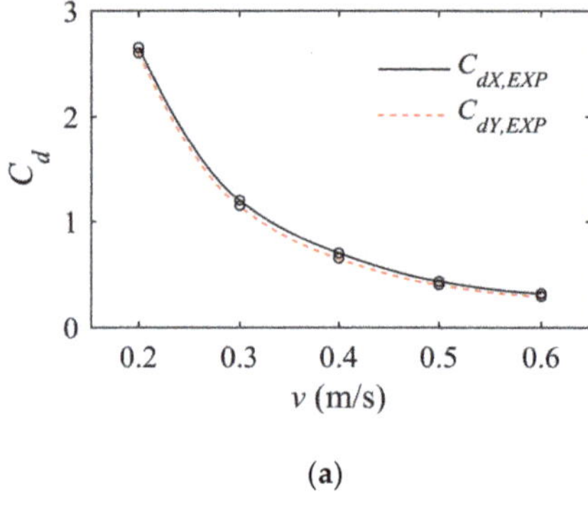

(a)

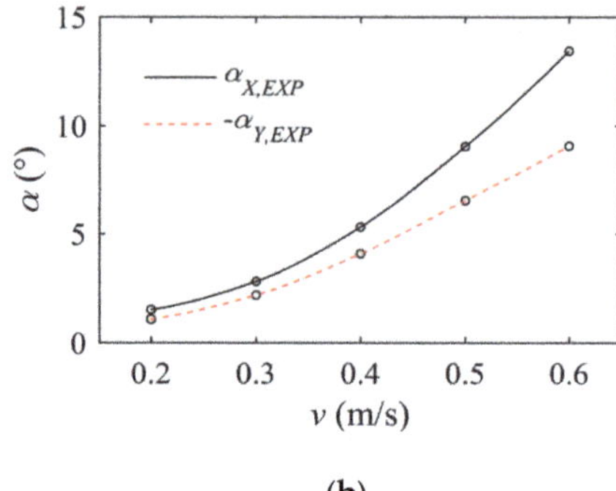

(b)

Figure 6. Comparison of average drag coefficients and offset angles: (**a**) Drag force coefficient variation for X and Y-direction towing, (**b**) Offset angle variation for X and Y-direction towing.

The vertical offset angle increased with velocity, the average offset angle in the X-direction towing shows a clear non-linear variation, but the maximum offset angle shows an almost linear increase. In stable towing states, the maximum average offset angle $\alpha_{X,EXP}$ = 13.430° at 0.6 m/s, and the minimum offset angle $\alpha_{Y,EXP}$ = 1.063° at 0.2 m/s. The maximum difference between the average offset angles in the X and Y-direction tests ($\alpha_{X,EXP}$ and $\alpha_{Y,EXP}$) was 32.4%, and the minimum was 22.4%. As shown in Figure 6, the difference in the drag force coefficients was very small. However, there was an obvious difference between offset angles in the X and Y-direction and $\alpha_{Y,EXP}$ was less than $\alpha_{X,EXP}$. Thus Y-direction towing could be a better choice if the offset angle limit is more important for towing operations, especially for high speed towing.

Figure 7 shows a series of images of the attitude of the model at different times when the speed is 0.4 m/s. Other cases show the same phenomenon. The blue arrow indicates the towing direction. In Figure 7a, it can be seen that the longitudinal of the model is parallel with the velocity, and at this time the offset angle is the largest. As shown in Figure 7b, the model starts to rotate when the offset angle decreases. Then, as shown in Figure 7c,d, the model clearly starts yaw motion, and the drag force and offset angle gradually stabilize. Finally, the model maintains the attitude as shown in Figure 7e: the longitudinal is almost vertical to the towing direction, and the heading angle changes repeatedly over a small range. Due to the asymmetry in the model shape, in the transition from acceleration to stable towing, the point of application of fluid forces moves with the fluctuation of the towing velocity and the change of the attitude. This forms an unbalanced hydrodynamic force that makes the heading angle of the model unstable. Finally, the heading angle of the model changes until the hydrodynamic force is symmetrically distributed again, and the model is stable and moves forward with an almost fixed heading angle.

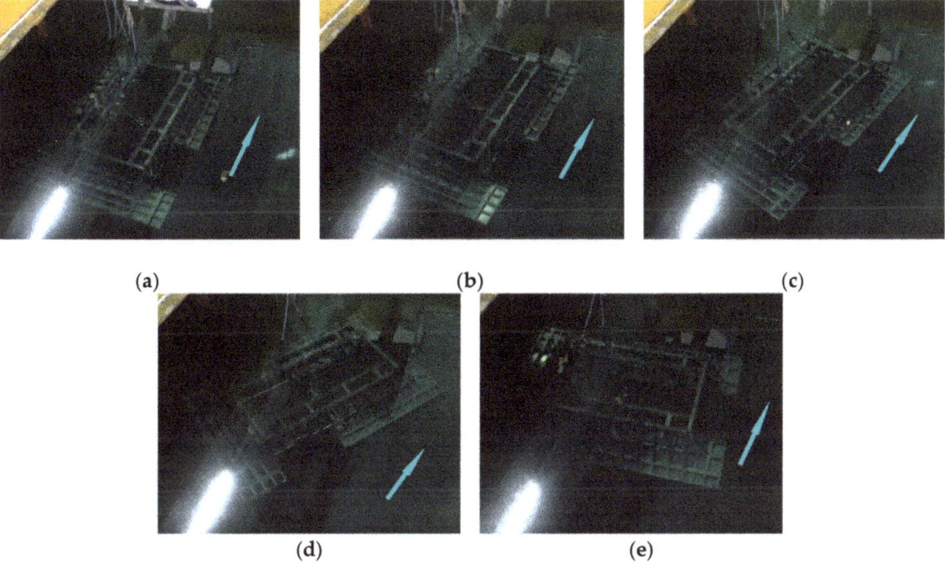

Figure 7. Change in attitude of the model in the towing experiment: (**a**) offset angle is at a maximum, (**b**) offset angle decreasing, (**c**) transitional phase, (**d**) transitional phase, and (**e**) stable.

3. Numerical Simulations

To verify the experimental results presented above, a series of CFD simulations, with both X-direction and Y-direction towing were carried out. Furthermore, a detailed study of the flow field was made to reconstruct the vortex patterns. In this section, details of the numerical setup are provided using the X-direction towing test as an example.

3.1. Mathematical and Numerical Models

Each case was simulated using the Reynolds-Average Navier-Stokes equation (RANS) method. Numerical simulations used the CFD software STAR-CCM+ 13.04, developed by SIEMENS in Berlin, Germany. The governing equation was modeled using RANS [20]:

$$\frac{\partial \rho}{\partial t} + \nabla(\rho \, \overline{\mathbf{v}}) = 0 \tag{6}$$

$$\frac{\partial}{\partial t}(\rho \, \overline{\mathbf{v}}) + \nabla(\rho \, \overline{\mathbf{v}} \otimes \overline{\mathbf{v}}) = -\nabla \, \overline{p}\mathbf{I} + \nabla(\mathbf{T} + \mathbf{T}_t) + \mathbf{f_b} \tag{7}$$

where ρ is the fluid density, $\overline{\mathbf{v}}$ and $\overline{p}$ are the mean velocity and pressure, respectively, $\mathbf{I}$ is the identity tensor, $\mathbf{T}$ is the viscous stress tensor, $\mathbf{f_b}$ is the resultant of the body force, and $\mathbf{T}_t$ is Reynolds stress tensor.

The finite volume method was employed to discretize the governing equations with the Segregated Flow Solver. The RANS equation was solved using the Semi-Implicit Method for Pressure Linked Equations (SIMPLE) pressure-velocity coupling algorithms to decouple pressure and velocity. To provide closure of the governing equations, the Eddy viscosity model was introduced to the model in terms of the mean flow quantities. The $k - \varepsilon$ model was chosen. This is a two-equation model that solves transport equations for the turbulent kinetic energy k and the turbulent dissipation rate ε to determine the turbulent eddy viscosity. The specified equations are described as follows:

$$\frac{\partial}{\partial t}(\rho k) + \nabla \cdot (\rho k \overline{\mathbf{v}}) = \nabla \cdot \left[\left(\mu + \frac{\mu_t}{\sigma_k} \right) \nabla k \right] + P_k - \rho(\varepsilon - \varepsilon_0) + S_k \tag{8}$$

$$\frac{\partial}{\partial t}(\rho\varepsilon) + \nabla\cdot(\rho\varepsilon\bar{\mathbf{v}}) = \nabla\left[\left(\mu + \frac{\mu_t}{\sigma_\varepsilon}\right)\nabla\right] + \frac{1}{T_e}C_{\varepsilon 1}P_\varepsilon - C_{\varepsilon 2}f_2\rho\left(\frac{\varepsilon}{T_e} - \frac{\varepsilon_0}{T_0}\right) + S_\varepsilon \tag{9}$$

where μ is the dynamic viscosity, and σ_k, σ_ε, $C_{\varepsilon 1}$, and $C_{\varepsilon 2}$ are model coefficients. P_ε and P_k represent production terms. The damping function is represented by f_2, while S_k and S_ε are user-specified source terms. The larger-eddy time-scale $T_e = k/\varepsilon$. The relationship between the specific time-scale T_0, the model coefficient C_t, the kinematic viscosity v, and the ambient turbulence value ε_0 is defined by:

$$T_0 = max\left(\frac{k_0}{\varepsilon_0}, C_t\sqrt{\frac{v}{\varepsilon_0}}\right) \tag{10}$$

The RANS method and $k - \varepsilon$ model have been successfully used in ocean engineering and provide a good compromise between robustness, computational cost, and accuracy [21].

The force on the surface along direction vector $\mathbf{n}_f$ is computed as:

$$\mathbf{F} = \sum_f\left(\mathbf{f}_f^{pressure} + \mathbf{f}_f^{shear}\right)\cdot\mathbf{n}_f \tag{11}$$

where $\mathbf{f}_f^{pressure}$ and $\mathbf{f}_f^{shear}$ are the pressure and shear force vectors, respectively, on the surface face f. The pressure force vector $\mathbf{f}_f^{pressure}$ along direction vector $\mathbf{a}_f$ on the surface face f is related to the face static pressure p_f and the reference pressure p_{ref} according to:

$$\mathbf{f}_f^{pressure} = \left(p_f - p_{ref}\right)\mathbf{a}_f \tag{12}$$

The sheer force vector on the surface face f is computed as:

$$\mathbf{f}_f^{shear} = -T_f\cdot\mathbf{a}_f \tag{13}$$

3.2. Boundary Conditions

To simulate the vorticity field around the model, a calculation domain was first established for the whole body. In this study, the domain size shown in Figure 8 was adopted. It can be seen that the domain extends for 2 L in front of the overset, 4 L behind the overset, 1.5 L to the side, 1 L below the overset, and 1.1 L above the overset. The distance between the boundaries of the calculation domain and module is more than 2 L in the longitudinal direction and more than L in the lateral and vertical direction when the module has its maximum moving range [22]. The flow is stable between the module and the inlet surface and does not cause backflow within 2 L [23].

The boundary conditions are specified as follows: the model is considered as a moving boundary, and a no-slip condition is imposed on the model surface, symmetry conditions are used for the top, bottom, and side boundaries, at the velocity inlet, the flow velocity is defined as the tested velocity in each case, and at the pressure outlet, the initial hydrostatic pressure is defined as constant.

In this study, we used a spherical joint coupling to replace towropes. The spherical joint restricts the relative motion of the two bodies to a pure rotation about the joint position. A relative translation of the two rigid bodies is not allowed. The spherical joint coupling is often called a ball-and-socket joint. The motion of the spherical joint is the same as for towropes.

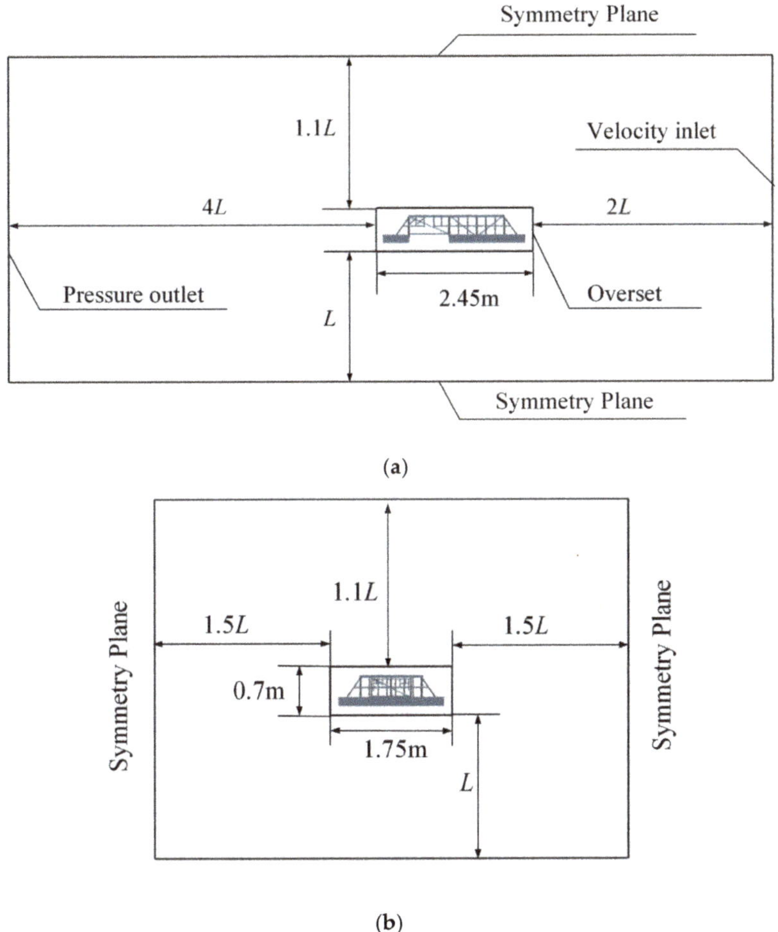

Figure 8. Computational domain and boundary conditions: (**a**) Side view of the computational domain, (**b**) Front view of the computational domain.

3.3. Mesh Generation

Because the model has a large movement, a dynamic mesh was adopted. As shown in Figure 9, a trimmed cell mesh was used for the discretization of the whole domain. The mesh generation was conducted carefully to ensure computational accuracy. The computational domain was separated into three regions: stationary, transition, and overset [24]. The target size of the mesh was 0.025 m in the overset and transition region to avoid half grids and equates to $B/60$ and nearly 0.01 L. The minimum size was 2 mm, which can precisely describe the structure of the module. To stabilize the calculation on the interfaces, the boundaries of the overset region were set above four layers from the module. The area of the transition mesh was large enough to include the range of motion, and the boundaries were more than 10 layers from the boundaries of the overset region, even when the offset angle approached the maximum. The model moves with the overset region in the transition region using the overset mesh and Dynamic Fluid Body Interaction (DFBI) solver. The stationary and transition regions did not change during the dynamic mesh process [25]. The total number of cells in the grid was 3,200,593. This fine mesh size provides a good distribution of most of the variables around the model.

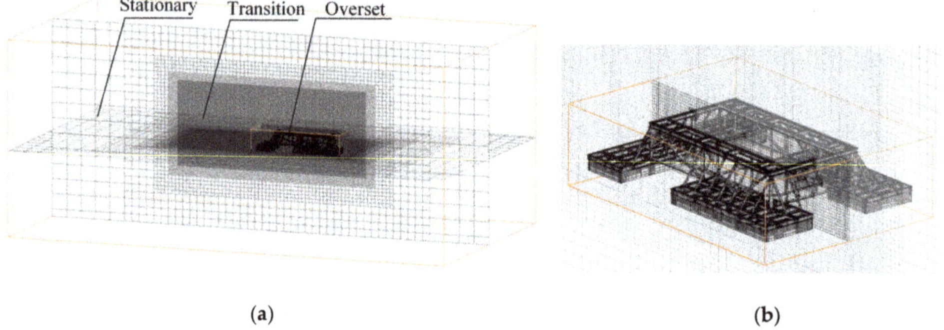

(a)   (b)

Figure 9. Computational mesh: (**a**) Mesh arrangement in the computational domain, (**b**) Mesh arrangement on the model.

4. Result and Discussion

4.1. Validation of Numerical Prediction

Figure 10 shows the calculated drag force coefficient and the vertical offset angle in comparison with experimental measurements for each case. The drag coefficients decreased non-linearly with velocity. The agreement was reasonable overall for all cases, and the relative error was usually less than 10%. For X-direction towing, the discrepancy between the drag force coefficient from the experimental data ($C_{dX,EXP}$) and numerical data ($C_{dX,CFD}$) was moderately high, especially for Case 2, where the relative difference was 5.4% (the absolute error was only 0.064). In other cases, the relative errors are less than 5%. The vertical offset angle in the X-direction tests ($\alpha_{X,EXP}$) and in simulations ($\alpha_{X,CFD}$) showed good agreement at the highest speeds: their relative difference was lower than 6.4%. The maximum error was 13.1%. However, the maximum absolute error in Cases 1 and 2 was just 0.284°. For Y-direction towing, the relative errors between $C_{dY,EXP}$ and $C_{dY,CFD}$ in the numerical simulations and the experiments all varied between 1.0% and 3.2%, which meets the needs of engineering applications. The maximum discrepancy of 0.056 occurred in Case 6. The maximum deviation between the offset angle in experiments ($\alpha_{X,CFD}$) and simulations ($\alpha_{Y,CFD}$) was 8.4% and the average was 5.8%. Although the average error of the offset angle in Y-direction towing appears large, the numerical difference was only about 0.218°, on average.

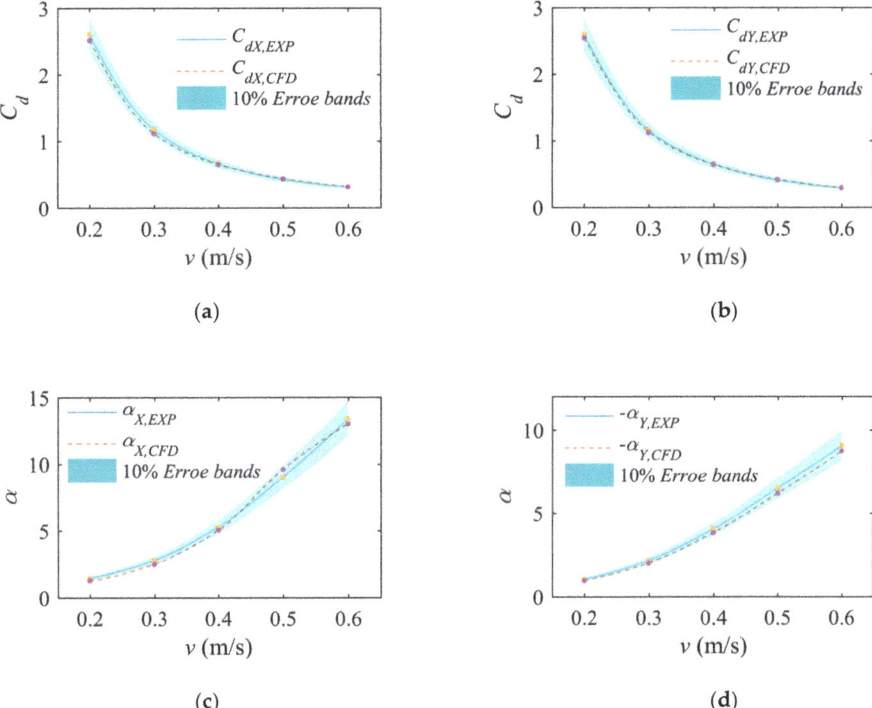

Figure 10. Drag coefficient and offset angle variations with velocity in experiments and simulations: (**a**) The variation of drag coefficient with towing speed in X-direction towing, (**b**) The variation of drag coefficient with towing speed in Y-direction towing, (**c**) The variation of offset angle with towing speed in X-direction towing, (**d**) The variation of offset angle with towing speed in Y-direction towing.

The comparison of the attitude of the numerical and physical model in a stable towing state is shown in Figure 11. The towing direction is along the X axial of the coordinate system in the Figure. The longitudinal of the model is almost perpendicular to the towing direction and shows the same phenomenon as in Figure 11b. Therefore, the validity of the numerical method is proven.

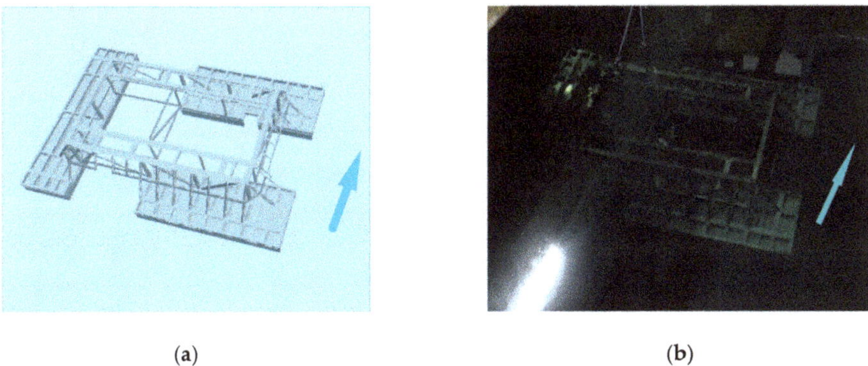

Figure 11. Stable towing state in Case 4: (**a**) Attitude of the module in the simulation, (**b**) Attitude in the experiment.

4.2. Detailed Vorticity Field Analysis

Figure 12 illustrates cross-sections of the flow field where the local velocity component along the negative of the X axial is equal to the speed of Cases 1–5. The cross-sections are colored according to vorticity magnitude. To clearly and carefully describe the flow, two-color bars were adopted. As is shown in Figure 12, there were two main vortexes that appeared in Case 1: one series of the vortex was generated at the top corner in the inlet section of the model (hereafter called vortex A), and another arose at the lower end (hereafter called vortex B). Both vortex A and B occurred mainly along the upper edge of the steel structure, while there was a little vortex at the end of the model in the outlet section. In Case 2, the main vortexes are still vortex A and B, but the distribution is larger than in Case 1. In addition, there were some vortexes being created in the middle of the model behind the steel pipe in the inlet section (hereafter called vortex C). For the intermediate towing velocities (Cases 3 and 4), in addition to vortex A and B, another series of vortexes was generated at the top corner in the outlet section (hereafter called vortex D). For the high towing velocity (Case 6), the distribution of vortex A, B, and C was larger than in the other cases, and another series of vortexes was generated at the lower end of the model in the outlet section (hereafter called vortex E). The reason for the occurrence of vortexes D and E is that the increasing trim angle caused an increase in the area of the incident flow surface in the outlet section. The arrows indicate the direction of the flow.

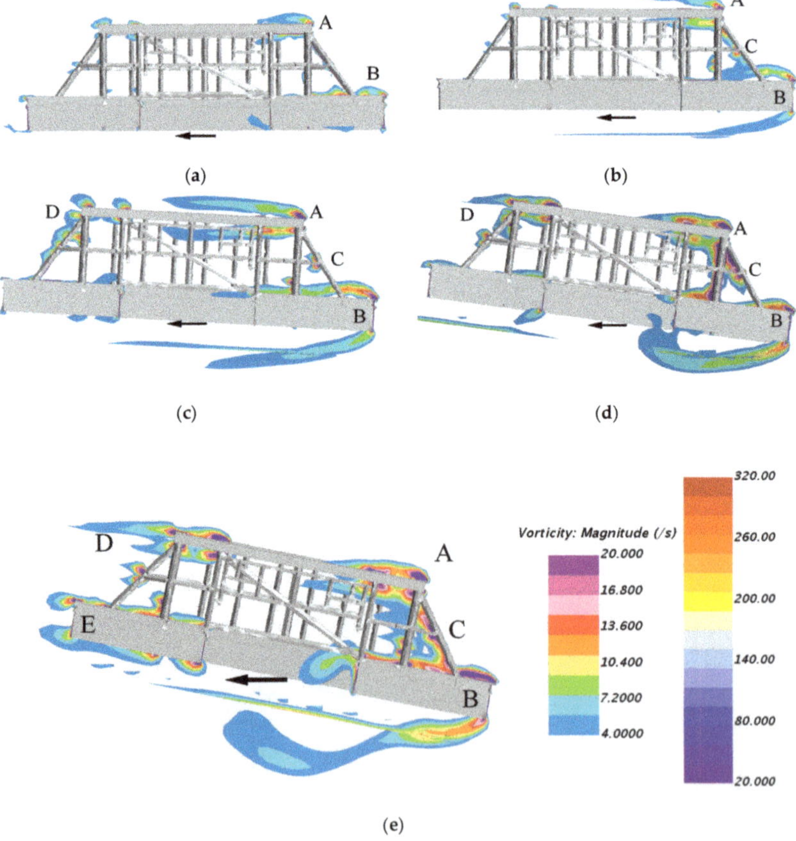

Figure 12. Cross-sections of the vortex flow field colored according to vorticity magnitude for X-direction towing: (**a**) Case 1, $\alpha_{X,CFD} = 1.295°$, (**b**) Case 2, $\alpha_{X,CFD} = 2.510°$, (**c**) Case 3, $\alpha_{X,CFD} = 5.073°$, (**d**) Case 4, $\alpha_{X,CFD} = 9.634°$, and (**e**) Case 5, $\alpha_{X,CFD} = 13.038°$.

According to the discussion above, the highest number of vortices and the largest distribution of the vortex field occur at high velocities. Therefore, top and bottom views of the vorticity visualization on the surface of the model in Case 5 are shown in Figure 13. The arrow indicates the direction of flow. As illustrated in Figure 13a, some of the high vorticity areas (over 214 Hz) arise at the upper surface on the truss in the outlet section, while others appear in the inlet section of the skirt plates, steel pipes, and at the front of the model. As shown in Figure 13b, there are three main areas with vorticity above 128 Hz under the perforated plates, and they all occur around the holes in the inlet section in rectangular spaces. Even though there are no large areas with high vorticity, the vorticity around the holes was over 192 Hz.

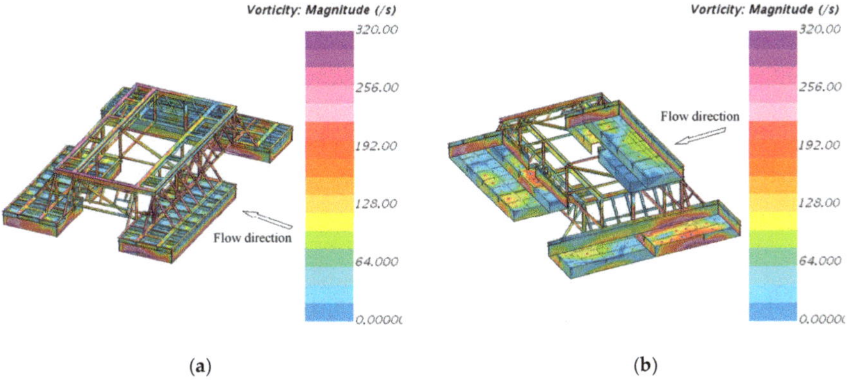

Figure 13. Vorticity visualization on the surface of the model in Case 5: (**a**) The top view of the flow field, (**b**) Lower view of the flow field.

For the Y-direction towing simulations, as shown in Figure 14, vortex A and B appeared in Case 6 to Case 10, and their distribution became wider and larger with the increase in velocity. Vortex C appeared mostly in Cases 7 to 10 and extended from the inlet section to the middle of the model. What is different from Case 1 is that vortex C was found in Case 6 and had a larger extent because there were more complex structures than for the X-direction towing at the inlet section. Vortex D occurred in Cases 8, 9, and 10. What is different from the X-direction towing cases is that the distribution of the vortex was narrower but longer for the same towing speed. Vortex E was not obvious in Y-direction towing cases because the longer perforated plate induced lower vorticity. The existence of the vortexes is the reason for the oscillation in the experimental data. The arrows indicate the direction of the flow.

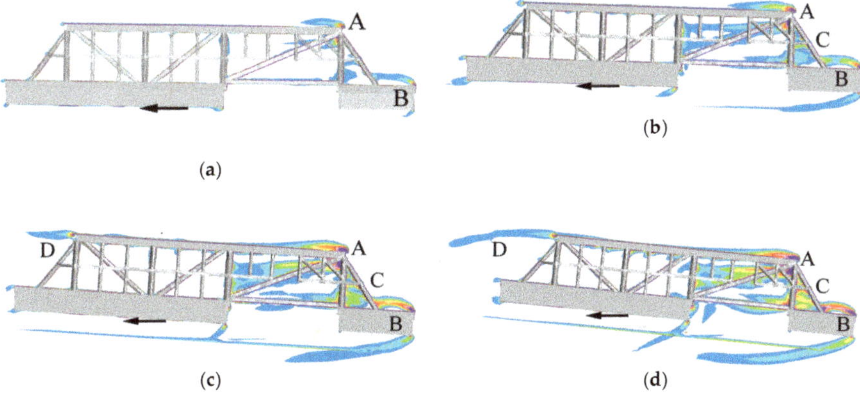

Figure 14. *Cont.*

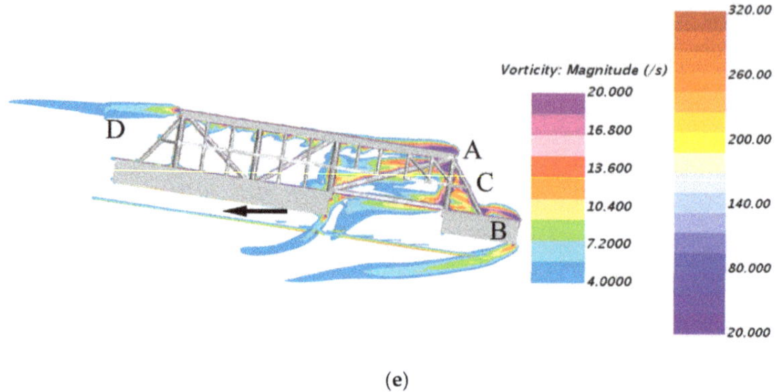

(e)

Figure 14. Cross-sections of flow field colored according to vorticity magnitude for Y-direction towing: (**a**) Vortex field in Case 6, $\alpha_{Y,CFD} = 0.974°$, (**b**) Vortex field in Case 7, $\alpha_{Y,CFD} = 2.018°$, (**c**) Vortex field in Case 8, $\alpha_{Y,CFD} = 3.838°$, (**d**) Vortex field in Case 9, $\alpha_{Y,CFD} = 6.218°$, (**e**) Vortex field in Case 10, $\alpha_{Y,CFD} = 8.770°$.

The vorticity visualization on the surface of the model in Case 10 for top and bottom views is shown in Figure 15. In terms of the direction of flow, the vortex distribution in Y-direction towing is similar to that in Case 6. The difference is that the vortex distribution in Case 10 is nearly symmetrical because of an almost symmetrical model. It is interesting that there are no large areas with a vorticity above 106 Hz, and that there are only two large areas with a vorticity of over 64 Hz. The areas with a vorticity of over 64 Hz are in the outlet section under the perforated plates.

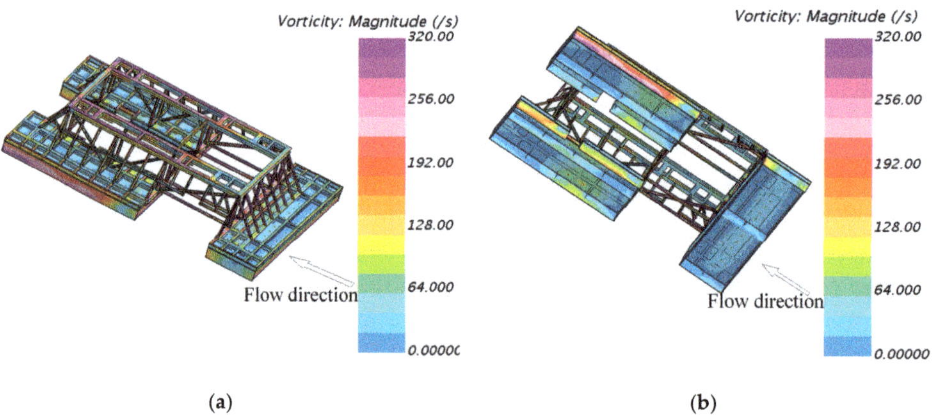

(a)   (b)

Figure 15. Vorticity visualization on the surface of the model in Case 10: (**a**) Top view of the flow field, (**b**) Bottom view of the flow field.

4.3. Detailed Flow Patterns Analysis

According to the analysis in the previous section, the flow field in the inlet section is more complex than that in the outlet section. Therefore, the detail of the flow pattern for the inlet section is needed. A more detailed set of visualizations is provided in Figures 16–19. The streamline inlet line is located near the holes above the perforated plate in the inlet section and is perpendicular to the direction of velocity. In Figure 16, there are two main observed flow patterns: pattern A and pattern B. The streamlines of pattern A, originating from the inlet, propagate according to a recirculating path below the perforated plate induced by external water flow from the velocity inlet surface. The recirculating path is mainly formed by the reflection of the skirt panel in pattern A. The streamlines of pattern B propagate from the inlet line to the middle of the model and then flow below the perforated plate because of the effect of the perforated plate, and finally converge on the streamlines of pattern A. The proportion of pattern A streamlines increased with velocity, and the recirculating path appeared in almost the same place. The recirculating path was not obvious in Case 1 because it was mainly caused by the flow through the holes on the perforated plate at a low velocity. As the towing velocity increased, more streamlines flowed over the perforated plate and were reflected by the skirt panel in the outlet section. The streamlines of pattern B become denser in low-velocity cases, and the streamlines flowed through the top of the perforated plate in Cases 1 and 2 because the trim angle was small and the streamlines were not significantly hindered by the perforated plate, especially in Case 1.

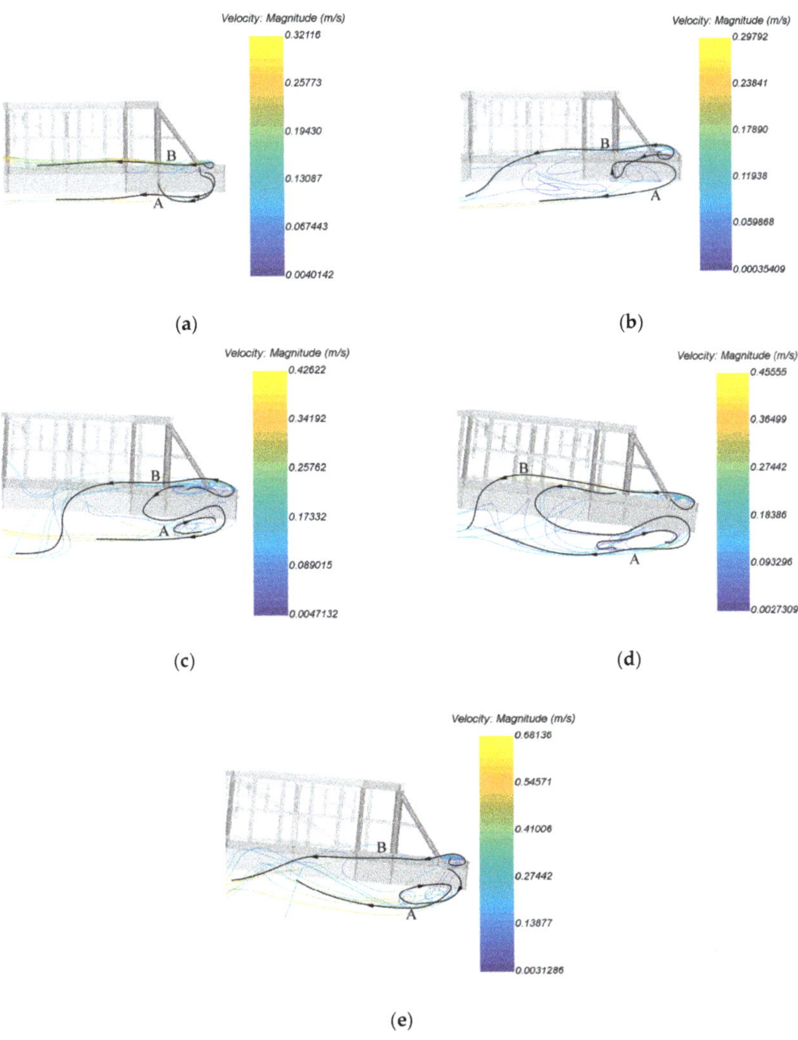

Figure 16. Flow patterns in the inlet section for X-direction towing cases: (**a**) Flow patterns in Case 1, (**b**) Flow patterns in Case 2, (**c**) Flow patterns in Case 3, (**d**) Flow patterns in Case 4, (**e**) Flow patterns in Case 5.

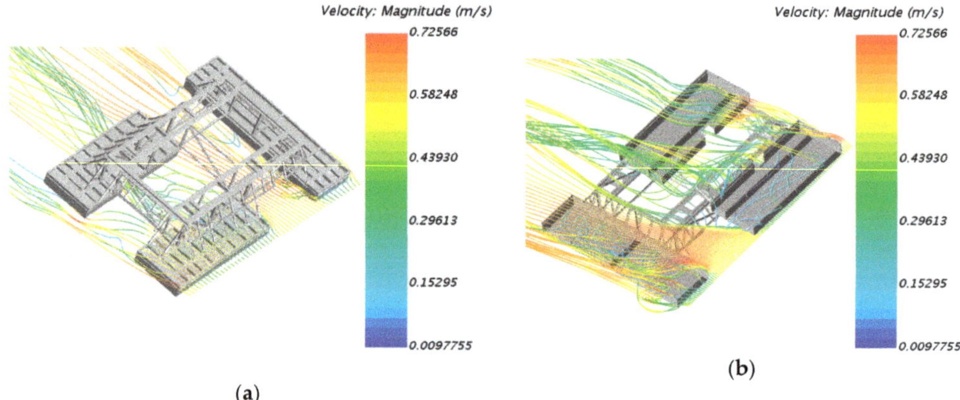

Figure 17. Plan view of streamlines from the inlet section in Case 6: (**a**) top of the flow field, (**b**) bottom of the flow field.

Figure 17 shows plan views of streamlines from the inlet section in Case 6. The streamlines in front of the model flow through the perforated plate and speed up, and are then separated by the steel pipes and flow below the perforated plate in the outlet section, and finally divide into two parts to converge with streamlines from far-field. Some of the front streamlines form large recirculating paths under the perforated plate and are forced to the back of the model. The streamlines at the back of the model flow through the side of the skirt plate to the bottom and form another recirculating path. Some of the streamlines flow down the perforated plate and speed up to converge with streamlines from far-field.

Figure 18 shows inlet flow patterns for Y-direction towing cases. The propagation of streamlines in patterns A and B was similar to that in X-direction cases. The difference is that the recirculating path clearly appears either in low-speed or high-speed cases. This occurs because the length between skirt panels in the Y-direction is larger than in the X-direction, and because there are more holes in the inlet section in the Y-direction. The reason why another distinct recirculating path arises at the middle of the model in Cases 7 and 8 is that the streamlines reflected by the skirt panel can only form an obvious recirculating path in the outlet section at the highest speeds, but the first recirculating path moves back when the speed is too high. Hence, two parts of the recirculating path combine together and are continuously distributed from the inlet section to the middle of the model in Case 10.

Figure 19 shows the perspective view of streamlines from the inlet section in Case 10. As with X-direction towing, the streamlines in front of the model flow through the perforated plate and speed up, are separated by the steel pipes, and then flow below the perforated plate in the outlet section. What is different in the Y-direction towing cases is that the streamlines were divided into three parts, which converge with streamlines from the far-field. The streamlines on the side flow down to the perforated plate to form a recirculating path and flow to the middle of the model. The streamlines in the middle of the model flow to the side because of the flow induction on the side of the model after acceleration, which causes little direct streamline flow through the middle of the model.

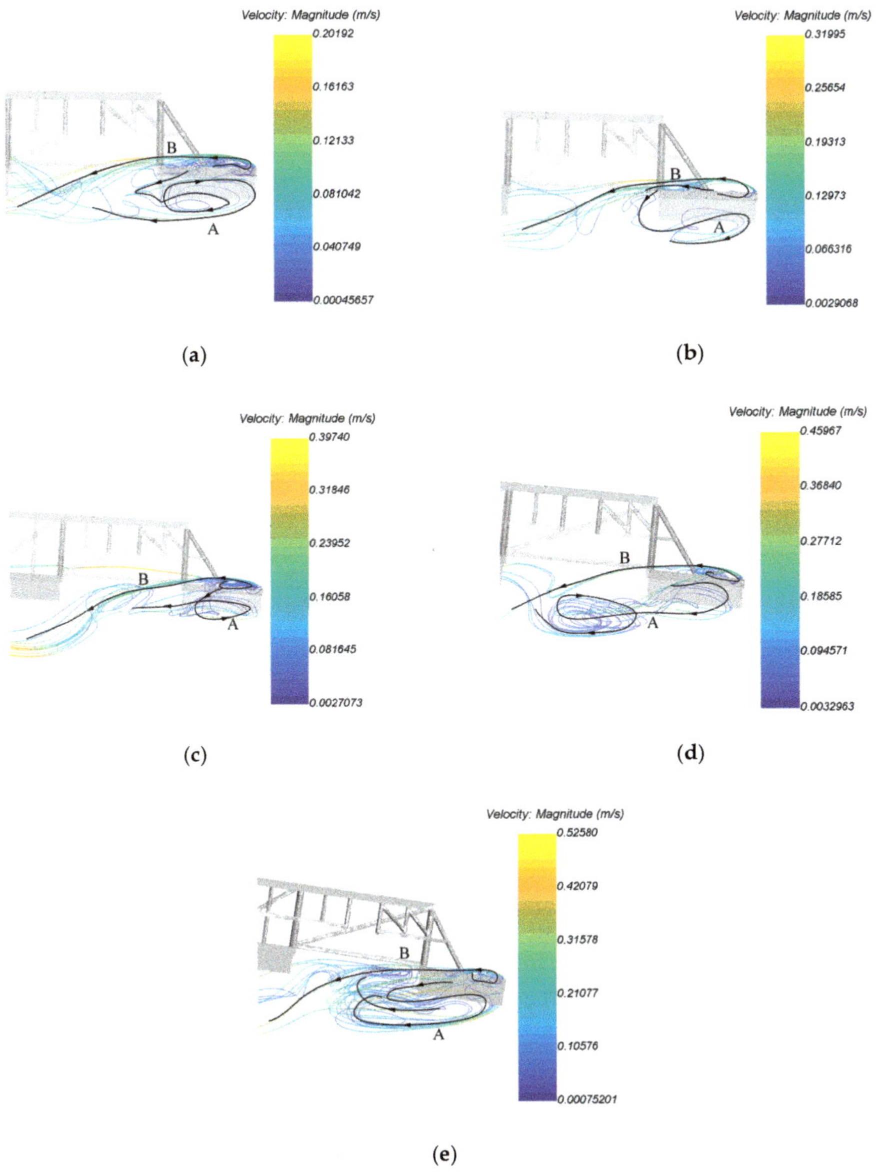

(**a**)

(**b**)

(**c**)

(**d**)

(**e**)

Figure 18. Flow patterns in the inlet section for the Y-direction towing cases: (**a**) Case 6, (**b**) Case 7, (**c**) Case 8, (**d**) Case 9, and (**e**) Case 10.

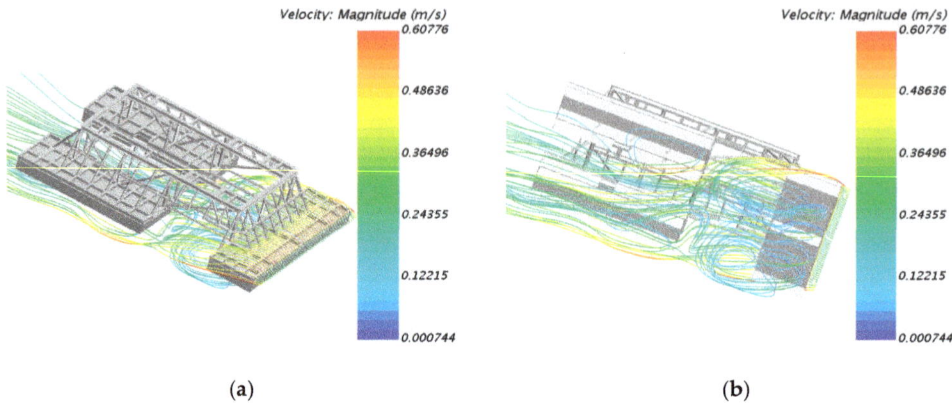

(**a**) (**b**)

Figure 19. Perspective view of streamlines from the inlet section in Case 10: (**a**) top view of the flow field, (**b**) bottom view of the flow field.

4.4. Resistance Component Analysis

Once the total drag force has been obtained, the frictional and pressure resistance components can be calculated. The RANS method calculates the hydrodynamic drag using an integral over the wetted surface [26]. The frictional, pressure and total resistance coefficients are calculated as follows [19]:

$$C_f = F_f / \frac{1}{2}\rho A U^2 \tag{14}$$

$$C_{pv} = F_{pv} / \frac{1}{2}\rho A U^2 \tag{15}$$

$$C_t = F_t / \frac{1}{2}\rho A U^2 \tag{16}$$

As shown in Figure 20 by the dotted line, the frictional, pressure, and total resistance all increase nonlinearly with velocity. In the X-direction, the frictional resistance F_{fx} increases slowly at high speeds and is very small compared with pressure resistance F_{pvx}. The proportion of F_{fx} to F_{pvx} is 0.69% in the case of 0.3 m/s. The reason for the small ratio of frictional resistance to pressure resistance is that a lot of vortexes appear behind the module, as shown in Figure 12. These vortexes are created by the separation of the boundary layer and the increase in the net pressure difference, and hence the increase in F_{pvx}. The distribution of vortexes increases in the high-speed towing cases and reduces the increase in F_{fx} induced by velocity. In the Y-direction, the frictional resistance F_{fy} is greater than that in the X-direction at the same towing speed because the vortexes are thinner. This is caused by the lag of the separation of the boundary layer. Furthermore, the pressure resistance is less than in the X-direction because of the decrease in the net pressure difference. Since the pressure resistance is the main component of total resistance, the total resistance in the X-direction is greater than in the Y-direction [27].

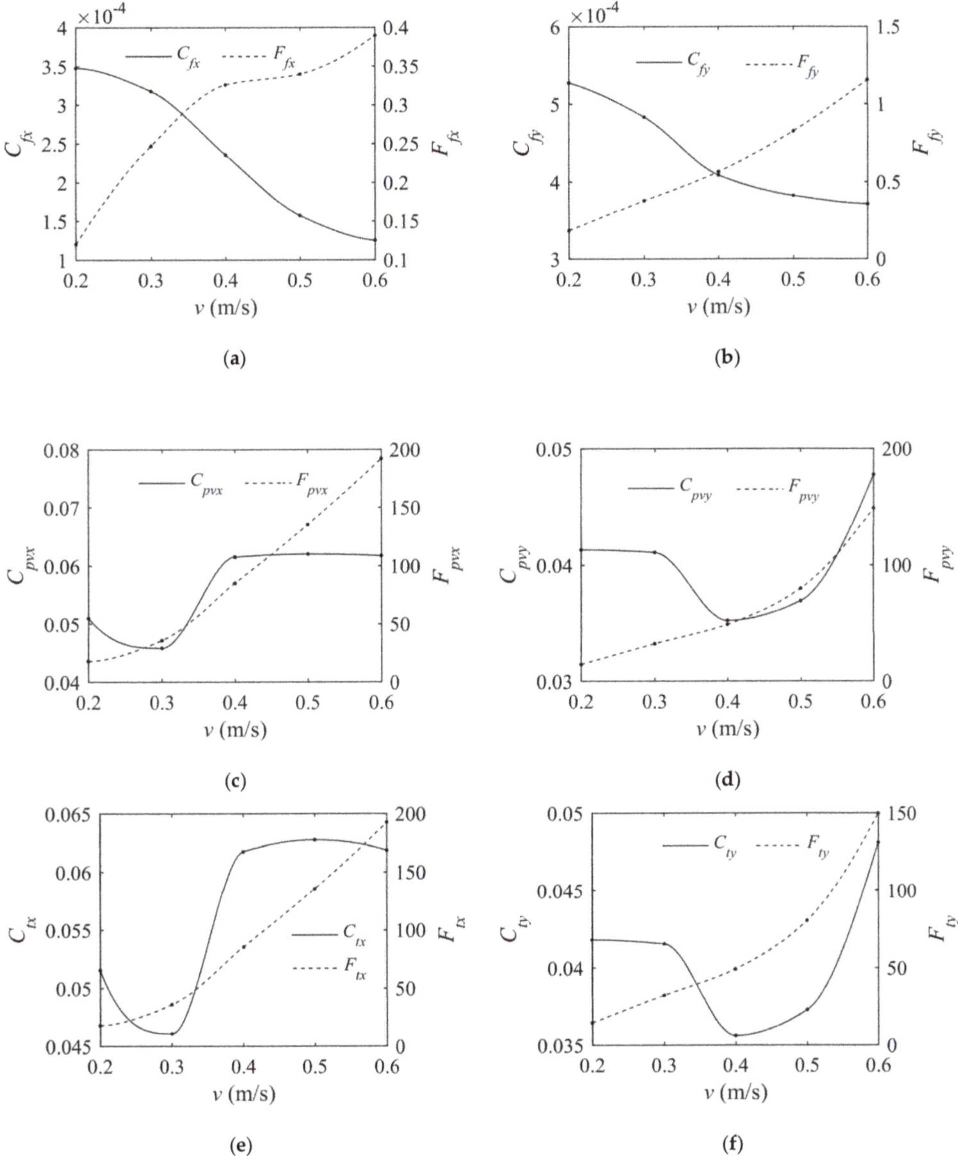

Figure 20. The variation of the resistance coefficients: (**a**) frictional resistance coefficient in the X-direction, (**b**) frictional resistance coefficient in the Y-direction, (**c**) pressure resistance coefficient in the X-direction, (**d**) pressure resistance coefficient in the Y-direction, (**e**) total resistance coefficient in the X-direction, (**f**) total resistance coefficient in the Y-direction.

As shown in Figure 20a,b in solid lines, the frictional drag coefficients in the X-direction (C_{fx}) and the Y-direction (C_{fy}) decreased as the velocity increased. C_{fy} was larger than C_{fx} for the same towing velocity, and the maximum relative difference (66.3%) was evident at a towing velocity of 0.6 m/s. In contrast, the pressure drag coefficient in the X-direction (C_{pvx}) was larger than in the Y-direction (C_{pvy}). The maximum relative difference (74.8%) appeared at a towing velocity of 0.4 m/s. Initially, C_{pvx} reached its minimum at 0.3 m/s. The reason for this is that the Re number is greater than 3×10^5

and laminar flow becomes turbulent before the separation of the laminar boundary layer. The pressure resistance induced by the net pressure difference increased slightly when the turbulent boundary layer separated. Therefore, C_{pvx} clearly decreased, indicating the same phenomenon but with a different threshold for the Re number as a cylinder [28]. With increased towing speed, C_{pvx} increased to around 0.062 when the Re number was greater than 9.6×10^5 because the location of the boundary layer separation stops moving forward. In the Y-direction, the turbulent boundary layer separated after 0.3 m/s, and the C_{pvy} reached its minimum at the speed of 0.4 m/s. However, C_{pvy} still increased because the location of the boundary separation was still moving forward. Finally, C_{pvy} reaches its maximum (0.0477) at 0.6 m/s. The variation in the total drag coefficient (C_t) was similar to that of C_{pv} because of the larger pressure drag component.

5. Conclusions

The present study initially used a physical model to investigate the drag force and vertical offset angle of the towrope of a subsea module, and to analyze the underwater attitude and flow field characteristics. It is interesting to note that the drag force and offset angle are smaller in lateral direction towing cases. The difference between the offset angles of longitudinal and lateral direction cases is more obvious than the difference between the drag force coefficients. These characteristics were then reproduced in a numerical simulation.

The numerical simulation results for the drag force coefficient and offsets angle were in agreement with the physical model results. Most of the differences in the drag force coefficients and offset angles were less than 10%. Thus, it was shown that the CFD solver, numerical methods, and computing grids can be applied for the purpose of accurate and efficient drag force and offset angle estimation in relation to underwater towing operations.

By analyzing the detailed vorticity fields around the module, we found that there were three kinds of vortexes in the inlet section that appeared in both the longitudinal and lateral direction towing cases. However, the distribution of vortexes in the lateral direction was narrower than in the longitudinal direction. In addition, the detailed flow field in the inlet section was analyzed: two main flow patterns were found and were broadly similar for longitudinal direction and lateral direction towing. Streamlines flow through holes and form recirculating paths in high-speed cases, and another recirculating path appears in middle-speed cases in lateral-direction towing because of the longer length between the skirt panels. Further analysis was provided of 3D flow in the longitudinal direction and lateral direction towing cases at the highest speed. The phenomena presented by fluid flow is in agreement with the experimental data. Thus, the CFD method used in this paper can reflect the flow detail of the module.

Finally, the frictional, pressure, and resistance coefficients were studied. The distribution of vortexes affects resistance significantly. There were greater frictional resistance and less pressure resistance in the lateral direction, ultimately leading to less total resistance. According to the comparison of the pressure coefficients, the turbulence boundary layer separates at greater Reynolds numbers, and the location of the separation was further forward in the lateral direction than in the longitudinal direction. Thus, it is clear that resistance coefficients need to be assessed when considering different towing directions.

In general, lateral-direction towing is more effective for towing operations of the module than longitudinal-direction towing. In the lateral direction, the offset angle and drag force at low speeds are smaller. The drag force and its oscillation increase slightly with the towing speed. The resistance is smaller, and this reduces potential damage to the structure. Attention should be paid to a sudden increase in drag force and offset angle when increasing the velocity of tugboats, even in low-speed towing operations. Towing operations also place restrictions on the attitude of the module, especially in the starting state, because the heading angle of the module will change from the initial state.

Author Contributions: This paper is the result of collaborative teamwork. R.G. wrote the paper, Y.Z. and L.Y. reviewed and edited the text, Y.Z. analyzed the data, and Z.W. obtained the resource data. All authors approved the manuscript.

Funding: This research was funded by the National Natural Science Foundation of China (Grant No. 51809067), the Fundamental Research Funds for the Central Universities (Grant Nos. 3072019CFM0101 and 3072019CF0102), and the National Science and Technology Major Project of China (Grant No. 2016ZX05057020).

Acknowledgments: The authors thank Chiahsing Liu at SIEMENS for his technical support in the simulation setting.

Conflicts of Interest: The authors declare no conflict of interest.

References

1. Nam, B.W.; Kim, N.W.; Hong, S.Y. Experimental and numerical study on coupled motion responses of a floating crane vessel and a lifted subsea manifold in deep water. *Int. J. Naval Archit. Ocean Eng.* **2017**, *9*, 552–567. [CrossRef]
2. Ramos, R.V.; Camponogara, E.; Martins, F.F.V.J. Design optimization of oilfield subsea infrastructures with manifold placement and pipeline layout. *Comput. Chem. Eng.* **2018**, *108*, 163–178. [CrossRef]
3. Neto, E.D.A.; Jacovazzo, B.M.; Correa, F.N.; Jacob, B.P. Numerical evaluation of a subsea equipment installation method designed to avoid resonant responses. *Appl. Ocean Res.* **2019**, *88*, 288–305. [CrossRef]
4. Olsen, T.A. Subsurface Towing of Heavy Module. Master's Thesis, Norwegian University of Science and Technology, Trondheim, Norway, 6 January 2011.
5. Risoey, T.; Mork, H.; Johnsgard, H.; Gramnaes, J. The pencil buoy method—A subsurface transportation and installation method. *Offshore Technol. Conf.* **2007**, *30*, 1–7. [CrossRef]
6. Yamazu, T.; Haraguchi, K.; Inoue, T.; Arai, K. Development of a Highly Flexible and High-Resolution Deep-Towed Streamer. *J. Mar. Sci. Eng.* **2019**, *7*, 254. [CrossRef]
7. Jacobsen, T. Subsurface Towing of a Subsea Module. Master's Thesis, Norwegian University of Science and Technology, Trondheim, Norway, 6 June 2010.
8. Kang, W.H.; Zhang, C.; Yu, J. Stochastic extreme motion analysis of jack-up responses during wet towing. *Ocean Eng.* **2016**, *111*, 56–66. [CrossRef]
9. Zhang, P.; Peng, Y.; Ding, H.; Hu, R.; Shi, J. Numerical analysis of offshore integrated meteorological mast for wind farms during wet towing transportation. *Ocean Eng.* **2019**, *188*, 106271–106283. [CrossRef]
10. Ding, H.; Hu, R.; Le, C. Towing Operation Methods of Offshore Integrated Meteorological Mast for Offshore Wind Farms. *J. Mar. Sci. Eng.* **2019**, *7*, 100. [CrossRef]
11. Rattanasiri, P.; Wilson, P.A.; Phillips, A.B. Numerical investigation of a fleet of towed AUVs. *Ocean Eng.* **2014**, *80*, 25–35. [CrossRef]
12. Wu, J.; Ye, J.; Yang, C.; Chen, Y.; Tian, H.; Xiong, X. Experimental study on a controllable underwater towed system. *Ocean Eng.* **2005**, *32*, 1803–1817. [CrossRef]
13. Go, G.; Ahn, H.T. Hydrodynamic derivative determination based on CFD and motion simulation for a tow-fish. *Appl. Ocean Res.* **2019**, *82*, 191–209. [CrossRef]
14. Sun, F.J.; Zhu, Z.H.; LaRosa, M. Dynamic modeling of cable towed body using nodal position finite element method. *Ocean Eng.* **2011**, *38*, 529–540. [CrossRef]
15. Wu, J.; Chwang, A.T. Experimental investigation on a two-part underwater towed system. *Ocean Eng.* **2001**, *28*, 735–750. [CrossRef]
16. Wu, J.; Chwang, A.T. Investigation on a two-part underwater manoeuvrable towed system. *Ocean Eng.* **2001**, *28*, 1079–1096. [CrossRef]
17. Wu, J.; Chwang, A.T. A hydrodynamic model of a two-part underwater towed system. *Ocean Eng.* **2000**, *27*, 455–472. [CrossRef]
18. Jacobsen, T.; Leira, B.J. Numerical and experimental studies of submerged towing of a subsea template. *Ocean Eng.* **2012**, *42*, 147–154. [CrossRef]
19. Elkafas, A.G.; Elgohary, M.M.; Zeid, A.E. Numerical study on the hydrodynamic drag force of a container ship model. *Alex. Eng. J.* **2019**. [CrossRef]
20. STAR-CCM+ Users' Guide Version 13.04. In *CD-Adapco, Computational Dynamics-Analysis & Design*; Application Company Ltd.: Melville, NY, USA, 2018.

21. Zhang, X.; Lin, Z.; Mancini, S.; Li, P.; Li, Z.; Liu, F. A Numerical Investigation on the Flooding Process of Multiple Compartments Based on the Volume of Fluid Method. *J. Mar. Sci. Eng.* **2019**, *7*, 211. [CrossRef]

22. Jiang, Y.; Sun, H.; Zou, J.; Hu, A.; Yang, J. Experimental and numerical investigations on hydrodynamic and aerodynamic characteristics of the tunnel of planing trimaran. *Appl. Ocean Res.* **2017**, *63*, 1–10. [CrossRef]

23. Gao, T.; Wang, Y.; Pang, Y.; Chen, Q.; Tang, Y. A time-efficient CFD approach for hydrodynamic coefficient determination and model simplification of submarine. *Ocean Eng.* **2018**, *154*, 16–26. [CrossRef]

24. Pan, Y.; Zhang, H.; Zhou, Q. Numerical Prediction of Submarine Hydrodynamic Coefficients using CFD Simulation. *J. Hydrodyn.* **2012**, *24*, 840–847. [CrossRef]

25. Saranjam, B. Experimental and numerical investigation of an unsteady supercavitating moving body. *Ocean Eng.* **2013**, *59*, 9–14. [CrossRef]

26. Maki, K.J.; Broglia, R.; Doctors, L.J.; Di Mascio, A. Numerical investigation of the components of calm-water resistance of a surface-effect ship. *Ocean Eng.* **2013**, *72*, 375–385. [CrossRef]

27. Molin, B. *Hydrodynamique Des Structures Offshore*, 1st ed.; National Defense Industry Press: Beijing, China, 2012; pp. 100–103.

28. Cantwell, B.; Coles, D. An experimental study of entrainment and transport in the turbulent near wake of a circular cylinder. *J. Fluid Mech.* **1983**, *136*, 321–374. [CrossRef]

Journal of
Marine Science and Engineering

Article

Numerical Investigation into Freak Wave Effects on Deepwater Pipeline Installation

Pu Xu [1,*], Zhixin Du [1] and Shunfeng Gong [2]

[1] College of Civil Engineering, Fuzhou University, Fuzhou 350116, China; 15259143923@163.com
[2] Institute of Structural Engineering, Zhejiang University, Hangzhou 310058, China; sfgong@zju.edu.cn
* Correspondence: puxu@fzu.edu.cn

Received: 18 January 2020; Accepted: 3 February 2020; Published: 14 February 2020

Abstract: Freak waves are an extreme marine environment factor in offshore structure design and become a potential risk, particularly for laying oil-gas pipelines in deep waters. The objective of this study was to reveal the freak wave effects on dynamic behaviors of offshore pipelines for deepwater installation. Thus, a dedicated finite element model (FEM) for deepwater pipeline installation by the S-lay method was developed with special consideration of freak waves. The FEM also took pipelay vessel motions, pipe–stinger roller interactions, and the cyclic contacts between the pipeline and seabed soil into account. Real vessel and stinger data from an actual engineering project in the South China Sea were collected to obtain an accurate simulation. Moreover, an effective superposition approach of combined transient wave trains and random wave trains was introduced, and various types of freak wave trains were simulated. Extensive numerical analyses of a 12 inch gas pipeline being installed into a water depth of 1500 m were implemented under various freak wave conditions. The noticeable influences of freak waves on the pipeline and seabed responses were identified, which provides significant awareness of offshore pipelines for deepwater installation design and field operation monitoring.

Keywords: offshore pipeline; installation simulation; deepwater; freak wave; S-lay method

---

1. Introduction

Freak waves occur unexpectedly far out at sea with remarkably large wave heights and are deemed to be an extreme marine environment condition. The irregular distribution of freak wave heights does not comply with the basic law of Rayleigh distribution for normal ocean waves, especially in deep waters. The unique feature of freak waves makes it difficult for marine structural engineers to sufficiently consider the huge wave loads in the design stage. In the past, plenty of tremendous accidents, including shipwrecks and massive destruction of offshore structures, have been caused by the great impact of freak waves [1,2]. These accidents have produced a striking warning on the potential risk of freak waves for lay barge and offshore structures and have attracted wide attention on the investigation into freak-wave-induced structural responses.

The offshore pipeline is a representative type of marine structure that is widely utilized for crude oil and natural gas transportation from subsea well sites to surface processing facilities. In recent years, the great demand for energy resources has facilitated the expansion of oil-gas exploitation into deepwater areas. The S-lay approach is one of most common methods of deepwater pipeline installation to the sea floor owing to its excellent adaptability and workability [3]. In this pipelay technique, numerous section pipes with designed lengths are welded and inspected on the operating lines of the vessel. The qualified pipeline is drawn by the tensioners and slides over the stinger to arrive at the seabed. The overall pipeline is characterized as an S-shaped curve and divided into two regions, as displayed in Figure 1. The upper curved section of the pipeline from the tensioner to the lift-off point (LOP) is denoted as the overbend, and the suspended section from the LOP to the seabed is known as the sagbend. The whole process of laying the pipeline generally takes months or even longer periods, and is more likely to encounter the occurrence of freak waves. Therefore, to assess the influences of freak waves on the dynamic behaviors of S-laying pipelines in deep waters is highly significant for the purpose of pipelay design and operation safety.

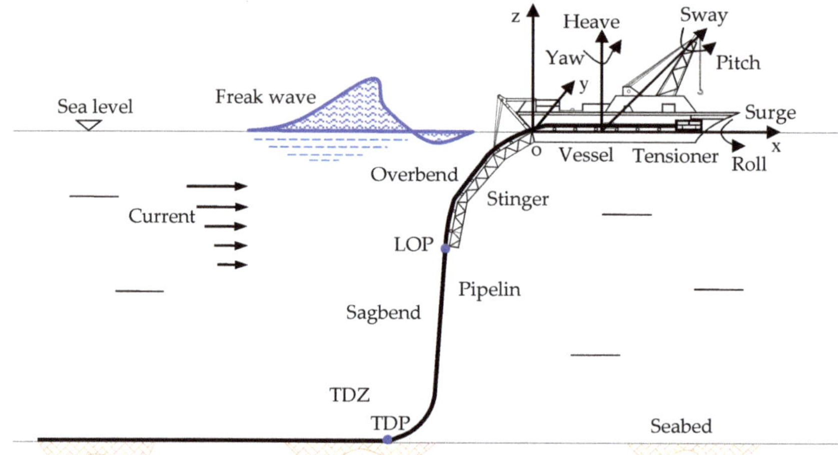

Figure 1. Deepwater pipeline installation by the S-lay method under exposure to freak waves.

A reasonable wave generation method is necessary to simulate freak waves and explore their impact on marine structures. A numerical technique is extensively employed due to its convenient simulation and good repeatability. Davis and Zarnick [4] initially proposed a wave focusing method to simulate freak waves by controlling the focalized time and space. Subsequently, Baldock et al. [5] applied the technique to accumulate numerous water waves and produce a huge transient wave group. Fochesato et al. [6] superposed wave trains of different directions to generate three-dimensional (3D) freak waves. Zhao et al. [7,8] presented four focusing models to generate the freak wave trains and numerically simulated the wave effect on a floating structure. Liu et al. [9] developed a modified phase modulation approach to focus wave trains with the specified phase and obtained the precise wave spectrum that coincided well with the target results. Hu et al. [10] employed a probability-based superposition method to calculate the generation probability of a freak wave. Recently, Tang et al. [11] improved the phase modulation model to generate a freak wave and investigated its effect on the dynamic responses of the Floating Production Storage and Offloading (FPSO) and Single Point Mooring (SPM) system. Pan et al. [12] conducted extensive tests to observe the great differences in cylinder motion responses under irregular waves and freak waves and obtained the important influencing factors. The wave focusing and superposition techniques were demonstrated to effectively generate freak wave trains, which provided a prior simulation of the wave impact on pipeline installation in deepwater areas.

Under the excitation of surface waves and vessel motions, the laying pipelines usually exhibit sophisticated non-linear, dynamic responses. The complicated S-lay problems mainly result from large pipeline deflections, pipe material plasticity, hydrodynamic loads, and boundary interactions. These non-linear features could cause some difficulties for analytical approaches and experimental tests to obtain accurate and comprehensive simulations. As a consequence, numerical techniques are preferred for modeling systematical behaviors of offshore pipelines in the S-lay process [13–15]. Gong et al. [16] and Gong and Xu [17] developed a full FEM on the basis of OrcaFlex to simulate the structural behaviors of pipeline installation and explored the influence of normal sea states on the pipeline responses. Ivić et al. [18,19] established a pipeline laying model by the use of non-linear elastic beam elements to analyze the static behaviors of the S-laying pipe and formulated a specialized optimization method for the pipeline laying operation. Xie et al. [20] investigated the dynamic loading history of laying pipelines in light of a test-verified FEM and confirmed obvious pipeline plastic deformations resulting from the S-lay operation. Cabrera-Miranda and Paik [21] quantified the probabilistic distribution of loads on a marine riser and observed the highly random characteristic of the loads to aid in the determination of nominal design values. Wang et al. [22] pointed out the probable underestimation of pipeline dynamic behaviors for practical engineering and built a real-time installation monitoring system to predict on-site pipeline responses. Recently, Liang et al. [23,24] presented a refined FEM to take account of the complex surface contact behaviors of overbend pipes and reproduced the pipe laying process of a deep S-lay case in the laboratory. Kim and Kim [25] employed the FEM-based linear beam element to present an efficient, linearized, dynamic analysis approach for pipeline installation design. The aforementioned studies usually adopted normal random waves as the input ocean conditions to calculate the dynamic responses of laying pipelines. The neglect of the freak wave effect could result in the inadequacy of pipeline installation design for field operation safety.

The objective of this study was to thoroughly investigate the freak wave effects on the dynamic responses of offshore pipelines during deepwater S-lay installation. A new extended FEM with particular consideration of freak waves was developed on the basis of our previous model [16] for S-lay pipelines. This model took the induced vessel motions, pipe–stinger roller contacts, and pipe–seabed soil interactions into account. The real vessel, stinger roller, and seabed soil data from an actual engineering project in South China Sea were collected to obtain an accurate simulation. Furthermore, an effective superposition technique was employed to generate freak waves by combining transient wave trains and random wave trains. The insertion of various freak wave trains into the S-lay FEM was then implemented to carry out a large number of numerical analyses of a 12 inch gas pipeline being installed into a 1500 m water depth. Finally, the influences of the freak wave energy ratio coefficient, focusing location, phase range, and peak value were sufficiently assessed on the pipeline and seabed responses. The dynamic amplification factors (DAFs) of the axial tension, bending moment, von Mises stress, longitudinal strain, pipeline embedment, and seabed resistance are discussed in detail in relation to pipeline installation design and field operation safety.

2. Deepwater Pipeline Installation Simulation

A reasonable FEM for S-lay system was presented by Gong et al. [16] to explore the random wave effects on the dynamic behaviors of deepwater pipeline installation. This model, developed within the framework of OrcaFlex [26], was validated with acceptable accuracy and effective applicability by an actual engineering case of S-laying pipelines. In this study, a new extension of the FEM was implemented to consider the freak waves with wave-induced pipeline behaviors, pipe–stinger roller contacts, pipe–seabed soil interaction, and pipelay vessel motions, as displayed in Figure 1. The following section presents a concise description of the main features of the deepwater pipeline installation model by the S-lay technique.

2.1. Pipeline Model

In the FEM of the S-lay system, the entire pipeline, from the tensioner to the sea floor, was discretized into a sequence of mass nodes connected together by massless line segments, as displayed in Figure 2. The local xyz-frames of references for the node and line segment were established, and the mechanical properties of the pipe weight, buoyancy, drag force, and so on for each half-segment were concentrated on its adjacent node. At either side of the node, two rotational springs and dampers were employed to model the bending stiffness and damping of the line segment. At the center of the line segment, an axial spring with a damper was utilized to represent its axial stiffness and damping, and a torsional spring with a damper was applied to characterize its torsional stiffness and damping. For the detailed calculation derivation of the tension force, bending moment, and torque moment, one can refer to the literature [16], and their expressions are given by

$$T_e = T_w(\varepsilon) + (1 - 2v) \cdot (P_oA_o - P_iA_i) + EA_{nom} \cdot \xi \cdot (dL/dt)/L_0 \tag{1}$$

$$M_2 = M_b(\kappa_2) + \varsigma \cdot (d\kappa_2/dt) \tag{2}$$

$$T_t = T_{or}(\phi/L_0) + \zeta \cdot (d\phi/dt) \tag{3}$$

where T_e and T_w are the effective tension and wall tension relating to the axial strain; P_i and P_o are the internal pressure and external pressure; A_i and A_o are the internal and external cross-section areas; v is the Poisson's ratio; EA_{nom} is the nominal axial stiffness defined at zero strain; L is the instantaneous length of the line segment; L_0 is the unstretched length of the line segment; M_b is the bending moment relating to the curvature κ_2; T_{or} is the torque moment relating to the twist angle φ; and the damping coefficients ξ, ς and ζ separately represent axial, bending, and torsional effects of structural damping.

The oil–gas pipelines installed in deep waters comprise carbon–manganese steel with a distinct yield point and some plastic deformation capability. The pipeline material features were simulated by the J2 flow theory of plasticity performance with isotropic strain hardening. The Ramberg–Osgood model [27] was applied to represent the non-linear stress and strain relationship of the adoptive X65 line pipe, which could be expressed as

$$\varepsilon(\sigma) = \sigma/E + B(\sigma/\sigma_y)^n \tag{4}$$

where σ_y is the effective yield stress, E is the elastic modulus, and B and n are the coefficient and the power exponent of the constitutive model.

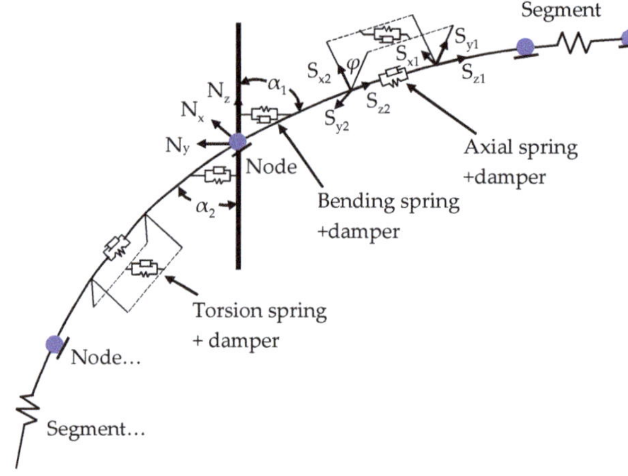

Figure 2. Node and line segment model of the S-lay pipeline.

2.2. Pipe–Stinger Roller Interaction

In the overbend, the sections of the pipeline were continuously supported by 10 roller boxes that were regularly spaced and settled on the articulated stinger, which was 75 m in length, as illustrated in Figure 3. The stinger with three sections of truss structure was collected from the actual design for the Hai Yang Shi You (HYSY) 201 pipelay vessel [28]. A group of pipe segments was employed to simulate the stinger's geometrical and mechanical properties. The clashing contacts between the pipeline and stinger rollers would vary with the vessel motions. Before the calculation of pipe–stinger roller interactions, an inspection had to be implemented to confirm whether the pipeline was in contact with the roller. If a mutual interaction was identified, the contact force was calculated and applied to the pipe and the roller, which was given by

$$F_r = [1/(1/k_1 + 1/k_2)] \times [d - (r_1 + r_2)] \tag{5}$$

where k_1 and k_2 are the contact stiffness of the pipe and the roller, d is the shortest separation distance of the center lines between them, and r_1 and r_2 are the corresponding radii.

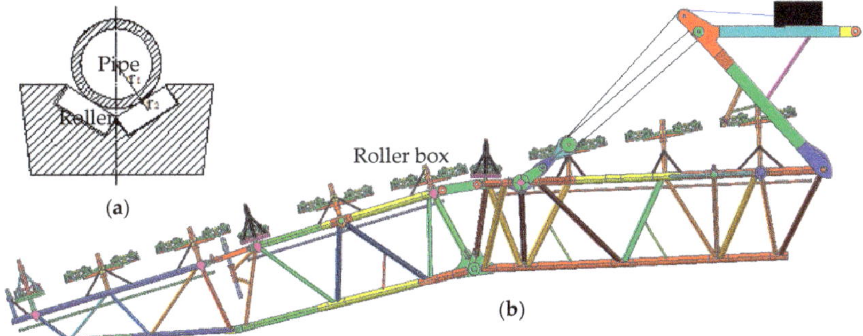

Figure 3. Schematic diagram of the deepwater S-lay stinger: (**a**) Pipe–roller interaction; (**b**) articulated stinger of the Hai Yang Shi You (HYSY) 201 vessel.

2.3. Pipe–Seabed Soil Interaction

At the touchdown zone (TDZ), the laying pipeline was freely supported by the seabed soil, which was liable to be trenched and remolded under dynamic installation in deep waters. In the vertical plane, the dynamic features of the cyclic pipe–seabed interactions were simulated by a non-linear hysteretic soil model with hyperbolic secant stiffness formulations [29], as shown in Figure 4. In this model, four types of pipe–soil penetration modes were applied to characterize the cyclic variations of the pipeline periodic embedment into the seabed. For the not-in-contact pattern, the seabed resistance $P(z)$ is naturally zero. For other three patterns, including initial penetration, uplift, and repenetration, the relationships between the seabed resistance $P(z)$ and the penetration z non-linearly vary in hysteretic cycles with the incessant shift of the penetration modes.

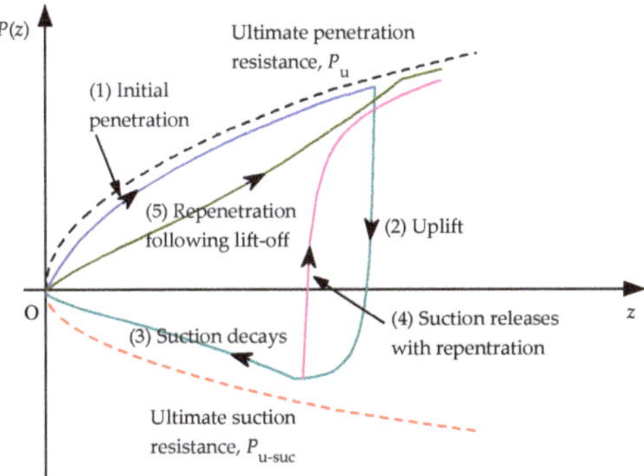

Figure 4. Non-linear hysteretic seabed soil model [29].

In another view of the horizontal plane, the lateral and axial pipe–seabed interactions were simulated by the modified Coulomb friction model, which expressed the lateral resistance and axial friction force with the deflection as a bilinear equation, as illustrated in Figure 5. When the lateral displacement y varies from $-y_{\text{breakout}}$ to $+y_{\text{breakout}}$, the linear friction force is given by $F_y = -k_s A y$, in which k_s refers to the seabed shear stiffness and A represents the contact area. When the lateral displacement y exceeds the range between $-y_{\text{breakout}}$ and $+y_{\text{breakout}}$, the friction force of the pipe is equal to $\mu P(z)$, where $P(z)$ is the vertical seabed resistance and μ is the soil friction coefficient. This model could effectively avoid the discontinuous nature of the friction force at zero lateral displacement and was conveniently implemented in the numerical program [30].

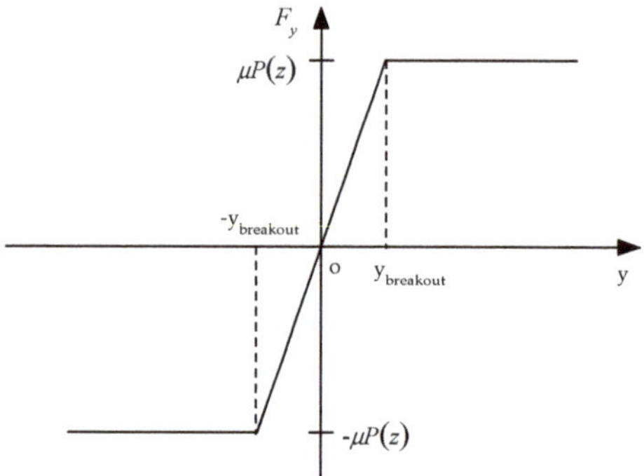

Figure 5. Modified Coulomb friction model.

2.4. Pipelay Vessel Motions

According to the geometrical features of the HYSY 201 vessel, a pipelay vessel model was built with a length of 204.65 m and breadth of 39.2 m. The wave frequency motion of the vessel was simulated by use of the displacement response amplitude operators (RAOs), which define the vessel motion responses for each degree of freedom (DoF) to one specified wave direction and wave period. Considering the six DoFs of vessel motions (surge, sway, heave, roll, pitch, and yaw), the motion response spectra of the vessel at the stinger base were derived in light of the RAOs of the HYSY 201 vessel, as illustrated in Figure 6. Almost all six DoFs of the vessel motion responses were greatly noticeable for the quartering seas, in which the heave motion was comparatively remarkable for all seas. It is noted that the slow drift motion of the vessel was restrained to be very small by the advanced dynamic positioning system [31], so it was not taken into consideration in the following analyses due to its small effect on the pipeline dynamic behaviors.

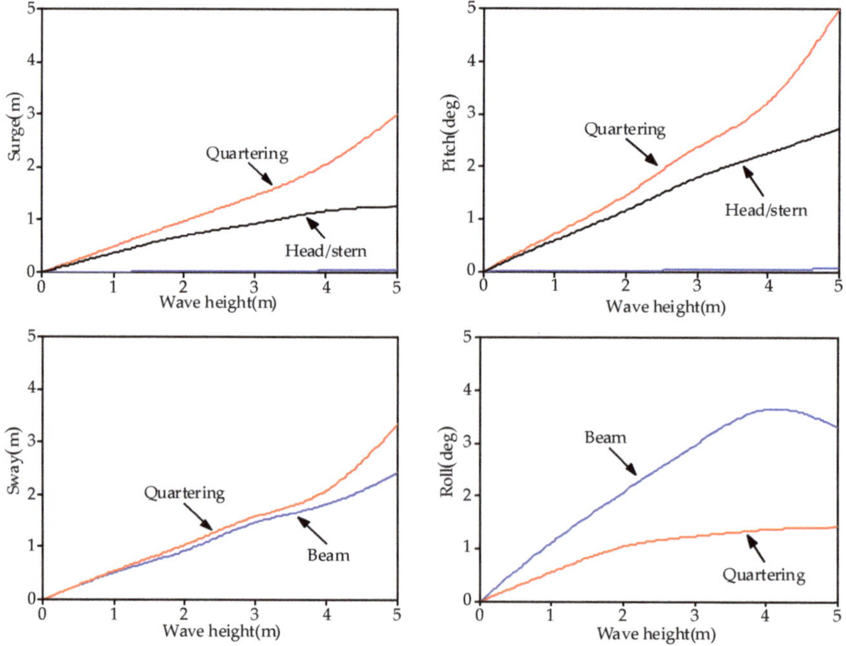

Figure 6. *Cont.*

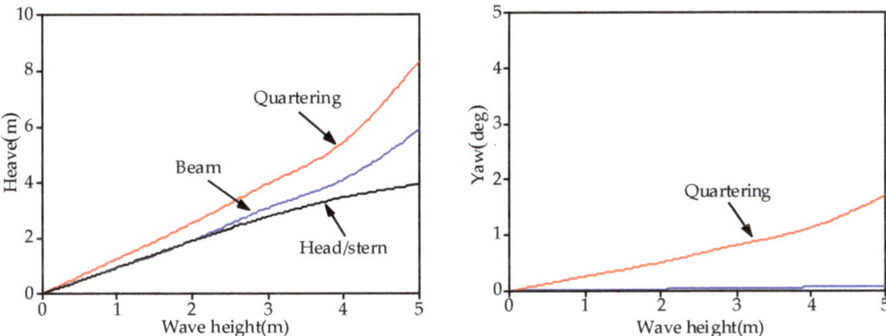

Figure 6. Motion response spectra of the HYSY 201 pipelay vessel at the stinger base.

3. Freak Wave Generation

Based on a large amount of ocean observations and laboratory tests, a great many generation models of freak waves have been developed to investigate the phenomenon of rogue wave impact [32,33]. In contrast with the non-linear model, the linear superposition model is simply understood by offshore structure engineers and can be rapidly simulated by researchers. It is also noted that during deepwater installation, the pipeline usually experiences large tension forces of the tensioners, and the influence of hydrodynamic forces induced by non-linear wave factors is very small. A time history train of freak waves must be inserted into the developed installation FEM for the dynamic analysis of an S-lay pipeline. Therefore, the linear superposition technique was employed to generate the freak wave trains.

3.1. Linear Superposition Approach

In the linear superposition method, freak waves are described as a combination of transient wave trains with random wave trains for different energy proportions. The transient waves were simulated by the wave focusing model which converges the wave energy of a certain number of wave components at a specified position at the assigned time. The random waves were deemed to be a stationary stochastic process of dispersed energy. The standard JONSWAP spectrum was employed to represent the random sea states in the South China Sea. The generation formula of freak waves can be expressed as

$$\eta(x,t) = E_{\mathrm{p}1}\sum_{i=1}^{N}\left[\int_{f_{i-1}}^{f_i} 2S(f)df\right]^{\frac{1}{2}}\cos[k_i(x-x_{\mathrm{p}}) - \omega_i(t-t_{\mathrm{p}})] + E_{\mathrm{p}2}\sum_{i=1}^{N}\left[\int_{f_{i-1}}^{f_i} 2S(f)df\right]^{\frac{1}{2}}\cos[k_ix - \omega_it + \phi_i] \quad (6)$$

where $E_{\mathrm{p}1}$ and $E_{\mathrm{p}2}$ are the energy ratio coefficients of transient waves and random waves; the spectral density function is $S(f) = \alpha g^2/(16\pi^4 f^5)\exp\left[-1.25(f/f_m)^{-4}\right]\gamma^\beta$; α is the spectral energy coefficient; and g is the gravitational constant. $\beta = \exp\left[-(f-f_m)^2/(2\tau^2 f_m^2)\right]$, in which τ is the spectral width parameter; f_m is the peak frequency; γ is the peak enhancement factor; N is the number of wave components; k_i, ω_i, and ϕ_i are the wave number, angular frequency, and phase lag of the ith wave component; and x_{p} and t_{p} are two constants separately representing the focusing position and time of transient waves.

3.2. Case Study

Before the generation of freak waves, there should be a clear mathematical definition. The popularly acceptable criterion for freak waves was adopted in this study, which defines the maximum wave height to be more than two times its significant wave height. Based upon the ocean statistics in the South China Sea, the significant wave height, $H_s = 2.0$ m, and the peak period, $T_p = 8.7$ s, were specified for the JONSWAP spectrum, as illustrated in Figure 7. The corresponding peak frequency, $f_m = 0.115$ Hz, and the spectral energy coefficient, $\alpha = 0.002$, were calculated along with the peak enhancement factor, $\gamma = 3.3$. Besides, the spectral width parameter σ was varied with the value of wave frequency. If $f \leq f_m$, $\tau = 0.07$; otherwise, $f > f_m$, $\tau = 0.09$.

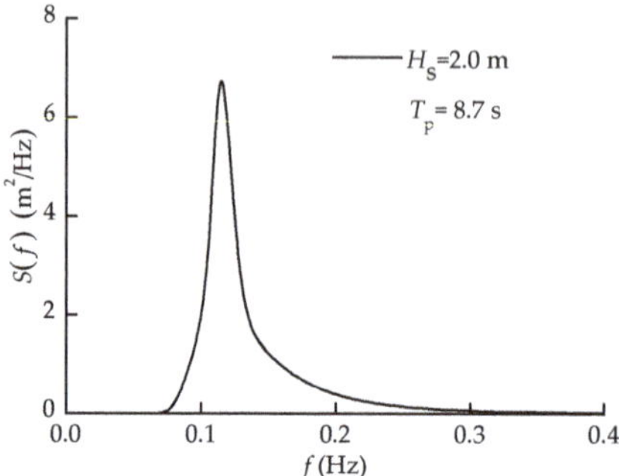

Figure 7. JONSWAP spectrum for $H_s = 2.0$ m.

The selected wave spectrum was discretized into 900 components by use of the equal energy approach. These wave components were then gathered in the numerical flume to constitute a sequence of transient wave trains and random wave trains. The energy proportions for both wave trains were set as $E_{p1} = 0.4$ and $E_{p2} = 0.6$, and the distribution range φ of the phase lag was taken as 1.1π. The wave focusing time and position were set at $t_p = 1000$ s and $x_p = 0$ m. Figure 8 illustrates a part of the time history trains of generated freak waves, and the beginning simulation time point was shifted to 750 s with the duration of 500 s so as to cover the maximum wave height in the globe time domain. It can be observed that the freak wave amplitude suddenly surged to a great wave crest of 5.1 m at the focusing time of 1000 s. The maximum wave height attained 7.8 m, which is 3.9 times larger than its significant wave height. The wave time history trains properly reflect the basic characteristic of freak waves in the ocean sea and satisfy the wave amplitude criterion for its definition.

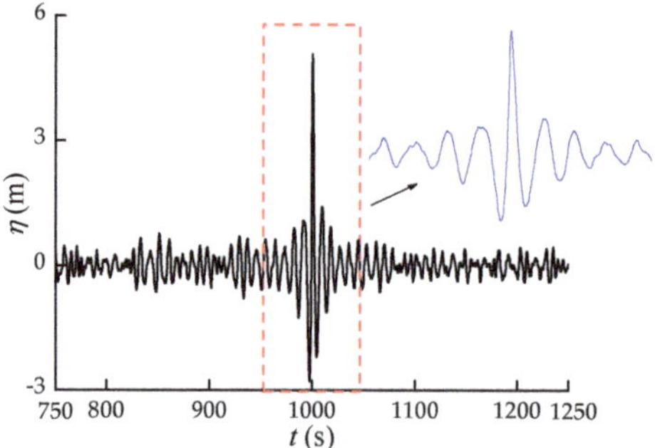

Figure 8. Time history trains of the freak wave.

3.3. Sensitive Analysis

The time history trains of freak waves are crucial for the investigation into their effects on the dynamic behaviors of S-laying pipelines. The sensitive analyses of generation factors for the freak wave train must be conducted. Four input parameters, including the wave energy ratio coefficient, focusing position, phase range, and peak value, were selected for the wave simulation by the linear superposition technique. Plenty of freak wave trains were obtained, and a group of represented trains with the duration of 500 s are illustrated in Figure 9. Significant differences in the wave crest and wave trough at the middle time were observed under different initial conditions.

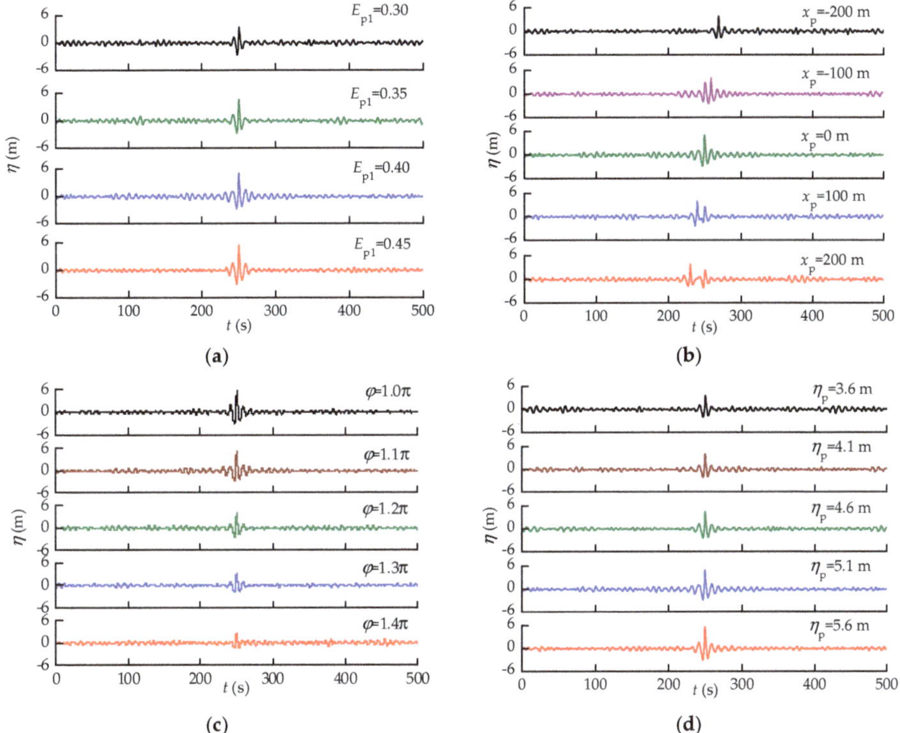

Figure 9. Time history trains of freak waves under different input conditions: (**a**) energy ratio coefficient; (**b**) focusing position; (**c**) phase range; (**d**) wave peak value.

Figure 10 displays the maximum wave heights of freak wave trains with the variation of the initial input parameters. It can be seen that the maximum wave height linearly increases with the increase of the energy ratio coefficient E_{p1}, which denotes the energy proportion of the transient wave in the freak wave trains. When the focusing location occurred from the distance $x_p = -200$ m to $x_p = 200$ m, the maximum wave height firstly augmented and then reduced, and the crest value attained 7.86 m at the point $x_p = 0$ m. As the phase range added from 1.0π to 1.4π, the maximum wave height gradually decreased. Oppositely, along with the augmentation of the wave peak value from 3.6 to 5.6 m, the maximum wave height linearly magnified.

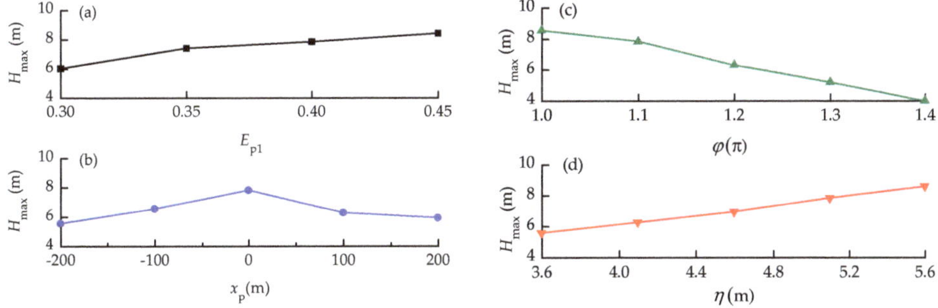

Figure 10. Variation of the maximum freak wave height with different input conditions: (**a**) energy ratio coefficient; (**b**) focusing position; (**c**) phase range; (**d**) wave peak value.

4. Numerical Implementation of Pipeline Installation under Freak Waves

4.1. Pipelay Parameters

According to a practical engineering case, a 12 inch pipeline was installed into a water depth of 1500 m in the Liwan3-1 (LW3-1) gas field in the South China Sea. The laying pipeline parameters listed in Table 1 were adopted, which included the outer diameter D, wall thickness t'_p, steel pipe density ρ_p, elastic modulus E, Poisson's ratio v, effective yield stress σ_y, thickness t_c and density ρ_c of the corrosion coatings, weight per unit length in air w_a, and submerged weight per unit length w_s.

Table 1. Laying pipeline parameters.

D (mm)	t'_p (mm)	ρ_p (kg/m^3)	E (MPa)	v	σ_y (MPa)	t_c (mm)	ρ_c (kg/m^3)	w_a (N/m)	w_s (N/m)
323.9	23.8	7850	2.07×10^5	0.3	448	3.0	950	1754.9	917.2

X65 material grade was used for the steel pipe, whose stress–strain relationship curve is displayed in Figure 11 on the basis of the Ramberg–Osgood model. The non-linear relationship between the bending moment and curvature of the steel pipe shown in Figure 12 was obtained by use of the hysteretic bending model, which gave a precise simulation of the bending state of the overbend pipeline under the cyclic clashing contacts with stinger rollers.

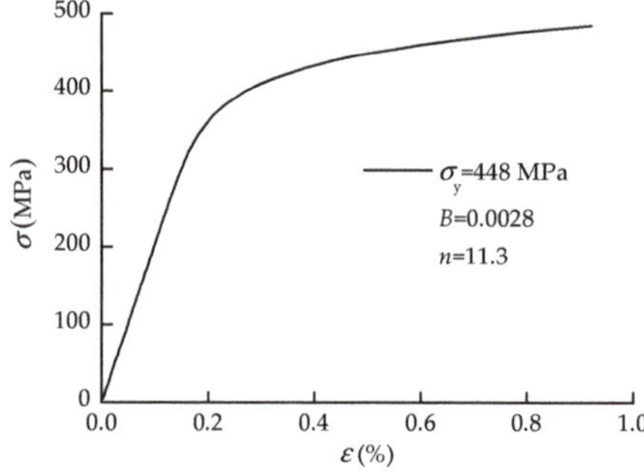

Figure 11. Stress–strain curve for the Ramberg–Osgood model.

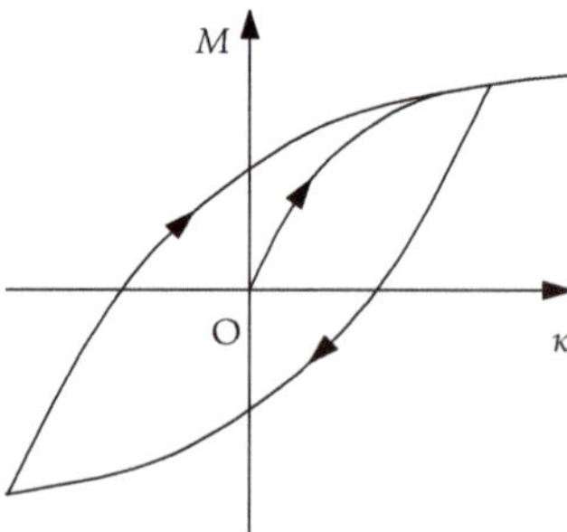

Figure 12. Non-linear hysteretic moment–curvature relationship.

The vertical distribution of the current speed among various water depths is illustrated in Figure 13 in light of the field measurement. The current direction was set as 0° in line with the pipelay heading. The ocean current was considered as a two-dimensional steady flow in the vertical plane. The hydrodynamic loads were calculated by means of Morison's equation [34] and are given by $F_d = (\Delta \cdot a_w + C_a \cdot \Delta \cdot a_r) + 0.5 \cdot C_D \cdot \rho_w \cdot A \cdot v_r|v_r|$, in which Δ is the mass of fluid displaced by the pipe,

α_w is the fluid acceleration relative to the earth, C_a is the added mass factor, α_r is the fluid acceleration relative to the pipe, C_D is the drag coefficient, ρ_w is the density of sea water, A is the drag area, and v_r is the fluid velocity relative to the pipe. For the hydrodynamic calculation, C_a was taken as 1.0, and the C_D for the axial and the normal directions was assumed to be 0.024 and 1.2, respectively.

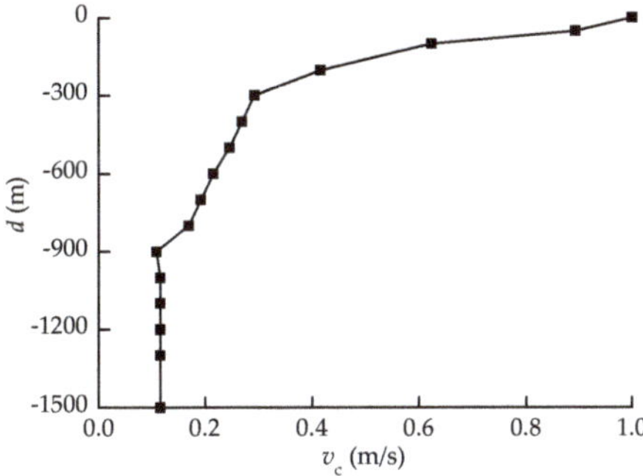

Figure 13. Current speed distribution with various water depths.

With regard to the non-linear hysteretic soil model applied in this study, a group of seabed soil parameters was selected to describe the basic features of soft clay in the deep water, as listed in Table 2. The ultimate penetration resistance $P_u(z)$ and the nominal bearing capacity factor $N_c(z/D)$ are non-linearly related to the penetration z and are given by [29]

$$P_u(z) = N_c(z/D) \cdot S_u(z) \cdot D \tag{7}$$

$$N_c(z/D) = a \cdot (z/D)^b \tag{8}$$

where the soil undrained shear strength refers to $S_u(z) = S_{u0} + S_{ug}z$, in which S_{u0} is the mudline shear strength and S_{ug} is the shear strength gradient, and a and b are the non-dimensional penetration factors. The saturated soil density ρ_{soil} and normalized maximum stiffness K_{max} were taken for the soft clay; other soil model parameters for different penetration patterns were specified as the defaults, including the suction ratio f_{suc} and the decay factor λ_{suc}, repenetration coefficient λ_{rep} and soil buoyancy factor f_b. In addition, the seabed soil friction coefficient μ and shear stiffness k_s were adopted for the simulation of axial and lateral pipe–seabed interactions [35].

Table 2. Seabed soil model parameters.

S_{u0} (kPa)	S_{ug} (kPa/m)	ρ_{soil} (t/m³)	a	b	K_{max}	f_{suc}	λ_{suc}	λ_{rep}	f_b	μ	k_s (kN/m³)
1.5	1.5	1.5	6.0	0.25	200	0.6	1.0	0.3	1.5	0.55	33.3

4.2. Calcultion Method

The dynamic calculation of laying pipeline responses induced by freak waves contained two modules: one was the S-lay model and another was the freak wave train. Firstly, a global S-lay model with the framework of OrcaFlex was established at a water depth of 1500 m. This model comprised the pipelay vessel, tensioner, stinger, pipeline, and seabed. Under the combined actions of self-weight, buoyancy, and internal forces, the equilibrium positions of the laying pipeline from the tensioner on the vessel via the stinger to the seabed were initially determined by utilization of the catenary technique. Subsequently, the hysteretic bending stiffness and the clashing mutual contacts of pipe-stinger rollers were further taken into account to obtain the full equilibrium configurations of the pipeline. The final static results of the S-lay system were taken as the initial values of the dynamic simulation.

Based upon the linear superposition technique, a series of time history trains of freak waves were obtained from the MATLAB program. These freak wave trains were then inserted into the developed S-lay model as the input conditions of extreme sea states. The geometric non-linearities of the laying pipeline, spatial variations of hydrodynamic forces, and clashing contacts were sufficiently incorporated in the simulation. The time domain calculations of the S-lay system under freak waves were conducted by use of the explicit dynamic integration approach. Besides, critical damping and target damping were utilized to cut down the spurious non-physical high frequency responses, and they were demonstrated to have little effect on the pipeline behaviors. Finally, the whole motion equations for the vessel and all line nodes were solved by iterative update of the forces and moments on the nodes and segments at each time step.

4.3. Time History Response of Pipelay Vessel Motions

The pipelay vessel motions are significant top excitation boundaries of the S-laying pipeline and cause dynamic responses. Under the generated freak wave trains shown in Figure 8 for the extreme quartering sea, the time history responses of six DoFs of pipelay vessel motions were calculated by use of displacement RAOs, as displayed in Figure 14. The heave motion among vessel translation responses was more prominent than the surge and the sway motion, and the pitch motion among vessel rotation responses was larger than the roll and the yaw motion. These results validate the above-mentioned response spectra of the pipelay vessel. Moreover, all six DoFs of the vessel motion response amplitudes abruptly rose and attained maximum values near the middle time of 250 s, which also reflects the basic time history characteristic of freak waves.

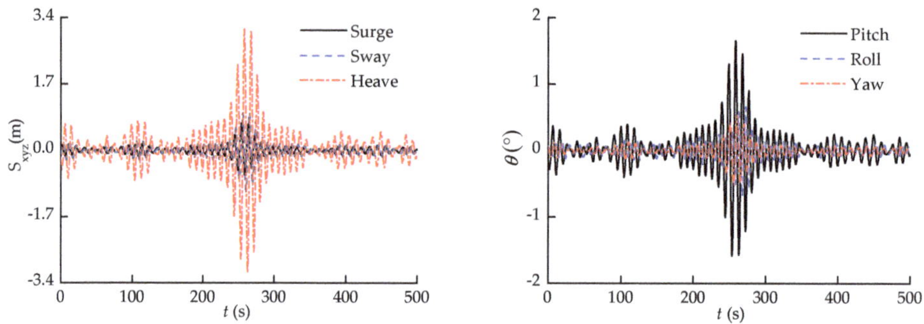

Figure 14. Time history responses of pipelay vessel motions under freak waves.

5. Results Analysis

5.1. Effect of the Wave Energy Ratio Coefficient

In the simulation of freak wave trains, the energy ratio coefficient directly dominates the energy proportion of transient waves and random waves. By selecting four energy ratio coefficients, $E_{p1} = 0.30$, 0.35, 0.40 and 0.45, a group of freak wave trains in Figure 9a was utilized as the input marine environment conditions for the dynamic analyses of the S-laying pipeline. The pipeline and seabed response results, which includes the axial tension, bending moment, von Mises stress, longitudinal strain, pipeline embedment, and seabed resistance, are illustrated in Figure 15.

As the energy ratio coefficient E_{p1} increased from 0.30 to 0.45, the axial tension of the overall pipeline noticeably rose, and its maximum value at the top end increased by 39.9% from 3512.57 to 4915.02 kN. The bending moments of the pipeline had some differences between the overbend and the sagbend, and the maximum results appeared at the last contact roller location in the overbend with a minor increase of 12.1% from 585.86 to 656.95 kN·m. In the sagbend, the maximum bending moment occurring near the touchdown point (TDP) had a prominent augmentation of 202.3% from 151.97 to

459.33 kN·m, and some bending moment crests along the touchdown pipeline formed as a result of pipe bending and soil softening. Under the combined axial tension, bending moment, and hydrostatic force, the von Mises stress of the pipeline increased to some extent and attained 20.8% of the maximum value from 461.27 to 557.10 MPa. Similarly, the maximum longitudinal strain of the pipeline rose by 23.9% from 0.285% to 0.353%. Moreover, the maximum pipeline embedment had a remarkable enlargement of 1139.5% from 0.043 ($0.130D$) to 0.533 m ($1.616D$), and the maximum seabed resistance grew by 199.8% from 1.837 ($2.701w_s$) to 5.507 kN/m ($8.097w_s$). It is demonstrated from these results that the energy ratio coefficient has an obvious effect on the pipeline dynamic behaviors and seabed resistance. Especially, when the E_{p1} reaches 0.45, the freak wave causes drastic dynamic responses of the pipeline and seabed interaction in the TDZ.

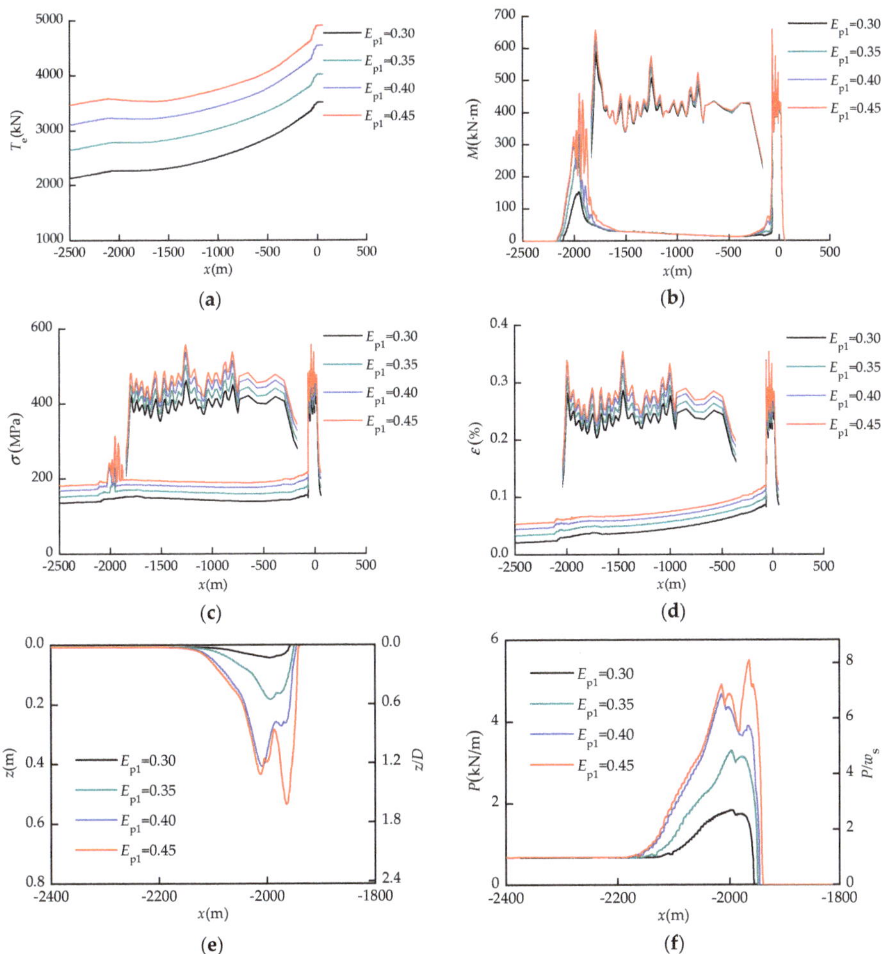

Figure 15. Dynamic responses of the S-laying pipeline under various freak wave energy ratio coefficients: (**a**) tension; (**b**) bending moment; (**c**) stress; (**d**) strain; (**e**) pipeline embedment; (**f**) seabed resistance.

5.2. Effect of the Wave Focusing Location

A noticeable characteristic of freak waves is the crest value appearing at the focusing position where the wave energy accumulates. To explore the influence of the wave focusing location on the dynamic responses of the laying pipeline, five focusing locations, $x_p = -200, -100, 0, 100, 200$ m, were assumed to simulate the freak waves shown in Figure 9b, which were separately taken as the input surface wave conditions to perform a dynamic analysis of pipeline installation. The pipeline and seabed response results illustrated in Figure 16 are the axial tension, bending moment, von Mises stress, longitudinal strain, pipeline embedment, and seabed resistance.

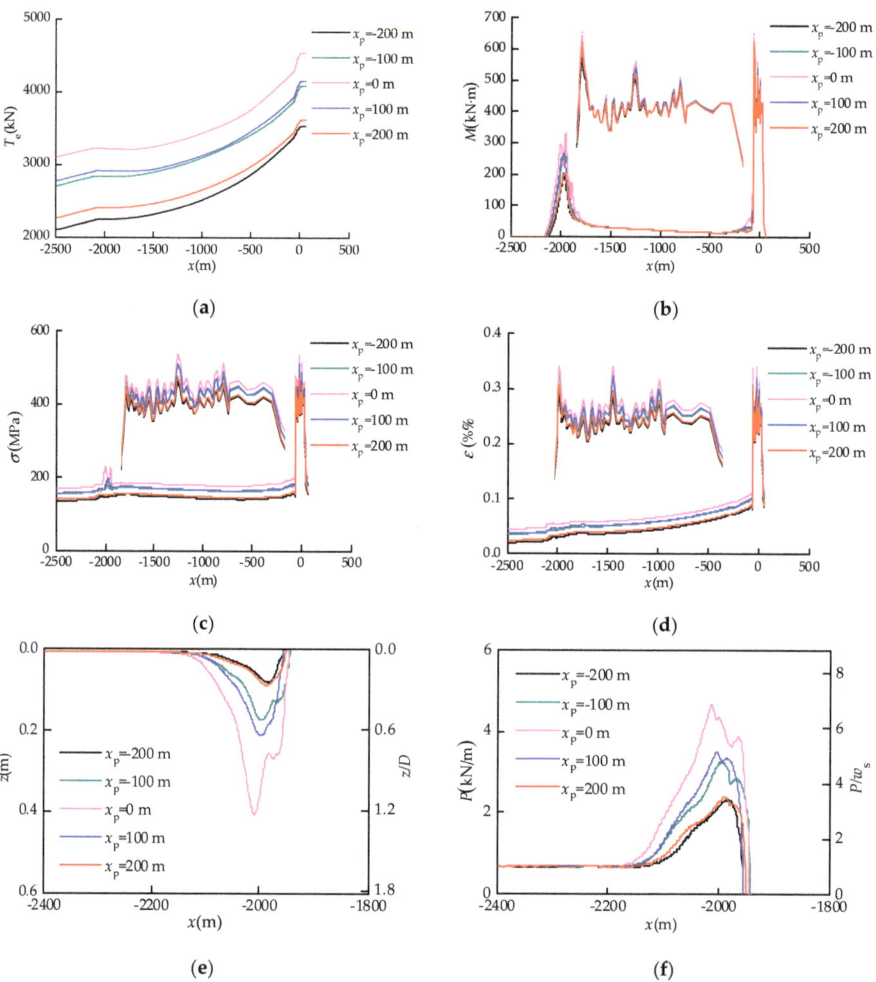

Figure 16. Dynamic responses of the S-laying pipeline under various freak wave focused positions: (**a**) tension; (**b**) bending moment; (**c**) stress; (**d**) strain; (**e**) pipeline embedment; (**f**) seabed resistance.

When the wave focusing location varied from −200 m to 200 m, the pipeline and seabed responses firstly rose and then dropped. The maximum responses occurred at the center position $x_\mathrm{p} = 0$ m with a peak axial tension of 4550.13 kN, bending moment of 654.13 kN·m, von Mises stress of 536.49 MPa, longitudinal strain of 0.340%, pipeline embedment of 0.406 m (1.231D), and seabed resistance of 4.685 kN/m (6.889w_s). The pipeline responses and seabed resistance at $x_\mathrm{p} = 100$ m and $x_\mathrm{p} = 200$ m were slightly larger than the corresponding results at $x_\mathrm{p} = -100$ m and $x_\mathrm{p} = -200$ m, for which the freak waves produced by the forward focusing location led to more prominent motion responses of the pipelay vessel. Evidently, the dynamic behaviors of the laying pipeline and seabed resistance are greatly influenced by the wave focusing location.

5.3. Effect of the Wave Phase Range

The wave phase range plays a significant role in generating freak waves and, to some degree, determines the wave height by controlling the phases of wave components in a specified region. Five groups of the wave phase range, $\varphi = 1.0\pi$, 1.1π, 1.2π, 1.3π and 1.4π, were selected to produce the freak wave trains shown in Figure 9c, and the influence of the wave phase range on the pipeline behaviors was explored by the combination of these wave trains with the developed S-lay FEM for time domain dynamic analyses. The pipeline and seabed response results on aspects of axial tension, bending moment, von Mises stress, longitudinal strain, pipeline embedment, and seabed resistance are displayed in Figure 17.

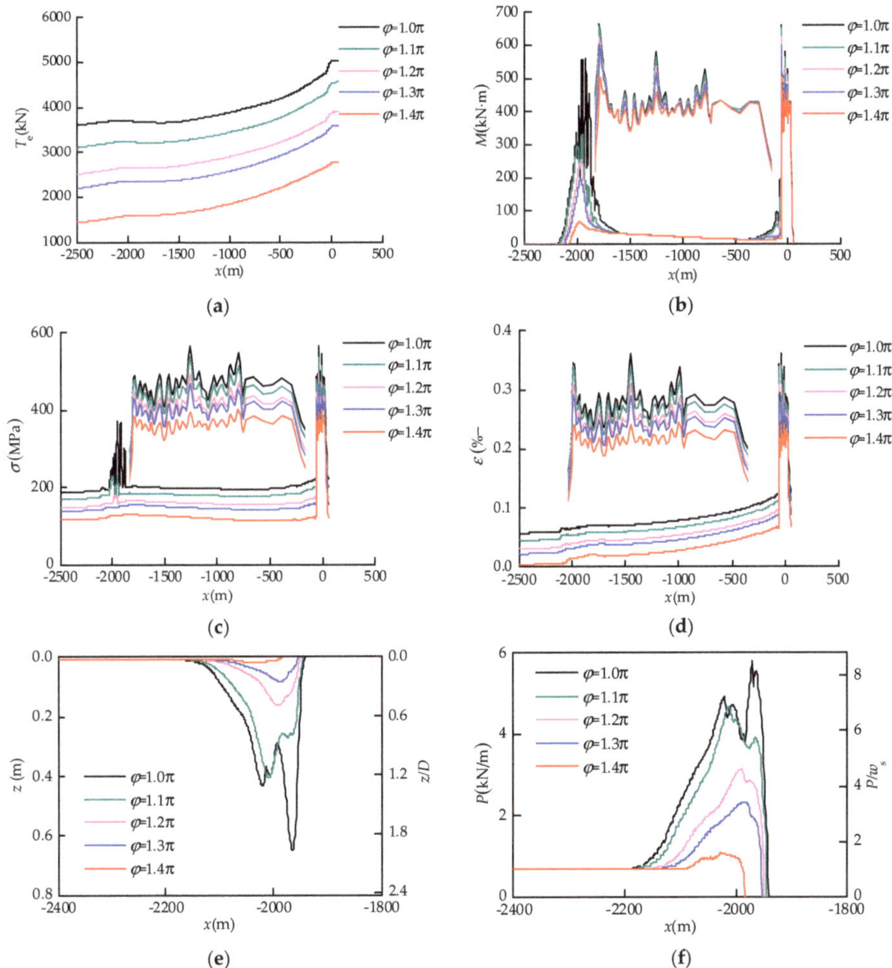

Figure 17. Dynamic responses of the S-laying pipeline under various freak wave phase ranges: (**a**) tension; (**b**) bending moment; (**c**) stress; (**d**) strain; (**e**) pipeline embedment; (**f**) seabed resistance.

With the increase of the wave phase range from 1.0π to 1.4π, all the pipeline and seabed responses had prominent decreases. The reductions in the maximum values were 45.0% for the axial tension, from 5034.62 to 2769.11 kN; 23.6% for the bending moment, from 661.99 to 506.08 kN·m; 29.0% for the von Mises stress, from 565.67 to 401.41 MPa; and 32.4% for the longitudinal strain, from 0.361%

to 0.244%. Moreover, the maximum pipeline embedment reduced by 97.2%, from 0.649 (1.967D) to 0.018 m (0.055D), and the maximum seabed resistance dropped by 81.1%, from 5.793 (8.518w_s) to 1.097 kN/m (1.613w_s). When the wave phase range was 1.0π, the bending moment and von Mises stress of the pipeline in the TDZ tremendously jumped to form some crests, as illustrated in Figure 17b,c. This phenomenon can be explained from the peak curves of pipeline embedment and seabed resistance shown in Figure 17e,f as the cyclic motions of the pipeline penetrating into and uplifting from the seabed, resulting in the softening and trenching of seabed soil, and the drastic pipe–seabed interactions induced by freak waves causing great pipeline flexural deflections. These results adequately demonstrate that the phase range of freak waves greatly influences the dynamic behaviors of the pipeline and the seabed, particularly for the entire axial tension and pipeline embedment as well as seabed resistance in the TDZ.

5.4. Effect of Wave Peak Value

Another noteworthy feature of freak waves is the peak value which generally represents the impact levels on offshore structures. In order to better understand the wave peak value effect on the dynamic behaviors of the S-laying pipeline, the five representative freak wave trains shown in Figure 9d were generated with their corresponding wave peak values of 3.6, 4.1, 4.6, 5.1, and 5.6 m. A time domain analysis of pipeline installation under these freak waves was conducted to obtain the pipeline and seabed response results, as shown in Figure 18.

Since the wave peak value gradually became larger from 3.6 to 5.6 m, all the pipeline and seabed responses showed an obvious increase. The axial tension of the overall pipeline became greater, and the maximum tension enlarged by 32.6%, from 3647.71 to 4838.41 kN. The bending moment of the pipeline mildly increased by 9.7% for its maximum value, from 602.24 to 660.69 MPa, in the overbend. Meanwhile, the bending moment in the sagbend had a great increase with the increment of its maximum value, attaining 103.2%, from 197.25 to 400.83 kN·m. Likewise, the maximum von Mises stress of the pipeline rose by 17.1%, from 471.98 to 552.47 MPa, and the maximum longitudinal strain of the pipeline rose by 19.5%, from 0.293% to 0.350%. Additionally, the maximum pipeline embedment and seabed resistance remarkably enlarged by 507.0% and 127.3%, respectively, from 0.086 (0.261D) to 0.522 m (1.582D) and from 2.387 (3.510w_s) to 5.426 kN/m (7.978w_s). Therefore, the increase in the freak wave peak value would result in great augmentation of pipeline behaviors and seabed resistance.

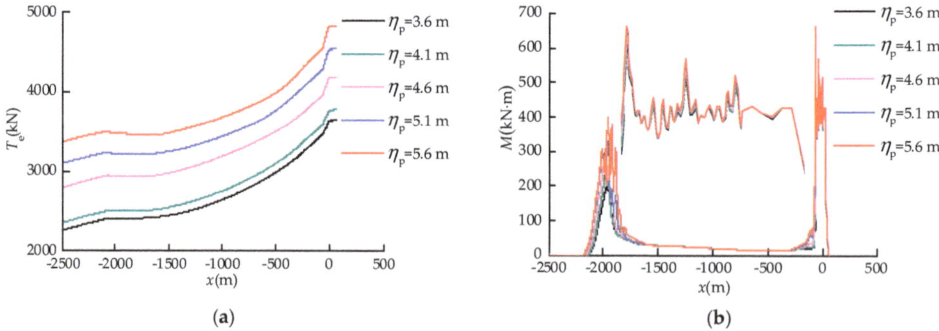

Figure 18. *Cont.*

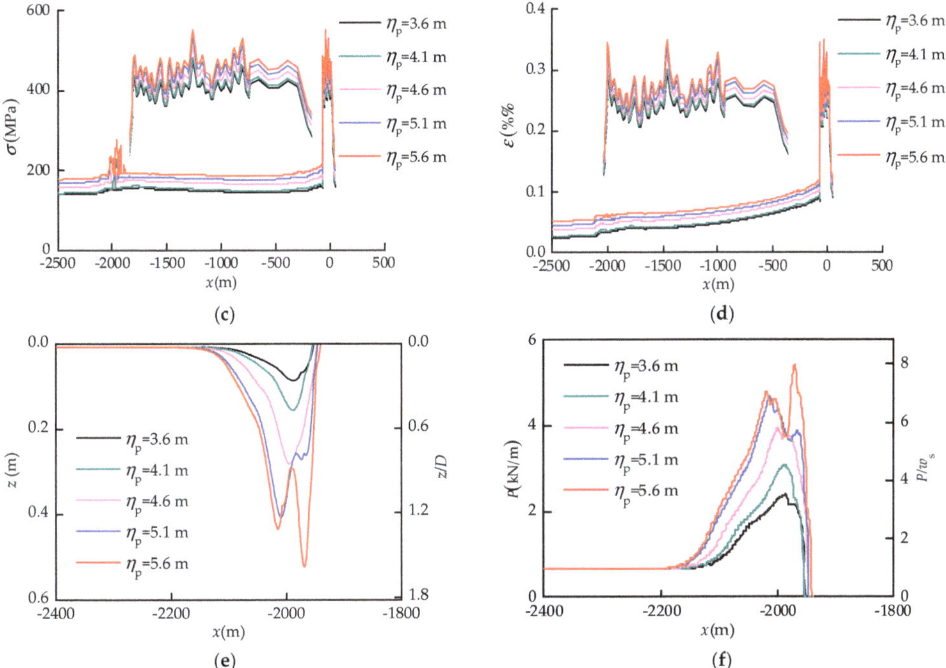

Figure 18. Dynamic responses of the S-laying pipeline under various freak wave peak values: (**a**) tension; (**b**) bending moment; (**c**) stress; (**d**) strain; (**e**) pipeline embedment; (**f**) seabed resistance.

6. Discussion and Implications

The deepwater S-lay FEM is a complicated, non-linear structural system, and dynamic response analysis of the laying pipeline under freak waves is difficult and time-consuming for marine structure engineers. A simplified technique was presented to estimate the pipeline dynamic response amplitudes by means of the dynamic amplification factors (DAFs), which are defined by the maximum responses relative to the corresponding static results. As a consequence, the dynamic response amplitudes of the laying pipeline can be easily determined if the static responses and DAFs are given.

As shown in Figure 19a, the pipeline and seabed DAFs in the parametric analyses were obtained with the variation of the energy ratio coefficient. Along with the increase of the energy ratio coefficient E_{p1}, the DAFs of axial tension, bending moment, stress, and strain gradually increased, in which the tension DAF was relatively prominent from 1.69 to 2.37. The pipeline embedment DAF largely rose from 3.75 to 46.41, and the seabed resistance DAF increased from 2.35 to 7.04. Figure 19b shows the variation in pipeline and seabed DAFs with the wave focused position, and all the DAFs firstly rose up and then dropped down. The maximum tension DAF reached 2.19, and the maximum pipeline embedment and seabed resistance DAFs reached 35.35 and 5.99, respectively. Moreover, the pipeline and seabed DAFs with the variation in the wave phase range are illustrated in Figure 19c. As the wave phase range increased, all of the DAFs reduced step by step. The DAF reductions were from 2.43 to 1.34 for axial tension, from 56.50 to 1.57 for pipeline embedment, and from 7.40 to 1.40 for seabed resistance. Oppositely, the pipeline and seabed DAFs enlarged stage by stage with the augmentation of the wave peak value, as displayed in Figure 19d, the DAF increment in axial tension was from 1.42 to 1.56, and the DAF increments in pipeline embedment and seabed resistance were from 7.49 to 45.44 and from 3.05 to 6.93. These obtained DAFs of the pipeline and seabed behaviors could offer intuitional knowledge for offshore pipeline engineers, which could be used to consider the freak wave effects in the initial design stage.

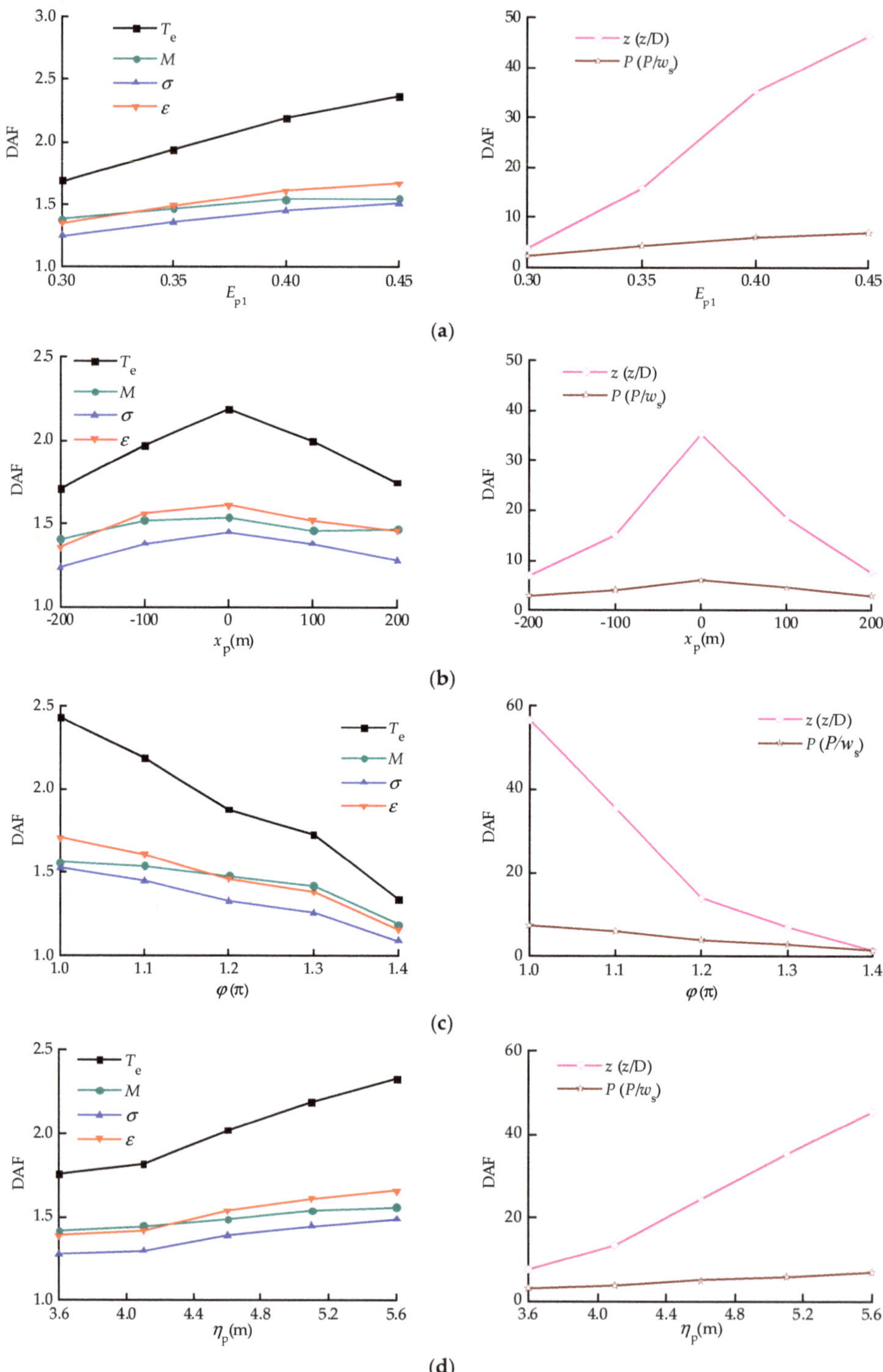

Figure 19. Effect of freak wave conditions on dynamic amplification factors of pipeline behaviors and seabed resistance: (**a**) energy ratio coefficients; (**b**) focused position; (**c**) phase range; (**d**) peak value.

7. Conclusions

This paper presented a profound investigation of freak wave effects on the dynamic responses of offshore pipelines for deepwater installation. For this purpose, an extended FEM of the S-lay system was developed in OrcaFlex with the particular consideration of freak waves. The linear superposition method of combined transient wave trains and random wave trains was applied to generate a series of freak wave trains. The wave induced pipelay vessel motions, pipe–stinger roller interactions in the overbend, as well as the cyclic contacts between the pipeline and seabed in the TDZ were also taken into account in the dynamic analysis of laying pipelines. The influences of the freak wave energy ratio coefficient, focusing location, phase range, and peak value on the pipeline and seabed behaviors were estimated in detail, and the DAFs of the axial tension, bending moment, von Mises stress, longitudinal strain, pipeline embedment, and seabed resistance were derived. Some significant conclusions were obtained as follows:

(1) The reasonable selection of wave parameters can effectively generate a variety of freak wave trains by the linear superposition model. The maximum heights of freak wave trains are obviously different with variations in the energy ratio coefficient, focusing position, phase range, and peak value. The freak wave trains could be steadily incorporated into the developed S-lay FEM to implement the dynamic analysis of deepwater pipeline installation.

(2) The energy ratio coefficient has a great influence on the generation of freak waves and the induced pipeline dynamic responses. With an increase in the energy ratio coefficient for transient waves, all the pipeline behaviors and seabed resistance remarkably increase. Especially, when the E_{p1} reaches 0.45, the interaction responses of the touchdown pipeline and seabed soil are drastically noticeable, which causes tremendous variation in the bending moment, von Mises stress, and pipeline embedment in the TDZ.

(3) The dynamic behaviors of the laying pipeline and seabed resistance are also strongly influenced by the wave focusing location. When the focusing wave is located at the center position $x_p = 0$ m, the responses of the offshore pipeline and seabed resistance are the most significant. Besides, the axial tension, pipeline embedment, and seabed resistance for the forward wave focusing location are slightly larger than the corresponding results for the negative wave focusing location.

(4) The phase range and peak value of freak waves were proven to be important influencing factors in the pipeline and seabed responses. As the wave phase range increases, the axial tension, bending moment, von-Mises stress, longitudinal strain, pipeline embedment, and seabed resistance as well as their DAFs remarkably decrease. On the contrary, when the wave peak value becomes larger, the pipeline behaviors and seabed resistance obviously augment.

Author Contributions: Conceptualization, P.X. and S.G.; methodology, P.X. and S.G.; software, P.X.; validation, P.X.; formal analysis, P.X. and Z.D.; investigation, P.X. and Z.D.; resources, P.X. and S.G.; data curation, Z.D.; writing—original draft preparation, P.X. and Z.D.; supervision, S.G.; funding acquisition, P.X. and S.G. All authors have read and agreed to the published version of the manuscript.

Funding: This research was funded by the National Natural Science Foundation of China (grant numbers 51809048, 51779223) and the Natural Science Foundation of Fujian Province, China (grant number 2018J05081).

Acknowledgments: The authors would like to thank the anonymous reviewers for their constructive comments and suggestions.

Conflicts of Interest: The authors declare no conflict of interest.

Nomenclature

A_i	internal cross-section area
A_o	external cross-section area
B	coefficient of the Ramberg–Osgood model
C_a	added mass coefficient
C_D	drag coefficient

d shortest separation distance of the center lines between the pipe and roller
D pipe outer diameter
E elastic modulus
E_{p1} energy ratio coefficient of a transient wave
E_{p2} energy ratio coefficient of a random wave
EA_{nom} nominal axial stiffness
f_m peak frequency
g gravitational constant
k_1 pipe contact stiffness
k_2 roller contact stiffness
k_i wave number of the ith wave component
K_{max} soil normalized maximum stiffness
k_s soil shear stiffness
L instantaneous length of a line segment
L_0 unstretched length of a line segment
M_b bending moment
n power exponent of the Ramberg–Osgood model
N number of wave components
N_c soil nominal bearing capacity factor
P_i internal pressure
P_o external pressure
$P_u(z)$ soil ultimate penetration resistance
r_1 pipe radius
r_2 roller radius
$S(f)$ spectral density function
S_{u0} soil mudline shear strength
S_{ug} soil shear strength gradient
t_c corrosion coating thickness
t_p wave focusing time
t'_p pipe wall thickness
T_e effective tension
T_w wall tension
T_{or} torque moment
w_a pipe weight per unit length in air
w_s pipe submerged weight per unit length
x_p wave focusing position
α spectral energy coefficient
γ peak enhancement factor
κ_2 curvature
μ soil friction coefficient
ν Poisson's ratio
ρ_c pipe corrosion coating density
ρ_p pipe density
ρ_{soil} saturated soil density
ρ_w sea water density
ω_i angular frequency of the ith wave component
ϕ_i phase lag of the ith wave component
σ_y effective yield stress
φ twist angle
τ spectral width parameter
ξ axial damping coefficient
ς bending damping coefficient
ζ torsional damping coefficient

References

1. Kjeldsen, S.P. Measurements of freak waves in Norway and related ship accidents. In Proceedings of the Royal Institution of Naval Architects International Conference-Design and Operation for Abnormal Conditions III, London, UK, 29–30 April 2004.

2. Slunyaev, A.; Didenkulova, I.; Pelinovsky, E. Rogue waves in 2006–2010. *Nat. Hazards Earth Syst. Sci.* **2011**, *11*, 2913–2924.

3. Bruschi, R.; Vitali, L.; Marchionni, L.; Parrella, A.; Mancini, A. Pipe technology and installation equipment for frontier deep water projects. *Ocean Eng.* **2015**, *108*, 369–392. [CrossRef]

4. Davis, M.C.; Zarnick, E.E. Testing ship models in transient waves. In Proceedings of the 5th International Symposium on Naval Hydrodynamics, Bergen, Norway, 10–12 September 1964; David Taylor Model Basin Hydromechanics Lab: Washington, DC, USA, 1964.

5. Baldock, T.E.; Swan, C.; Taylor, P.H. A laboratory study of nonlinear surface wave in water. *Philos. Trans. Math. Phys. Eng. Sci.* **1996**, *354*, 649–676.

6. Fochesato, C.; Grilli, S.; Dias, F. Numerical modeling of extreme rogue waves generated by directional energy focusing. *Wave Motion* **2007**, *44*, 395–416. [CrossRef]

7. Zhao, X.Z.; Sun, Z.C.; Liang, S.X. Efficient focusing models for generation of freak waves. *China Ocean Eng.* **2009**, *23*, 429–440.

8. Zhao, X.Z.; Ye, Z.T.; Fu, Y.N.; Cao, F.F. A CIP-based numerical simulation of freak wave impact on a floating body. *Ocean Eng.* **2014**, *87*, 50–63. [CrossRef]

9. Liu, Z.Q.; Zhang, N.C.; Yu, Y.X. An efficient focusing model for generation of freak waves. *Acta Oceanol. Sin.* **2011**, *30*, 19–26. [CrossRef]

10. Hu, Z.Q.; Tang, W.Y.; Xue, H.X. A probability-based superposition model of freak wave simulation. *Appl. Ocean Res.* **2014**, *47*, 284–290. [CrossRef]

11. Tang, Y.G.; Li, Y.; Wang, B.; Liu, S.X.; Zhu, L.H. Dynamic analysis of turret-moored FPSO system in freak wave. *China Ocean Eng.* **2016**, *30*, 521–534. [CrossRef]

12. Pan, W.B.; Zhang, N.C.; Huang, G.X.; Ma, X.Y. Experimental study on motion responses of a moored rectangular cylinder under freak waves (I: Time-domain study). *Ocean Eng.* **2018**, *153*, 268–281. [CrossRef]

13. Gong, S.F.; Chen, K.; Chen, Y.; Jin, W.L.; Li, Z.G.; Zhao, D.Y. Configuration analysis of deepwater S-lay pipeline. *China Ocean Eng.* **2011**, *25*, 519–530. [CrossRef]

14. Marchionni, L.; Alessandro, L.; Vitali, L. Offshore pipeline installation: 3-dimensional finite element modelling. In Proceedings of the 30th International Conference on Offshore Mechanics and Arctic Engineering, Rotterdam, The Netherlands, 19–24 June 2011.

15. O'Grady, R.; Harte, A. Localised assessment of pipeline integrity during ultra-deep S-lay installation. *Ocean Eng.* **2013**, *68*, 27–37. [CrossRef]

16. Gong, S.F.; Xu, P.; Bao, S.; Zhong, W.J.; He, N.; Yan, H. Numerical modelling on dynamic behaviour of deepwater S-lay pipeline. *Ocean Eng.* **2014**, *88*, 393–408. [CrossRef]

17. Gong, S.F.; Xu, P. The influence of sea state on dynamic behaviour of offshore pipelines for deepwater S-lay. *Ocean Eng.* **2016**, *111*, 398–413. [CrossRef]

18. Ivić, S.; Čanađija, M.; Družeta, S. Static structural analysis of S-lay pipe laying with a tensioner model based on the frictional contact. *Eng. Rev.* **2014**, *34*, 223–234.

19. Ivić, S.; Družeta, S.; Hreljac, I. S-Lay pipe laying optimization using specialized PSO method. *Struct. Multidiscip. Optim.* **2017**, *56*, 297–313. [CrossRef]

20. Xie, P.; Zhao, Y.; Yue, Q.J.; Palmer, A.C. Dynamic loading history and collapse analysis of the pipe during deepwater S-lay operation. *Mar. Struct.* **2015**, *40*, 183–192. [CrossRef]

21. Cabrera-Miranda, J.M.; Paik, J.K. On the probabilistic distribution of loads on a marine riser. *Ocean Eng.* **2017**, *134*, 105–118. [CrossRef]

22. Wang, F.C.; Chen, J.; Gao, S.; Tang, K.; Meng, X.W. Development and sea trial of real-time offshore pipeline installation monitoring system. *Ocean Eng.* **2017**, *146*, 468–476. [CrossRef]

23. Liang, H.; Yue, Q.J.; Lim, G.; Palmer, A.C. Study on the contact behaviour of pipe and rollers in deep S-lay. *Appl. Ocean Res.* **2018**, *72*, 1–11. [CrossRef]

24. Liang, H.; Zhao, Y.; Yue, Q.J. Experimental study on dynamic interaction between pipe and rollers in deep S-lay. *Ocean Eng.* **2019**, *175*, 188–196. [CrossRef]

25. Kim, H.S.; Kim, B.W. An efficient linearised dynamic analysis method for structural safety design of J-lay and S-lay pipeline installation. *Ships Offshore Struct.* **2019**, *14*, 204–219. [CrossRef]
26. Orcina. *OrcaFlex User Manual, Version 9.7a*; Orcina: Cumbria, UK, 2014.
27. Ramberg, W.; Osgood, W.R. *Description of Stress-Strain Curves by Three Parameters*; Technical Note, No. 902; National Advisory Committee for Aeronautics (NACA): Washington, DC, USA, 1943.
28. Wang, F.C.; Luo, Y.; Xie, Y.; Li, B.; Li, J.N. Practical and theoretical assessments of subsea installation capacity for HYSY 201 laybarge according to recent project performances in South China Sea. In Proceedings of the Annual Offshore Technology Conference, Houston, TX, USA, 5–8 May 2014; pp. 2696–2704.
29. Randolph, M.F.; Quiggin, P. Non-linear hysteretic seabed model for catenary pipeline contact. In Proceedings of the 28th International Conference on Ocean, Honolulu, HI, USA, 31 May–5 June 2009.
30. Gong, S.F.; Xu, P. Influences of pipe–soil interaction on dynamic behaviour of deepwater S-lay pipeline under random sea states. *Ships Offshore Struct.* **2017**, *12*, 370–387. [CrossRef]
31. Ai, S.M.; Sun, L.P.; Tao, L.B.; Yim, C.S. Modeling and simulation of deepwater pipeline S-lay with coupled dynamic positioning. *J. Offshore Mech. Arct. Eng.* **2018**, *140*, 051704. [CrossRef]
32. Slunyaev, A.; Pelinovsky, E.; Sergeeva, A.; Chabchoub, A.; Hoffmann, N.; Onorato, M.; Akhmediev, N. Super-rogue Waves in Simulations Based on Weakly Nonlinear and Fully Nonlinear Hydrodynamic Equations. *Phys. Rev. E Stat. Nonlinear Soft Matter Phys.* **2013**, *88*, 012909. [CrossRef]
33. Lu, W.Y.; Yang, J.M.; Fu, S.X. Numerical Study of the Generation and Evolution of Breather-type Rogue Waves. *Ships Offshore Struct.* **2017**, *12*, 66–76. [CrossRef]
34. Morison, J.R.; O'Brien, M.D.; Johnson, J.W.; Schaaf, S.A. The force exerted by surface waves on piles. *J. Pet. Technol.* **1950**, *2*, 149–154. [CrossRef]
35. White, D.J.; Cheuk, C.Y. Modelling the soil resistance on seabed pipelines during large cycles of lateral movement. *Mar. Struct.* **2008**, *21*, 59–79. [CrossRef]

Article

Underwater Pipeline Oil Spill Detection Based on Structure of Root and Branch Cells

Huajun Song [1,2,*], **Jie Song** [1] and **Peng Ren** [1,2]

1   College of Oceanography and Space Informatics, China University of Petroleum, Qingdao 266580, China;
     s17050647@s.upc.edu.cn (J.S.); pengren@upc.edu.cn (P.R.)
2   Key Laboratory of Marine Environmental Survey Technology and Application, Guangzhou 510000, China
*   Correspondence: huajun.song@upc.edu.cn

Received: 26 October 2020; Accepted: 5 December 2020; Published: 11 December 2020

Abstract: The existing oil spill detection methods mainly rely on physical sensors or numerical models cannot locate the spill position accurately and in time. To solve this problem, combining with underwater image processing technology, an unsupervised detection algorithm for oil spill in underwater pipelines is proposed for the first time. First, the oil spill region to be detected is regarded as the moving target, and the foreground detection algorithm is applied to the processed images. Then, the HSV (Hue, Saturation, Value) color space of the image is used to screen the oil spill region meeting the threshold requirements. Next, the bitwise of foreground mask and HSV mask into cells are divided. Finally, according to the characteristics of the oil spill image, false detection is eliminated by classifying cells three times. After qualitative and quantitative analysis, it is proved that the proposed algorithm can detect oil spill region accurately.

Keywords: object detection; underwater image; computer vision; oil spill

---

1. Introduction

There have been several oil spills at sea over the decades, e.g., the Gulf of Mexico (2010), the Timor Sea (2009), and the North Sea (1988). Although oil spills occur infrequently, they cause severe economic losses and a substantial adverse impact on the global environment [1]. Take the oil spill in the Gulf of Mexico in 2010 as an example: the accident caused eleven deaths and economic losses amounted to tens of billions of dollars. About 757,000 m^3 of oil spilled, which seriously threatened the marine ecological environment.

To provide technical support for environmental risk assessment, emergency plan formulation and emergency response, researchers have carried out numerous oil spill detection studies. The existing oil spill detection methods can be roughly divided into two directions according to detection location.

Above the sea surface, most of them are detected by synthetic aperture radar (SAR) satellite images or physical sensors. Dhont proposed the combined use of SAR and underwater gliders for oil seeps detection [2]. However, due to poor timeliness, leakage points cannot be found in time and the cost is high.

The other direction is underwater which mainly focuses on the numerical model research. Øistein regarded the discharges as terms of multiphase plumes and proposed a plume model based on Lagrange [3–5]. With image processing technology development, many target detection and tracking algorithms based on image features have been proposed. Compared with numerical models, vision-based target detection algorithms are more versatile. Some scholars have carried out image-based oil spill research. Bescond proposed a photoacoustic detection and monitoring method of oil spill [6]. Osman estimated the oil spill flow rate base on Wavelet in optical images [7]. Qasem implemented an underwater remotely operated vehicle (ROV) and identified oil spills by binarized difference images [8]. Oh proposed an underwater multispectral imaging system for environmental monitoring to identify different oil samples [9]. However, the current use of image processing technology is based on the image of the simulation experiment, with a simple image background, which is different from the actual situation, and the primary purpose is to use the oil spill image which spill region is known to calculate the relevant values, such as oil spill rate, oil spill amount and oil spill area.

With the development of machine learning, it has been widely used in image processing. Using convolutional neural networks (CNN), general target detection algorithms have achieved good results on object recognition datasets such as Pascal VOC (Visual Object Classes) [10], COCO (Common objects in context) [11], and ImageNet [12], which consist of a large number of pictures and their annotated documents. However, target detection algorithms based on machine learning often require massive amounts of datasets to extract features of the target. Commonly used training sets often reach the level of millions or even hundreds of millions of samples. However, underwater oil spill images are challenging to obtain and scarce in practice, which cannot meet the requirements of statistical-based detection model training. Therefore, the target detection methods based on machine learning are not suitable for underwater oil spill detection task. Feature extraction has to adopt traditional methods.

Considering the oil spill movement characteristics are significant, motion detection can be applied in detecting the oil spill. Although there are various motion detection methods, none achieve good results in dynamic monitoring scenarios. Especially due to a lot of interference in the underwater environment, such as seawater and marine life movement, underwater image motion detection's difficulty is further increased. Thus, false detection is easy to occur, and the detection results need to be processed to reject false detection.

Considering the maturity of the existing algorithms and the limitation of their application on underwater oil spillage detection, an unsupervised detection algorithm based on underwater oil spill motion and color characteristics is proposed. It integrates the characteristics of the oil spill image to eliminate false detection by using the proposed cell structure. It achieved high accuracy in experiments. The rest of this paper is as follows. Section 2 introduces the proposed algorithm and its specific implementation. Section 3 presents the detection results and evaluation of the proposed algorithm. Section 4 gives the conclusion of the paper.

2. Method

The proposed algorithm processes the video images collected by underwater cameras. Firstly, the underwater oil spill is regarded as the motion foreground, and the motion detection algorithm is used to detect it. The suspicious oil spill region is screened out by threshold in HSV color space too. Next, the two suspicious oil spill regions are united into the pre-detecting mask and the mask image is separated into cells, which can be classified as oil cells or others. The spatiotemporal context information is used in cells to eliminate missed targets and lock the oil leakage region. Finally, combining the plume characteristic of spilled oil, a structure of root and branch cell is proposed, and a cluster of cells belonging to the same root cell is regarded as the detected oil spill region.

Figure 1 is the flowchart of the proposed algorithm. Figure 2a shows the implement of the algorithm. Figure 2b shows the process of pre-detection. It unites motion detection and color screening in HSV space to generate a suspected oil spill region mask. Figure 2b–d are the single image feature detection based on cell structures. From left to right are cell divisions, root-branch classification, and root-branch stacks. Figure 2d is the process of eliminating false detection by introducing STC information into cells and the final detection result.

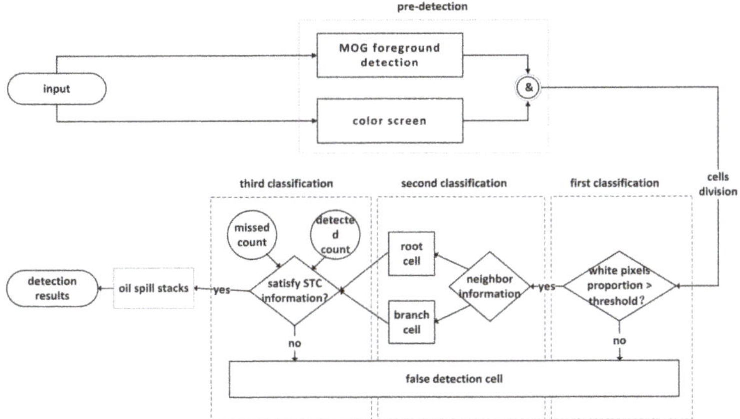

Figure 1. Algorithm flowchart.

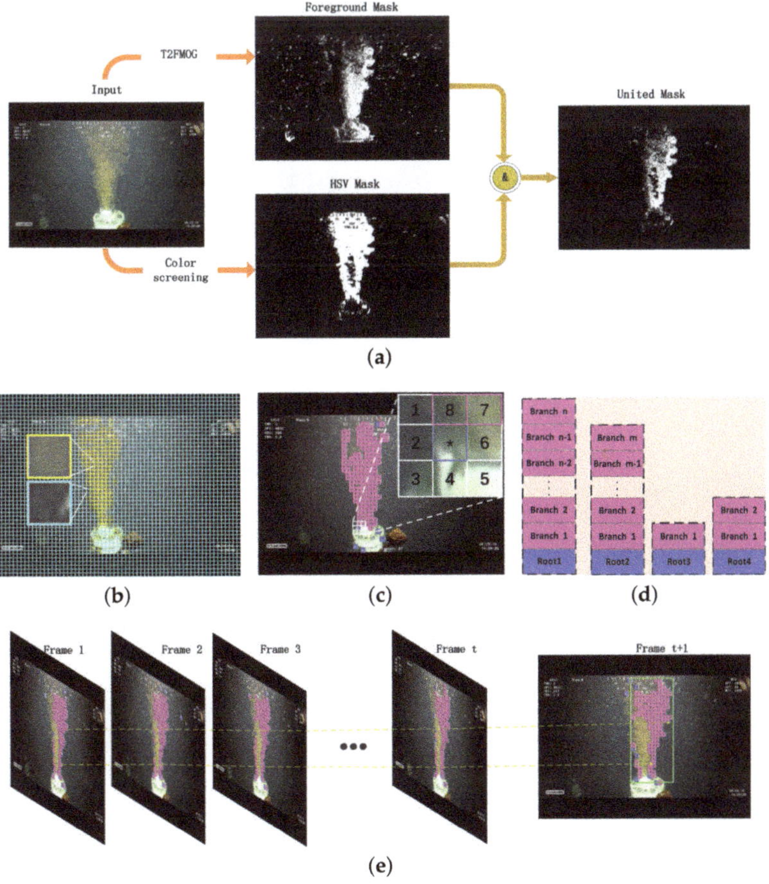

Figure 2. Algorithm implementation: (**a**) pre-detection; (**b**) cells division; (**c**) first classification; (**d**) second classification and (**e**) third classification.

In the past two decades, it is widely accepted that the progress of object detection has generally gone through two historical periods: "traditional object detection period (before 2014)" and "deep-learning-based detection period (after 2014)" [13]. Limited by the computing power and the lack of datasets, traditional target detection mostly uses handcrafted features, and then uses the classifier to classify the features. However, training a deep-learning-based model, such as SSD [14], Faster R-CNN [15], and YOLO [16], requires many data, thus oil spill detection tasks lacking underwater image information are not suitable for deep-learning-based detection models. To detect oil spill in underwater images, we use an unsupervised algorithm system combining multiple image features to achieve underwater oil spill detection.

2.1. Foreground Detection

Since the oil spill is continuously moving and changing, the underwater oil spill can be regarded as the moving target. The underwater oil spill detection can be regarded as the motion foreground detection task. Traditional motion detection algorithms include background modeling [17], clustering [18], image segmentation algorithm. However, it is difficult to detect an object in the dynamic background due to factors such as varying illumination, a sudden change in a scene, occlusion, and shadow [19]. Some researchers have applied machine learning methods to motion detection in recent years, such as FgSegNet series algorithms [20,21], which won first place in the CDNet2014 challenge [22]. However, these supervised algorithms are based on dataset training and have no universality, so they cannot apply to underwater oil spill detection.

However, looking at the literature, we can remark that there is often a gap between the current methods used in real applications and fundamental research [23]. After trials and contrasts, T2FMOG [24] based on a fuzzy mixture of Gaussians model is used in this paper, which has a higher tolerance to the background of slight motion and can produce ideal foreground detection effect in more complex underwater environment.

According to the uncertain mean vector or error, T2FMOG is divided into T2FMOG-UM and T2FMOG-UV. Due to the water flow in an underwater environment, the uncertainty variance is more likely to occur. Therefore, T2FMOG-UV can achieve better results. For RGB images, the multivariate Gaussian distribution with uncertain variance is as follows:

$$\eta(o, \mu, \overset{\sim}{\textstyle\sum}) = \frac{1}{\sqrt{(2\pi)^3 |\Sigma|}} \prod exp(-\frac{1}{2}\frac{(X_{t,c} - \mu_c)^2}{\sigma_c}) \tag{1}$$

in which $\sigma_c \in [\underline{\sigma}_c, \overline{\sigma}_c], c \in \{R, G, B\}$, each exponential component in Equation (1) is the Gaussian primary membership function (MF) with standard deviation. The upper MF is $\underline{h}(o) = f(o; \mu, \overline{\sigma})$ and the lower MF is $\overline{h}(o) = f(o; \mu, \underline{\sigma})$.

Between $\underline{h}$ and $\overline{h}$ is the footprint of uncertainty (FOU) as Figure 3. To control the intervals, a factor k_v introduced:

$$\underline{\sigma} = k_v\sigma, \overline{\sigma} = \frac{1}{k_v}\sigma, k_v \in [0.3, 1] \tag{2}$$

Our experiments verify that the result is more accurate when $k_v = 0.9$. After getting the T2FMOG-UV model of the image, we can detect the image foreground. Since the background is more present than moving objects and its value is practically constant, a background pixel corresponds to a high weight with a weak variance. Based on this prior knowledge, K Gaussian models are sorted by w_j/σ_j as a ratio. The first Gaussian distribution whose threshold exceeds T is regarded as the background distribution in Equation (3); other distributions are considered as foreground.

$$B = argmin_b(\sum_{i=1}^{b} w_{i,t} > T) \tag{3}$$

when a frame of the video comes in, use the log-likelihood and only consider the bilateral likelihood interval to test and classify each pixel:

$$H(X_t) = (\frac{1}{1/k_v{}^2 - k_v{}^2}) \frac{|X_t - \mu|^2}{2\sigma^2} \tag{4}$$

where μ and σ are the mean and the std of the original T1 MF without uncertainty. If $H(x_t) < k\sigma$, the pixel is attributed to Gaussian distribution. Therefore, the pixel test results can be divided into two cases:

1. If a pixel belongs to a certain Gaussian distribution, then the pixel belongs to the category represented by the Gaussian distribution.
2. If a pixel is not attributed to a Gaussian distribution, then the pixel is attributed to the motion foreground.

Furthermore, to ensure the accuracy of the background model, the parameters update according to Equations (5) and (6).

Case 1: Matched Gaussian distribution

$$w_{i,t+1} = (1 - \alpha)w_{i,t} + \alpha$$
$$\mu_{i,t+1} = (1 - \rho)\mu_{i,t} + \rho X_{t+1} \tag{5}$$
$$\sigma_{i,t+1}^2 = (1 - \rho)\sigma_{i,t}^2 + \rho(X_{t+1} - \mu_{i,t+1})(X_{t+1} - \mu_{i,t+1})^T$$

α is a fixed learning rate, $\rho = \alpha\eta(X_{t+1}, \mu_i, \Sigma_i)$. For unmatched distribution, μ and σ are invariant, $w_{j,t+1} = (1 - \alpha)w_{j,t}$.

Case 2: The final probability distribution K

$$w_{k,t+1} = Low\ Prior\ Weight$$
$$\mu_{k,t+1} = X_{t+1} \tag{6}$$
$$\sigma_{k,t+1}^2 = Large\ Initial\ Variance$$

Thus far, the motion foreground and the background model, which can be updated, are obtained by using T2FMOG. However, when considering factors such as water currents, marine life movement, and lens motion, there may be non-oil spilled parts in the foreground mask, so it is necessary to eliminate false detection by other methods.

2.2. Color Screening

In addition to the motion features, the oil spill region's color feature in an underwater environment is the most obvious. Therefore, the color threshold screening method is considered for screening the detected foreground further. However, in most videos, to display conveniently, the video pixel value is often in RGB format, which is designed according to the principle of color luminescence. The color is displayed by the excitation of red, green, and blue. Thus, it is not accurate enough for oil spill segmentation by using RGB color space. HSV is an intuitive color model, and its color parameters are hue, saturation, and lightness. Therefore, the videos can be converted to HSV space to screen the oil spill region under the foreground mask. The S and V channel calculation methods are: $S = V - min\{R, G, B\}$ amd $V = max\{R, G, B\}$, respectively. The channel calculation method of H is shown in Equation (7). When $H < 0$, $H = H + 360$.

$$H = \begin{cases} 60(G - B)/(V - min(R, G, B)) & if\ V = R \\ 60(B - R)/(V - min(R, G, B)) + 120 & if\ V = G \\ 60(R - G)/(V - min(R, G, B)) + 240 & if\ V = B \end{cases} \tag{7}$$

After obtaining the HSV model, the mask of the suspected oil spill region is obtained by threshold screening, and the screening method is as follows:

$$Mask_i = \begin{cases} 1, & \underline{T}_i \leq I_i \leq \overline{T}_i \\ 0, & others \end{cases} \tag{8}$$

where $\underline{T}_i$ and $\overline{T}_i$ are the lower and upper thresholds of each color channel, respectively. I_i is the pixel value on a single color channel in HSV space. The pixel value on mask is 1 only when the pixel values on three channels meet the threshold requirements; otherwise, it is 0.

A single frame mask of the suspected oil spill region can be obtained by performing a bitwise operation between the foreground detection mask and color screened mask. However, there is still false detection in practical applications caused by those moving objects with the same color as spilled oil. Further operations are needed to eliminate false detection after obtaining the detection mask of a single image.

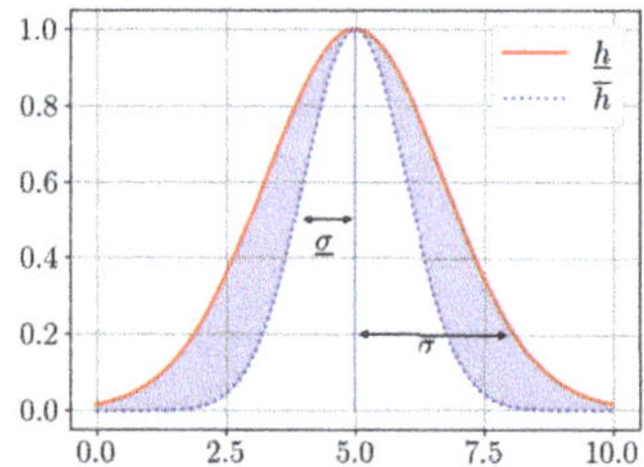

Figure 3. Gaussian membership function with uncertain variance.

2.3. False Detection Elimination

After watching more than 60 min of video containing oil spills, we found that the underwater oil spill image has two characteristics:

- Oil spillage is always produced from bottom to top.
- The movement of oil spillage occurs locally.

Based on these two characteristics, the root and branch cell structure and the STC method are proposed to eliminate false detection. The oil spill region is finally accurately screened out by classifying the cells three times.

Aiming at the first feature of underwater oil spill images, an algorithm of eliminating false detection based on cells with root and branch structures is proposed. First, the masked image is divided into S*S cells; then, the cells containing suspected oil spills are pre-classified and screened. The screening rule follows Equation (9), in which O_i is the category of the ith cell, W_i is the number of white pixels in the masked cell, and T is the set threshold, $0 < T < 1$. Figure 2b shows the screening diagram

$$Cell_i = \begin{cases} 1, W_i/S^2 \geq T \\ 0, W_i/S^2 < T \end{cases} \tag{9}$$

Then, the algorithm traverses the cells that have been classified as an oil spill from bottom to top and from left to right and secondarily classifies the pre-classified cells into root cells or branch cells. The classification method is shown in Figure 2c. The category of the cell is determined by the eight cells in the neighborhood. If there is no root cell or branch cell in the center cell neighbor positions 2–5, the center cell is classified as a root cell, and a new cell stack is established. If there is a root cell or branch cell in the center cell neighborhood, the center cell is pushed into the neighboring cell's stack. Figure 2d shows the second classification method.The second classification method is shown in Equation (10), in which $neighbor_i$ is the ith neighbor cell category.

$$Cell_i = \begin{cases} root, \sum_{i=2}^{5} neighbor_i > 0 \\ branch, \sum_{i=2}^{5} neighbor_i = 0 \end{cases} \qquad (10)$$

The movement of oil spillage occurs locally; that is, the movement amplitude of the oil spill in a short period is tiny, especially when the seawater movement is relatively gentle. The oil spill at the leakage point generally does not do translation movement. However, there is a diffusion movement at the edge part, and the internal motion amplitude is low or even relatively static. It is familiar to STC information, which models the statistical correlation between the low-level features (i.e., image intensity and position) from the target and its surrounding regions. Taking advantage of the oil spill movement characteristics, we propose an oil spill detection method that adds temporal and spatial context information to the cells. Due to the low amplitude of oil spill translation, the oil spill cell can always be detected within a certain period, and it cannot be detected for a long time when the target cell is an oil spill. The number of times each cell is detected is counted, and the cell is classified for the third time.

On the one hand, when the detection count is greater than the set value, the cell is confirmed to be an oil spill. On the other hand, when the continuous missed detection count is greater than a certain value, the detection count is set to zero, and the cell is regarded as a false detection for exclusion. The cell classification method of the tth frame is as follows:

$$Cell_t = \begin{cases} oil, \sum_{f=t-w}^{t} h_f > T_d \quad \& \quad \sum_{f=t-w}^{t} m_f < T_m \\ others, \sum_{f=t-w}^{t} m_f \geq T_m \end{cases} \qquad (11)$$

where w is the width of the counting window, h_f is the category of the fth frame determined by Equation (9), and the judgment rule of m_f is the opposite. T_d and T_m are detected and missed thresholds, respectively. The third classification process is shown in Figure 2c.

Finally, all cells are traversed again and popped up, taking the circumscribed rectangle of the graph formed by the cells in the same stack to form the detection boxes. The areas of the detection boxes are calculated, and the detection box with an area smaller than a certain threshold is regarded as invalid detection and excluded. The remaining detection boxes are used as the final detection result.

3. Results and Discussion

The dataset used in the experiment comes from underwater oil spill videos collected on the Internet, including some videos recorded by remote operated vehicle (ROV) during the Deep Horizon accidents. The videos were resized to a resolution of 640×480, and then cut into 1278 video frames.

Figure 4 shows the results of underwater motion detection [25]. The experimental results take the 60th and 215th frames of a video for comparison. Figure 4a shows the original frames. Figure 4b–d provided the detection results of several algorithms based on background modeling. It shows that the general background modeling algorithm cannot effectively distinguish the oil spill region and the seawater background, resulting in many missed detection. Figure 4e is the result of VIBE algorithm foreground detection, which can accurately detect part of the oil spill region, but there is still some missed detection. Figure 4f gives the detection results of the foreground detection algorithm used in this paper, which accurately detects oil spill movement in videos.

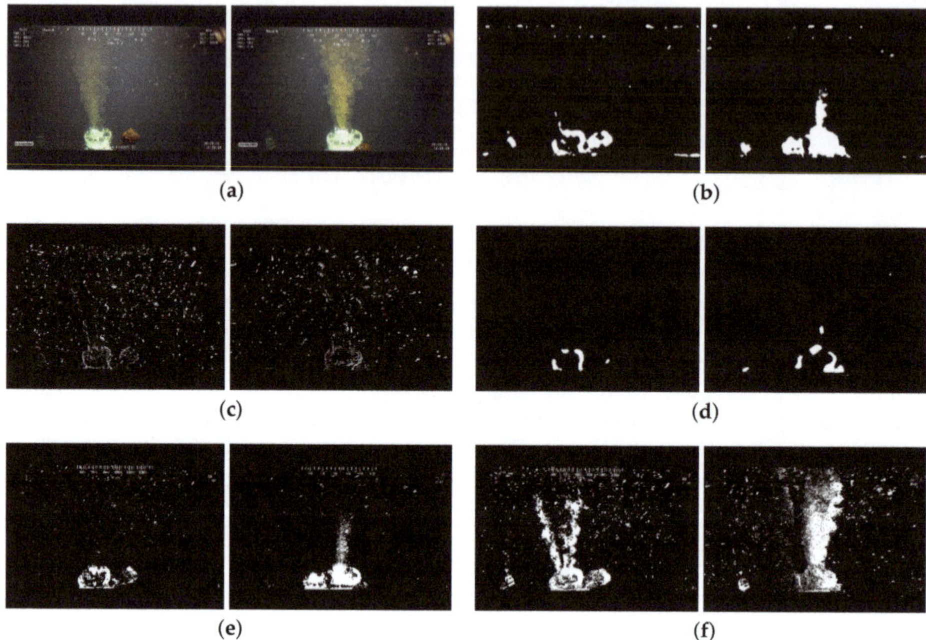

Figure 4. Foreground detection: (**a**) original frame; (**b**) LOBSTER [26]; (**c**) MOG2 [17]; (**d**) SuBSENSE [27]; (**e**) VIBE [28]; and (**f**) T2FGMG-UV.

Figure 5 shows the result of color threshold screening. Figure 5a is the original image, which is converted to HSV color space, as shown in Figure 5b, and Figure 5c shows the color mask obtained after threshold screening. The experimental results show that the algorithm can screen out accurate suspected oil spill region, but there is still false detection.

The mask of the suspected oil spill region is obtained by the motion mask bitwise and color mask, as shown in Figure 6a. Directly extracted detection results of the suspected oil spill region mask show that there are still some false detection boxes, as shown in Figure 6b. Figure 6c shows the cell stacks after removing false detection by using STC.

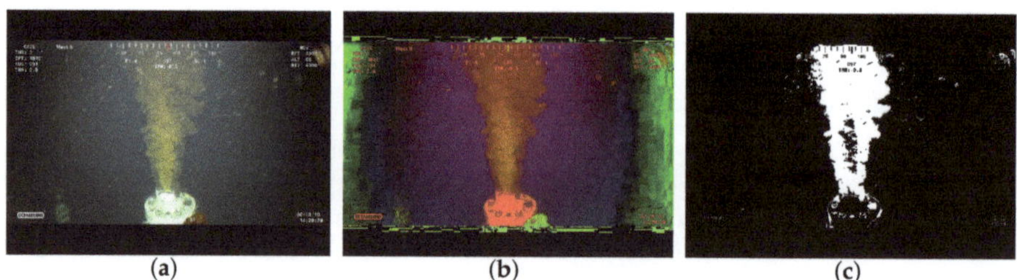

Figure 5. HSV threshold: (**a**) original image; (**b**) HSV image; and (**c**) color mask.

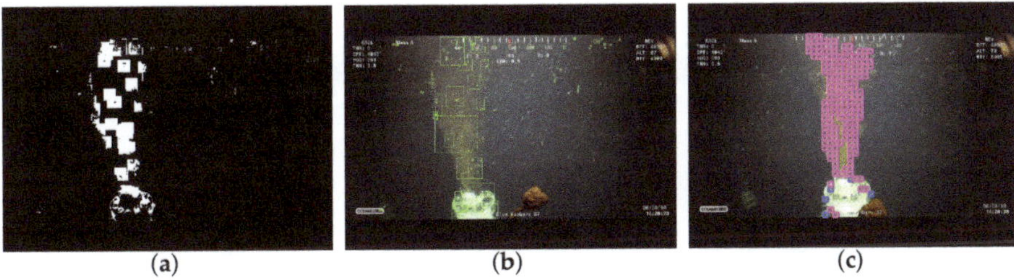

Figure 6. detection results: (**a**) fusion-mask; (**b**) pre-detection; and (**c**) cell stacks.

Figure 7a gives the proposed algorithm detection results. Compared to the frame difference in Figure 7b, our results locate the oil spill position and estimate the false detection. Figure 7c shows the yolov3 results. When the iteration number is low, there are many false detections, while, when the iteration number is high, the model is overfitting and it cannot detect oil spill.

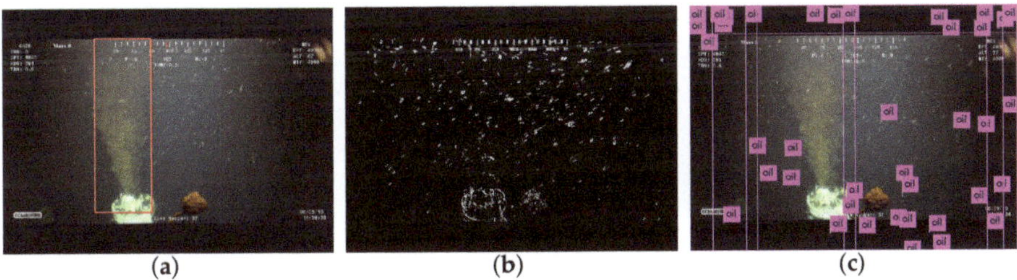

Figure 7. Results comparison: (**a**) ours; (**b**) frame difference [8]; and (**c**) yolov3 [29].

To evaluate algorithm results, we adopted the calculation method of VOC2010 [10]. It includes two import evaluation indicators: precision and recall. Their calculation methods are shown in Equations (12) and (13), respectively.

$$P = \frac{TP}{TP + FP} \tag{12}$$

$$R = \frac{TP}{TP + FN} \tag{13}$$

where TP, FP and TN mean the amount of true positive, false positive and false negative, respectively. They are decided by the intersection over union (IoU) threshold of the detected bounding box and ground truth.

To calculate average precision, the precision at each recall level r is interpolated by taking the maximum precision measured for a method for which the corresponding recall exceeds *r*. The interpolation method is shown in Equation (14). Then, we computed the average precision (AP) as the area under this curve (AUC) by numerical integration, as shown in Equation (15).

$$P_{interp(r)} = \max_{\tilde{r}:\tilde{r} \geq r} p(\tilde{r}) \tag{14}$$

$$AP = \sum_{r=0}^{r_{max}} P_{interp(r)} \tag{15}$$

The precision–recall (P-R) curve and average precision (AP) value of the test results are shown in Figure 8 and Table 1. The proposed algorithm achieves extremely high detection accuracy when the intersection over union (IoU) is set at 0.5. However, when the IoU is greater than 0.7, the accuracy of the proposed algorithm declines, which is due to the limitation of the algorithm caused by the motion detection algorithm and partial oil spill color is close to the color of seawater.

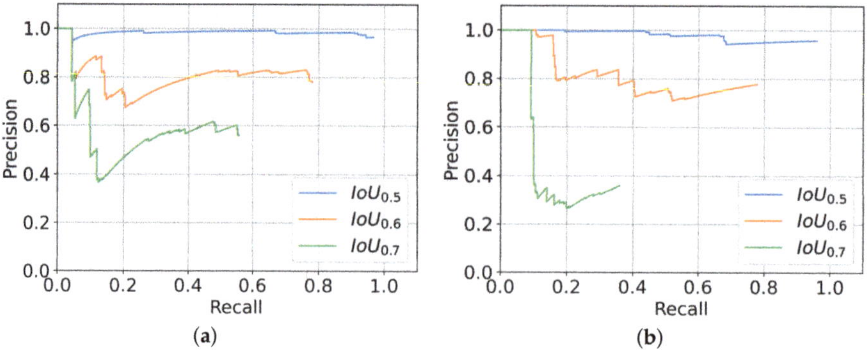

Figure 8. precision-recall graph: (a) Video 1; (b) Video 2.

Table 1. average precision.

Video	$AP_{0.5}$	$AP_{0.6}$	$AP_{0.7}$
1	0.958	0.662	0.369
2	0.943	0.659	0.192

4. Conclusions

This paper presents a method to detect oil spills by underwater images, which can be applied to fixed underwater cameras or remote operated vehicle (ROV). Compared with existing algorithms, it can process more complex real underwater images. It needs less hardware support, thus it can be integrated into an embedded system for independent use with ROVs in the future. It has massive potential for further treatment of the spill area too, such as calculation of spill speed and volume. It has the characteristics of low cost and good timeliness.

The experimental results show that the proposed algorithm can be applied to the alarm operation with less strict requirements for the oil spill position.

However, when the IoU is greater than 0.7, the limitations of the pre-detection part of the algorithm leads to a decline in the proposed algorithm's accuracy. Therefore, further optimization is needed in the detection applications with higher requirements for location and area.

Author Contributions: Conceptualization, H.S. and J.S.; methodology, H.S.; software, J.S.; validation, H.S. and J.S.; formal analysis, J.S. and P.R.; writing—original draft preparation, H.S. and J.S.; and writing—review and editing, J.S. and P.R. All authors have read and agreed to the published version of the manuscript.

Funding: This research was funded by National Natural Science Foundation of China grant number No. 61602517, the Fundamental Research Funds for the Central Universities(grant No. 18CX02109A), Science and Technology on Electronic Test & Measurement Laboratory Open fund (grant No. 6142001180514) and the Key Program of Marine Economy Development Special Foundation of Department of Natural Resources of Guangdong Province (GDNRC [2020]012)

Conflicts of Interest: The authors declare no conflict of interest.

Abbreviations

The following abbreviations are used in this manuscript:

HSV	Hue, Saturation, Value	STC	spatiotemporal context
CNN	convolutional neural networks	VOC	visual object classes
COCO	common objects in context	SSD	single shot multibox detector
YOLO	you only look once	T2FMOG	type-2 fuzzy mixture of Gaussian model
MF	membership function	FOU	footprint of uncertainty
RGB	red, green, blue	ROV	remote operated vehicle
P-R	precision-recall	AP	average precision
IoU	intersection over union	AUC	area under curve

References

1. Brkić, D.; Praks, P. Proper Use of Technical Standards in Offshore Petroleum Industry. *J. Mar. Sci. Eng.* **2020**, *8*, 555. [CrossRef]
2. Dhont, D.; Jatiault, R.; Lattes, P. Combined Use of SAR and Underwater Gliders for Oil Seeps Detection. In Proceedings of the 2019 IEEE International Geoscience and Remote Sensing Symposium (IGARSS 2019), Yokohama, Japan, 28 July–2 August 2019; IEEE: Piscataway, NJ, USA, 2019; pp. 7865–7868. [CrossRef]
3. Johansen, Ø. DeepBlow—A Lagrangian plume model for deep water blowouts. *Spill Sci. Technol. Bull.* **2000**, *6*, 103–111. [CrossRef]
4. Johansen, Ø.; Rye, H.; Cooper, C. DeepSpill—field study of a simulated oil and gas blowout in deep water. *Spill Sci. Technol. Bull.* **2003**, *8*, 433–443. [CrossRef]
5. Johansen, Ø. Development and verification of deep-water blowout models. *Mar. Pollut. Bull.* **2003**, *47*, 360–368. [CrossRef]
6. Bescond, C.; Kruger, S.E.; Levesque, D.; Brosseau, C. Photoacoustic detection and monitoring of oil spill. In *AIP Conference Proceedings*; AIP Publishing LLC.: Melville, NY, USA, 2019; p. 020025. [CrossRef]
7. Osman, A.; Ovinis, M.; Hashim, F.; Faye, I. Wavelet-based optical velocimetry for oil spill flow rate estimation. *Measurement* **2019**, *138*, 485–496. [CrossRef]
8. Qasem, F.; Susilo, T.; Said, S.; Alarbash, Z.; Hasan, M.; Jabakhanji, B.; Beyrouthy, T. Preliminary Engineering Implementation on Multisensory Underwater Remotely Operated Vehicle (ROV) for Oil Spills Surveillance. In Proceedings of the 2019 3rd International Conference on Bio-Engineering for Smart Technologies (BioSMART), Paris, France, 24–26 April 2019; IEEE: Piscataway, NJ, USA, 2019; pp. 1–5. [CrossRef]
9. Oh, S.; Lee, M.; Seo, S.; Roy, M.; Seo, D.; Kim, J. Underwater multispectral imaging system for environmental monitoring. In *OCEANS 2014-TAIPEI*; IEEE: Piscataway, NJ, USA, 2014; pp. 1–3. [CrossRef]
10. Everingham, M.; Gool, L.V.; Williams, C.K.I.; Winn, J.; Zisserman, A. The Pascal Visual Object Classes (VOC) Challenge. *Int. J. Comput. Vis.* **2010**, *88*, 303–338. [CrossRef]
11. Lin, T.Y.; Maire, M.; Belongie, S.; Hays, J.; Perona, P.; Ramanan, D.; Dollár, P.; Zitnick, C.L. Microsoft coco: Common objects in context. In *European Conference on Computer Vision*; Springer: Berlin, Germany, 2014; pp. 740–755. [CrossRef]
12. Deng, J.; Dong, W.; Socher, R.; Li, L.J.; Li, K.; Fei-Fei, L. Imagenet: A large-scale hierarchical image database. In Proceedings of the 2009 IEEE Conference on Computer Vision and Pattern Recognition, Miami, FL, USA, 20–25 June 2009; IEEE: Piscataway, NJ, USA, 2009; pp. 248–255. [CrossRef]
13. Zou, Z.; Shi, Z.; Guo, Y.; Ye, J. Object detection in 20 years: A survey. *arXiv* **2019**, arXiv:1905.05055.
14. Liu, W.; Anguelov, D.; Erhan, D.; Szegedy, C.; Reed, S.; Fu, C.Y.; Berg, A.C. Ssd: Single shot multibox detector. In *European Conference on Computer Vision*; Springer: Berlin, Germany, 2016; pp. 21–37. [CrossRef]
15. Ren, S.; He, K.; Girshick, R.; Sun, J. Faster R-CNN: Towards Real-Time Object Detection with Region Proposal Networks. *IEEE Trans. Pattern Anal. Mach. Intell.* **2017**, *39*, 1137–1149. [CrossRef]
16. Redmon, J.; Farhadi, A. Yolov3: An incremental improvement. *arXiv* **2018**, arXiv:1804.02767.
17. Zivkovic, Z. Improved adaptive Gaussian mixture model for background subtraction. In Proceedings of the 17th International Conference on Pattern Recognition (ICPR), Cambridge, UK, 23–26 August 2004; IEEE: Piscataway, NJ, USA, 2004; Volume 2, pp. 28–31. [CrossRef]
18. Zivkovic, Z.; Van Der Heijden, F. Efficient adaptive density estimation per image pixel for the task of background subtraction. *Pattern Recognit. Lett.* **2006**, *27*, 773–780. [CrossRef]

J. Mar. Sci. Eng. **2020**, *8*, 1016

19. Das, B.P.; Jenamani, P.; Mohanty, S.K.; Rup, S. On the development of moving object detection from traditional to fuzzy based techniques. In Proceedings of the 2017 2nd IEEE International Conference on Recent Trends in Electronics, Information & Communication Technology (RTEICT), Bangalore, India, 19–20 May 2017; IEEE: Piscataway, NJ, USA, 2017; pp. 658–661. [CrossRef]
20. Lim, L.A.; Keles, H.Y. Foreground segmentation using convolutional neural networks for multiscale feature encoding. *Pattern Recognit. Lett.* **2018**, *112*, 256–262. [CrossRef]
21. Lim, L.A.; Keles, H.Y. Learning multi-scale features for foreground segmentation. *Pattern Anal. Appl.* **2019**, 1–12. [CrossRef]
22. Wang, Y.; Jodoin, P.M.; Porikli, F.; Konrad, J.; Benezeth, Y.; Ishwar, P. CDnet 2014: An expanded change detection benchmark dataset. In Proceedings of the IEEE Conference on Computer Vision and Pattern Recognition Workshops, Columbus, OH, USA, 23–28 June 2014; pp. 387–394. [CrossRef]
23. Garcia-Garcia, B.; Bouwmans, T.; Silva, A.J.R. Background subtraction in real applications: Challenges, current models and future directions. *Comput. Sci. Rev.* **2020**, *35*, 100204. [CrossRef]
24. El Baf, F.; Bouwmans, T.; Vachon, B. Type-2 fuzzy mixture of Gaussians model: Application to background modeling. In *International Symposium on Visual Computing*; Springer: Berlin, Germany, 2008; pp. 772–781. [CrossRef]
25. Sobral, A. BGSLibrary: An opencv c++ background subtraction library. In Proceedings of the IX Workshop de Visao Computacional: Botafogo, Rio de Janeiro, Brazil, 3–5 June 2013; CRC Press: Boca Raton, FL, USA, 2013; Volume 2, p. 7. Available online: https://github.com/andrewssobral/bgslibrary (accessed on 10 December 2020). [CrossRef]
26. St-Charles, P.L.; Bilodeau, G.A. Improving background subtraction using local binary similarity patterns. In *IEEE Winter Conference on Applications of Computer Vision*; IEEE: Piscataway, NJ, USA, 2014; pp. 509–515. [CrossRef]
27. St-Charles, P.; Bilodeau, G.; Bergevin, R. SuBSENSE: A Universal Change Detection Method With Local Adaptive Sensitivity. *IEEE Trans. Image Process.* **2015**, *24*, 359–373. [CrossRef] [PubMed]
28. Van Droogenbroeck, M.; Barnich, O. ViBe: A disruptive method for background subtraction. *Backgr. Model. Foreground Detect. Video Surveill.* **2014**, 7–11. [CrossRef]
29. Zhao, X.; Wang, X.; Du, Z. Research on Detection Method for the Leakage of Underwater Pipeline by YOLOv3. In Proceedings of the 2020 IEEE International Conference on Mechatronics and Automation (ICMA), Beijing, China, 2–5 August 2020; IEEE: Piscataway, NJ, USA, 2020; pp. 637–642. [CrossRef]

Publisher's Note: MDPI stays neutral with regard to jurisdictional claims in published maps and institutional affiliations.

Journal of
Marine Science and Engineering

MDPI

Article

Investigation of the Structural Strength of Existing Blast Walls in Well-Test Areas on Drillships

Byeongkwon Jung [1], Jeong Hwan Kim [2] and Jung Kwan Seo [1,2,*]

[1]   Department of Naval Architecture and Ocean Engineering, Pusan National University, Busan 46241, Korea;
     bk47.jung@gmail.com
[2]   The Korea Ship and Offshore Research Institute, Pusan National University, Busan 46241, Korea;
     kjhwan1120@pusan.ac.kr
*   Correspondence: seojk@pusan.ac.kr; Tel.: +82-51-510-2415

Received: 15 June 2020; Accepted: 29 July 2020; Published: 4 August 2020

Abstract: Blast walls are installed on the topside of offshore structures to reduce the damage from fire and explosion accidents. The blast walls on production platforms such as floating production storage, offloading, and floating production units undergo fire and explosion risk analysis, but information about blast walls on the well-test area of drillship topsides is insufficient even though well tests are performed 30 to 45 times per year. Moreover, current industrial practices of design method are used as simplified elastically design approaches. Therefore, this study investigates the strength characteristic of blast wall on drillship based on the blast load profile from fire and explosion risk analysis results, as well as the ability of the current design scantling of the blast wall to endure the blast pressure during the well test. The maximum plastic strain of the FE results occurs at the bottom connection between the vertical girder and the blast wall plate. Based on the results, several alternative design applications are suggested to reduce the fabrication cost of a blast wall such as differences of stiffened plated structure and corrugated panels, possibility of changing material (mild steel), and reduced plate thickness for application in current industrial practices.

Keywords: drillship; blast wall; explosion; well-test areas; strength evaluation; non-linear structural analysis

1. Introduction

One of the most important trends in the development of the modern oil and gas industry is reorientation to the development of offshore oil and gas fields. The first offshore drilling rigs were constructed in 1934 on drilling shelves with stationary substructures [1]. The development of continental shelves has resulted in a variety of accidents with catastrophic consequences due to the lack of attention devoted to the identification of prevention and mitigation of possible risks and hazard.

The most serious accidents on drilling ships and platforms of various types (semi-submersible, submersible, mobile, stationary) between 1979 and 2015 are included in historical databases [2]. Well-known examples include "Piper Alpha" in the North Sea, "Deepwater Horizon" in the Gulf of Mexico and "SOCAR (The State Oil Company of the Azerbaijan Republic)" in the Caspian Sea. These accidents resulted from flooding, explosion and fire, and oil slippages. The distribution of accidents shows that a considerable proportion of accidents are due to personnel errors, such as process disturbance, improper pilotage, and improper berthing to offshore oil and gas facilities. According to statistical data, these emergency incidents are distributed by Fattakhova and Barakhnina [3]. Incidents such as fire, explosion and oil slippages occurred mostly during drilling work.

Drillships have the functional ability of semi-submersible drilling rigs but greater mobility (i.e., they are ship-shaped vessels), and they can move quickly under their own propulsion from drill site to another drill site, in contrast to semi-submersibles, jack-up barges, and platforms. A generally drillship comprises a hull, mud module, subsea control module, mud process module, drill floor, pipe and casing storage, and well-test module. One unusual characteristic is that the hull bottom has a large opening (so called "moon-pool") that allows the seabed to be drilled with a drill pipe and casing. A typical drillship has a length of 220 to 230 m, a breadth 36 to 42 m and a depth of 18 to 20 m, and drilling equipment on its topside. Its main advantage is its ability to perform drilling operations in very deep sea (up to 12 km), where general fixed platforms are not able to be drilled.

Among a drillship's various operations, the well-test operation is similar to a process module of a production offshore platform. Therefore, the safety design of a drillship should consider fires and explosions. For example, water-based drilling fluids (WBDF), which if used with proper additives such as nano-particles can be more effective than conventional oil-based drilling fluids and also can mitigate the firing problems for high pressure and high-temperature wells. Oil-based drilling fluids can be toxic to the marine species, if any incident of oil spills happens, such as the Gulf of Mexico, whereas, considered WBDF with nano-particles are non-toxic [4–6].

For those reasons, the well-test area on drillship installed blast walls for minimizing and preventing explosion damage surrounding modules. The structural design of the blast wall should consider the type of explosive load to which the structure will be exposed because of the numerous uncertainties inherent in an explosion. According to this reason, numerous investigations of blast walls on offshore platforms have been conducted to determine design methods and for structural evaluation though theoretical, numerical, and experimental approaches.

The literature includes many studies on the use of corrugated and stiffened walls on offshore platforms. A formal framework was suggested for blast walls with FE analysis [7]. That study provided not only a better understanding of the underlying assumptions required to justify blast wall explosive load detection behavior but also a coherent basis for specification and design for dynamic analysis of blast wall capabilities. In the presence of theoretical studies of blast walls, they proposed a detailed and reasonable FE model and conducted two analyses: traditional non-linear static analysis and non-linear time-path dynamic analysis. Vignjevic et al. [8] considered different internal reinforcements (C-section and corrugated) to improve the energy absorption properties of thin walled rectangular beams under uniaxial and biaxial deep bending collapse, for loading angles. Also, corrugated reinforcements showed a greater potential for increasing specific energy absorption, which was supported by investigating key geometric parameters, including corrugation angle, depth, and number using LS-DYNA [9] simulations experimentally validated. Sohn et al. [10] studied the role of a flat-plated stiffener on the structural characteristics of a blast wall on an offshore installation exposed to hydrocarbon explosions. Blast walls are generally installed in oil and gas production structures to minimize and prevent the damage from explosions. Kang et al. [11] suggested a blast loading application method. The uniformly distributed loading condition, predicted by explosion risk analysis, was applied in most previous analysis methods. However, the analysis methods related to load conditions are inaccurate because the blast overpressure around the wall tends to show a low level in open areas and a high level in enclosed areas. Syed et al. [12] suggested a method for analysis of offshore stainless-steel blast walls. Blast walls are mostly designed using simplified analytical techniques, such as the single degree of freedom method, in which global deformation or displacement is used as the primary response parameter. This study uses detailed non-linear Finite Element analysis to present realistic responses of offshore blast walls under various high impulsive pressure loads generated from accidental hydrocarbon explosions. The numerical models were also verified against past experiments results with similar steel blast panels. An extensive parametric study was conducted in which the verified FE models were used to construct pressure-impulse (P-I) diagrams for various deformation levels. Sohn and Kim [13] studied the structural response of corrugated blast walls depending on blast load pulse shapes. Hydrocarbon explosions are among the most hazardous events for workers on offshore platforms. Corrugated blast walls are typically installed to protect structures against explosion loads, but the profiles of real explosion loads differ depending on the congestion and confinement of the topside structures. Xiao et al. [14] suggested numerical prediction of blast wall effectiveness for structural protection against air blasts. The propagation of shock waves and their interaction with the blast wall were simulated in FE code (LS-DYNA [9,15,16]) to calculate the overpressure-time profile of the blast wall. The influence of structural flexibility on the effectiveness of the blast wall was considered by including a steel sheet ("canopy") at the top of the wall. These studies help to understand the behavior of an air blast interacting with blast walls and helps to identify the principal parameters for the design of such walls.

According to existing blast wall studies are well developed, validated, and suggested. However, limited studies have examined the blast wall in the well-test area on the topside of a drillship, likely because although floating offshore platforms (e.g., floating liquid natural gas and floating production storage and offloading units) have great exposure to the risk of fire and explosion accidents, accidents aboard drillships are expected to occur during well test operation. Moreover, current industrial practices of blast wall design on drillships, design method is simplified elastically design approaches. The design of the blast wall around the well-test area was evaluated according to the bending and shear stress of the supporting vertical girder using a static pressure (0.2 bar) and beam theory, and this method of evaluation may be over-designed when considering the blast pressure against a realistic blast load. Furthermore, only beam theory is applied to evaluate the blast wall, so it was not possible to evaluate the connection of the vertical girder and deck. Therefore, a blast wall on a drillship should be re-evaluated with advanced and/or developed methods for proper and accurate design.

2. Blast Wall of Well-Test Area on Drillship Topside

2.1. Drillship Topside Arrangement and Well-Test Area

To determine the target drillship topside arrangement and well-test area, an existing drillship was considered for this study as shown in Figure 1. The blast wall is placed between the well-test area and the ROV (Remotely Operated Vehicle) station on the port side. In addition, the lifeboat stations on either side of the drillship are located further aft than the well-test area and are partially shielded by the funnel casings.

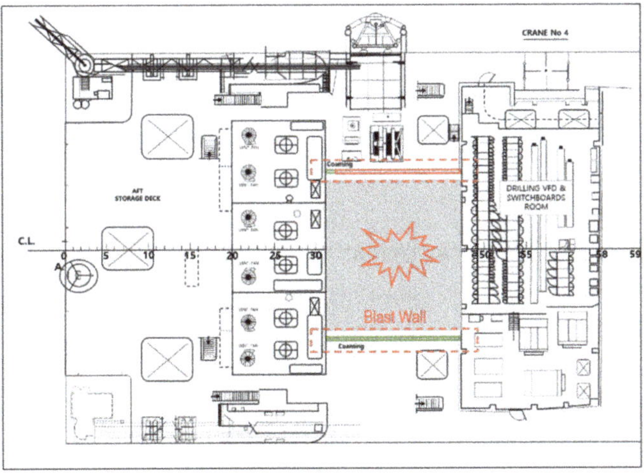

Figure 1. Typical well-test area arrangement on a drillship.

Well testing is normally performed approximately 30 to 45 times per year on a drillship [17]. The objectives of well testing are divided into to obtain and analyze representative samples of produced fluids, to measure the reservoir pressure and temperature, to determine the inflow performance relationship and deliverability, to evaluate the completion efficiency, to characterize well damage, and to evaluate work over and simulation treatments.

During well-test operation, oil or gas leakage is likely to result in an explosion accident. Therefore, to protect the equipment and structures on the topside, a blast wall is installed beside the expected ignition point. Figure 2 shows that the blast wall on the drillship topside is a stiffened panel structure that differs from a general blast wall and/or corrugated structure.

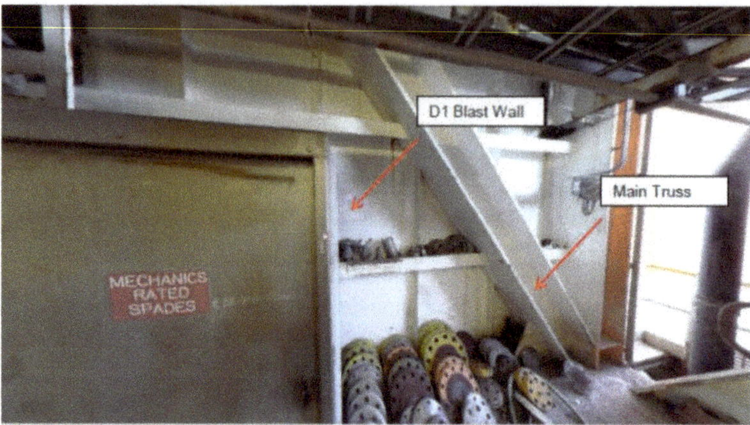

Figure 2. Example of a stiffened panel blast wall on drillship.

2.2. Characteristic of Explosion Load Profile

In industrial practice, simple beam theory has been applied to evaluate blast wall support with calculated blast pressure. The applied blast pressure is only considered static pressure without blast load profiles such as the duration time and the peak time.

In this study, to identify the load profile and scenarios of the well-test area, two specific studies were thoroughly reviewed. A previous experimental study [18] identified the characteristics of overpressure loads in an explosion, as shown in Figure 3a. A test module was fabricated for the explosion test that was similar to the processing module of a gas explosion in a floating liquid natural gas unit, and the explosion pressures were measured with a different geometrical effect, which is termed "porosity" and is given by Equation (1):

$$\text{Porosity} = 1 - \frac{\text{Volume of total structure}}{\text{Volume of total space}} \tag{1}$$

(**a**)

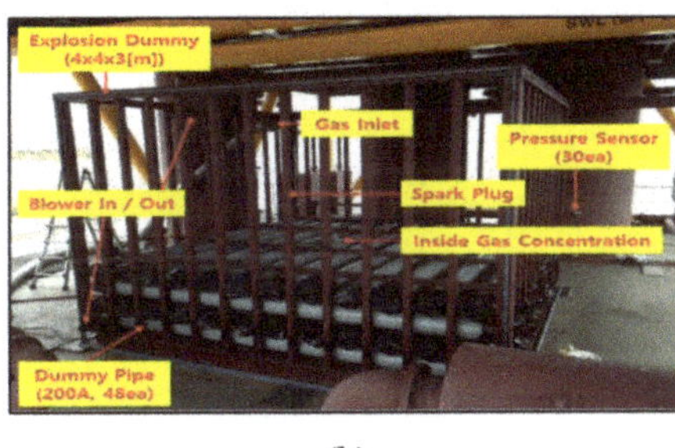

(**b**)

Figure 3. Full-scale test module for determination of explosion load profile. (**a**) Test module of a typical offshore topside. (**b**) Pipe rack structure for explosion.

The porosity was controlled by the number of pipes at the ignition point with a pipe rack structure for ignition, as shown in Figure 3b. Five scenarios were investigated with 0, 12, 24, 36, and 48 pipes. The ignition point was at the center of gas cloud.

As the experimental results [18] provide that the overpressure loads rose to a peak value within a very short period of rise time and decayed sharply in the experimental results. The rise time until the peak overpressure loads were reached differed depending on the considered structural congestion model. The rise time became shorter with increase in the degree of structural congestion. The overpressure loads fell into negative values compared with the ambient pressure and were recovered to the ambient pressure as the impact energy was released. In this result, porosity is an important factor in the overpressure in an explosion, and a structure with higher porosity has higher overpressure. The experimental results allow the explosion pressure profile to be categorized as "Arrival time", "Positive phase", and "Negative phase", respectively.

The general load profile of an explosion is quite complicated due to the explosion condition. The realistic characterization of blast pulse pressure action requires the pressure-time history to be traced, including the rise time, peak pressure, duration, and type of pressure decay. Blast pressure can be idealized as an impulsive loading that is characterized by peak pressure and duration time. To simplify the structural analysis, the time history of the panel load around its peak was idealized as a triangular impulse. The idealized blast load profiles can be placed into four categories, as shown in Figure 4. Those profiles were analyzed with three different domains: the impulsive domain ($\frac{\tau}{T} \leq 0.3$), the dynamic domain ($0.3 \leq \frac{\tau}{T} < 3$) and the quasi-static domain ($3 \leq \tau/T$). The structural behavior characteristics of corrugated blast walls and stiffened blast walls were simulated under various types of explosion loadings [5,8].

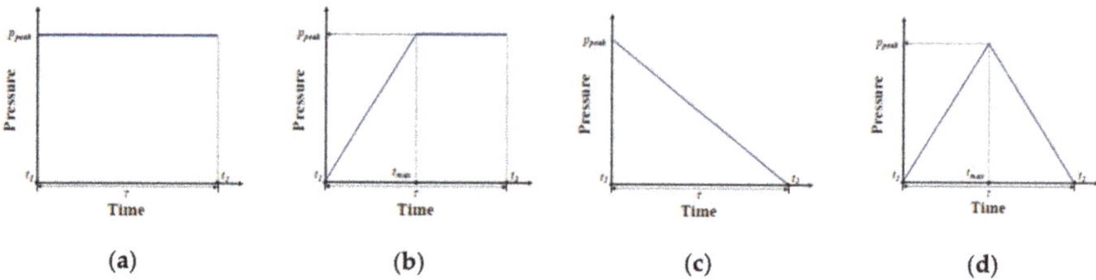

Figure 4. Common blast loading shapes [19]. (**a**) Rectangular. (**b**) Gradually applied. (**c**) Linear decay. (**d**) Triangular.

For selection of blast loading shapes, Sohn and Kim [13] provided that the effects of the shape of the load profile on the structural characteristic of blast walls show that the duration and peak pressure are not critical, but impulse is important in the impulsive domain. The rising time and the level of impulse have profound influences on the structural characteristics of blast walls in the dynamic domain. The results indicate that the structural characteristics of blast walls depend on the amount of applied momentum rather than the peak pressure or duration time. Furthermore, loads with linear decay and triangular loads have the same impulse value, but the ductility ratio differs greatly in the dynamic domain while remaining very similar in the impulse domain. In other words, the rising time is not critical in the impulse domain, but it is very important in the dynamic domain.

In fact, the current blast wall structure on drillship is designed with the simply linear beam theory approach, therefore the structure is assumed as a rigid body, and the stresses are concentrated at the bottom end of the vertical supports. According to these investigations, this study of blast load profile assumed a triangular load profile of positive phase and effect of negative phase.

2.3. Method of Strength Assessment of a Blast Wall in a Well-Test Area

An applicable method for the strength assessment of a blast wall in a well-test area suggested both fire/explosion analysis and structural analysis, as shown in Figure 5. Fire and explosion risk analysis is performed to evaluate accidental loading scenario hazards to define potential accidents that may occur to the facility during its lifetime from installation to decommissioning and assess the corresponding risk exposure. Potential accidents are defined as scenarios involving a ship collision, a dropped object, a fire, or a blast that introduces risk to personnel, the environment, or the facility.

To estimate the fire/explosion risk from the well-test area, certain operational days of well testing are assumed and included in the overall risk picture in the general oil/gas industrial risk practices. For advanced and quantitative risk evaluation, details of the equipment and piping and instrumentation diagrams for the well-test unit to be installed on the drill ship should be conducted with available date and calculation methods.

A typical well-test unit can be used to assess several components (valves, flanges, pie sections, etc.) for use as the basis of leak frequency assessment. Leak rates and frequencies are also calculated for each component, failure mode and type of medium with generic data or statistical methods.

The well-test unit has been divided into two segments such as gas and oil. Components that contain both oil and gas are allocated to the gas segment. This allocation is done due to the difficulties in predicting the amount of gas or oil that will be released during a leak (which depends on the gas/oil ratio in the reservoir to be tested). This is a conservative approach and is applied for components upstream of the separator from the drill floor, which is included in the gas segment.

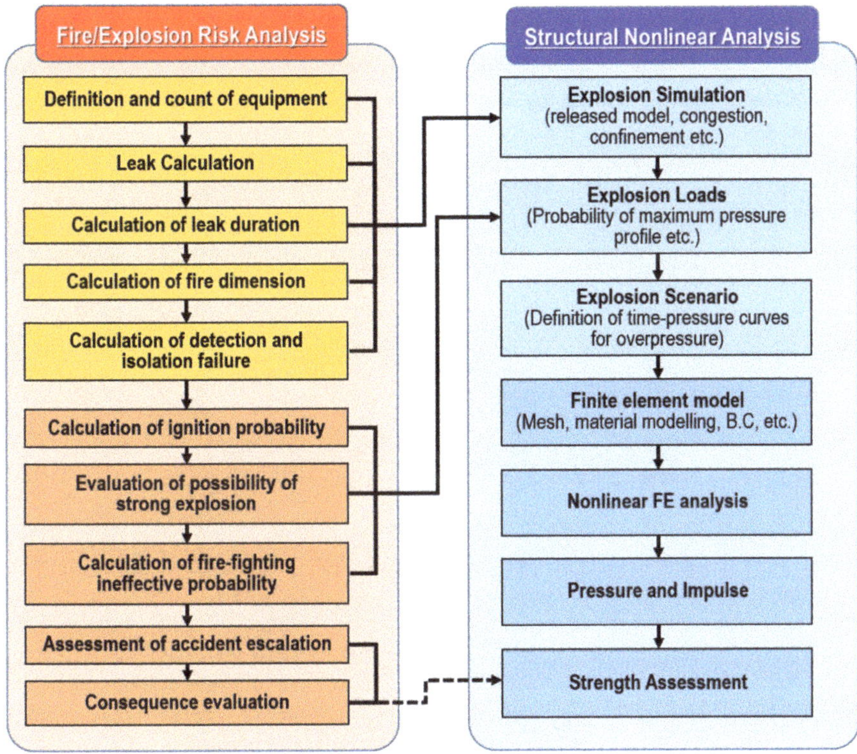

Figure 5. Flow chart risk and structural analyses of this study.

To evaluate the strength of the blast wall in this study, the explosion load requires for non-linear structural analysis. In addition, the design of the blast wall may be difficult because it requires predicting the type of explosive load to which a structure may be exposed due to the numerous uncertainties inherent to an explosion. Generally, oil/gas industries, classification societies and international/national organizations are used to idealized random explosion loads with symmetric triangular shapes because these well match an actual explosion loading with respect to attributes such as total impulse, peak pressure, rising time, and pulse duration. Therefore, the explosion load profile can be defined via fire/explosion risk analysis. Figure 5 (right side) shows the structural response characteristics of blast walls with various explosion loading shapes determined via non-linear dynamic FE analysis (procedure of structural non-linear analysis).

2.3.1. Calculation of Explosion Load Profile

The well-test area and adjacent areas have been assessed for potential fire and explosion accidents during well testing in current industrial practice [20]. The flow chart in Figure 5 presents the steps in the analysis. The ballast wall strength of the explosion load depends on factors such as the fuel type, congestion, confinement, and relative position of the ignition point within the gas cloud.

To determine the blast effect and prediction of explosion pressure, DNV-Phast [21] is used with the Baker–Strehlow–Tang model, which uses assumed parameters (material reactivity: low; flame expansion: 1.5; obstacle density: high; ground reflection factor: 2.0).

The affected volume ($30\,\text{m}^3$) is considered the confined volume within the gas cloud. The maximum explosion load can be read at around 5.0 m from the gas cloud and the ignition point. According to the analysis result, the explosive pressure based on the leak category calculation is shown in Table 1.

The arrangement of the well test and adjacent area in Figure 6 shows that the blast wall in the well-test area is located approximately 10 m from the ignition point; therefore, the peak pressure on the blast wall can be assumed via linear interpolation with the pressure value 5 m from the ignition point and the distance of the peak pressure 0.2 bar. Various peak pressures were also selected to evaluate the blast wall for load scenarios (0.2, 0.3, 0.6, 1.0, 1.5, 2.0, 2.5 and 3.0 bar).

Table 1. Overview of major explosion accidents.

Leak Category	Mass Used for Explosion (kg)	Maximum Explosion Load (Barg, at 5 m)	Distance to 0.2 Barg (m)	Peak Pressure on Blast Wall (Barg)
Small	-	-	-	-
Medium	0.47	0.65	13	0.37
Major	8.44	3.0	33	2.5
Large	131.34	3.4	51	3.05

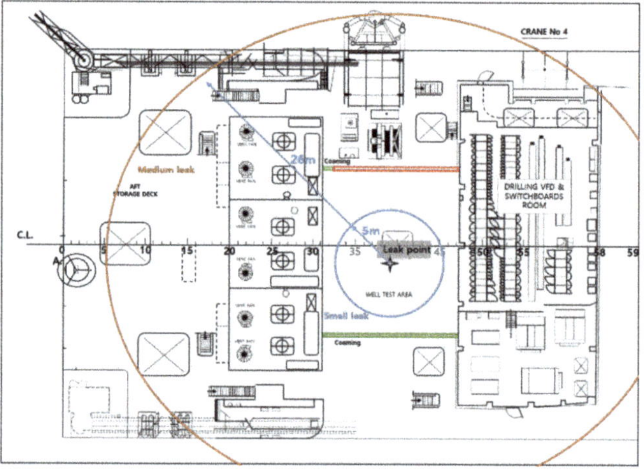

Figure 6. Arrangement of well test and adjacent area.

2.3.2. Applied Blast Load Scenarios

Various factors affect the duration time and thus the impulse of the blast pressure. The impulse is presented in Equation (2):

$$\text{Impulse (bar; s)} = \text{Peak pressure (bar)} \times \text{Duration time (s)} \tag{2}$$

The selected impulses are 0.001, 0.01, 0.02, 0.03, 0.04, 0.05 and 0.06 bar·s, and the duration time is calculated with Equation (2). Based on a review of the blast load profile study, the load applied in this study was a triangular loading pulse without the negative impulse, as shown in Figure 7. The considered load scenarios are presented with the selected peak pressures and durations in Table 2 and in Appendix A (Table A1. Load scenario of peak pressure at 0.3, 0.6, 1.0, 1.5, 2.0, 2.5 and 3.0 bar).

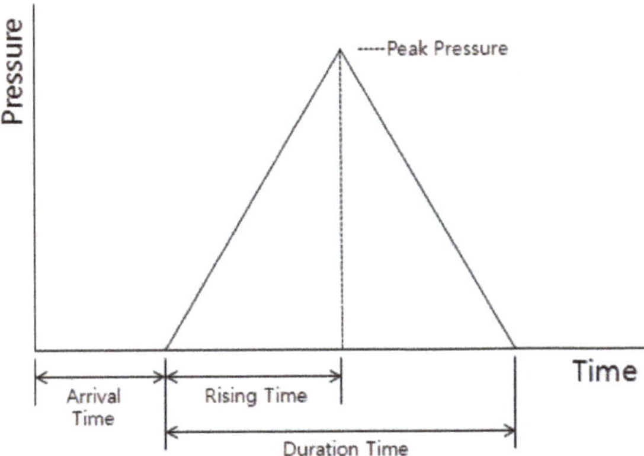

Figure 7. Idealized load profile in the analysis.

Table 2. Load scenario (0.2 bar).

Peak Pressure (bar)	Case Number *	Duration Time (s)	Peak Time (s)	Impulse (bar)
	P02M01	0.0100	0.0050	0.001
	P02M10	0.1000	0.0500	0.010
	P02M20	0.2000	0.1000	0.020
0.2	P02M30	0.3000	0.1500	0.030
	P02M40	0.4000	0.2000	0.040
	P02M50	0.5000	0.2500	0.050
	P02M60	0.6000	0.3000	0.060

* Note: The letter P is peak pressure, and the next two numbers are the value of peak pressure without a decimal point (0.2 bar indicated as 02); the letter M is impulse, and next two numbers are the value of impulse without a decimal point (0.001 bar indicated as 01).

2.4. Numerical Modelling of Blast Wall

Target Blast Wall

In this study, the commercial non-linear FE code (MSC/Nastran [22]) was applied to simulate and analyze highly non-linear structural characteristics by considering geometrical and material nonlinearities. Iso-parametric quadrilateral elements with four nodes and six degrees of freedom per node were used to analyze thin to moderately thick shell structures such as the corrugated steel plate elements [23–25]. Figure 8 shows the FE model of the blast wall in the well-test area on the drillship topside. A normal blast wall has a corrugated shape, but the blast wall in the well-test area is a stiffened panel. Details of the dimensions of the stiffeners and material properties are given below:

- Blast wall thickness: 10 mm (AH36)
- Vertical H-beam: H400 × 400 × 13 × 24 (AH36)
- Horizontal H-beam: H200 × 200 × 10 × 15 (AH36)
- Horizontal channel: C200 × 100 × 7.5 × 11 (AH36)

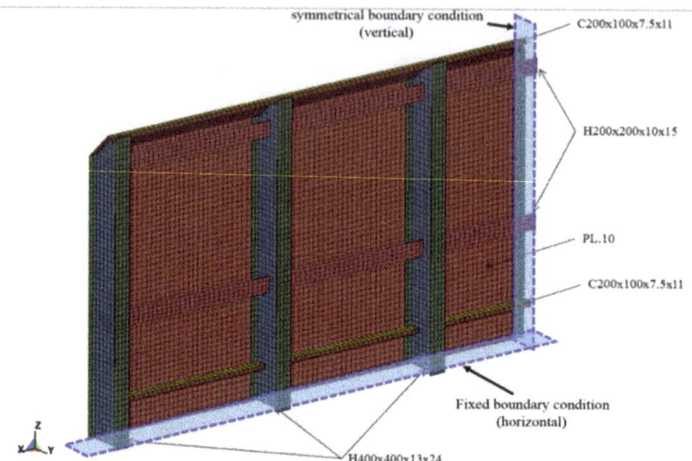

Figure 8. Scantling of blast wall and boundary conditions.

To represent the plastic behavior in a non-linear FE analysis, the material behavior obtained from the tensile tests [10]. Because elasto-perfectly plastic model was used for material model, only yield stress was converted to dynamic yield stress for FE simulations. In an impact load case such as explosion load, the effect of the strain-rate must be significant in the FE material model. The dynamic effect is the Cowper–Symonds equation, which follows as Equation (3), to consider the strain effect of material:

$$\frac{\sigma_{Yd}}{\sigma_Y} = 1.0 + \left(\frac{\dot{\varepsilon}}{C}\right)^{1/q} \tag{3}$$

where the coefficients C and q can be determined based on the test data [10]. The data indicate that the coefficients C and q are dependent on the material type, among other factors. The C and q values in this study were selected as 3200 and 5.0, respectively, for high-tensile steel and 40.4 and 5.0, respectively, for mild steel.

Figure 8 shows the boundary condition of the FE model. For the horizontal fixed condition, the vertical girders are welded on the deck and the blast wall edges are free, and for the vertical symmetrical boundary condition, the horizontal stiffeners and blast wall edge all have symmetrical boundary conditions (δ_x : free, δ_y : fixed, δ_z : free, γ_x : fixed, γ_y : free, γ_z : fixed) for the continued installation condition.

2.5. Analysis Results

Figure 9 shows typical results for the distribution of a von Mises stress plot and plastic strain plot of case P15M20 (peak pressure: 1.5 bar; impulse: 0.02 bar·s) for the target blast wall structure and the selected load scenarios in the previous sections. The maximum von Mises stress is observed at the connection of the vertical H-beam and the deck at 0.13 s, with a duration of 0.0333 s. A low stress level is calculated, which may be explained by the rigidity of the vertical H-beam (H400 × 400 × 13 × 24), which takes most of the load from the explosion.

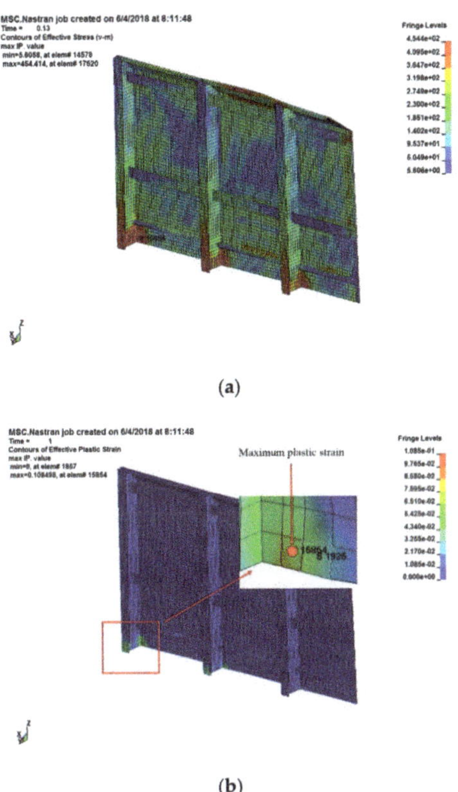

(a)

(b)

Figure 9. Stress and strain distribution of P15M20. (**a**)Von Mises stress. (**b**) Plastic strain.

Most of the blast wall plates do not reach the plastic strain limit compared with the three connections of the vertical H-beam and the deck. The furthest connection from the vertical symmetric boundary line has greater plastic strain than the other two connections. From this result, it can be concluded that the explosion load is concentrated on the three bottom connections (especially the connection furthest from the symmetric boundary) between the blast wall, and the deck for the blast wall has generally been designed with elastic criteria without consideration of energy absorption by deflection.

Figure 10a shows the relationship between plastic strain and the impulse curve of the blast wall at the maximum point. The plastic strain is increased when the peak pressure and the impulse are increased. The maximum and permanent deflection according to the peak pressure and impulse are summarized in Figure 10b,c. The maximum and permanent deflection have tendencies similar to the plastic strain, and the permanent deflection is slightly lower than the maximum deflection.

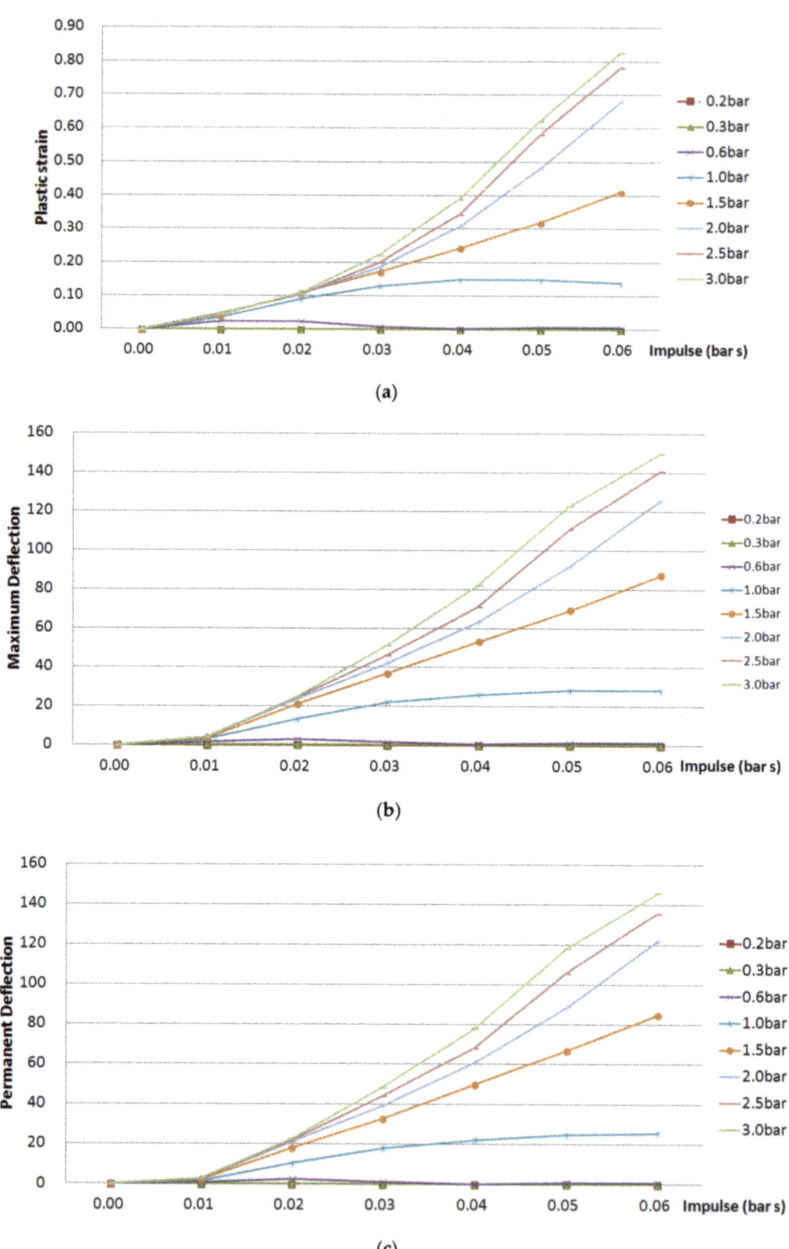

Figure 10. Relationship of structural response characteristics vs. impulse curve (at maximum point). (**a**) Plastic strain—Impulse curve for each peak pressure. (**b**) Maximum deflection vs. impulse. (**c**) Permanent deflection vs. impulse.

With peak pressure less than 1.0 bar, the permanent deflection was observed zero value, and with peak pressures greater than 1.0 bar, the maximum and permanent deflection are proportional to the peak pressure and impulse. However, in some ranges, the plastic strain and the maximum and permanent deflection are not proportional to the peak pressure and impulse in the range of below 1.0 bar of peak pressure and impulse between 0.01 bar and 0.03 bar.

In Figure 11, the plastic strain results are zoomed and compared to the ranges of below 1.0 bar of peak pressure and impulse between 0.01 bar and 0.03 bar. One is the current actual thickness of 10mm in industrial application, and the other is the blast wall thickness of 2.0 mm plate. The results show that the plastic strain is not proportional to impulse in the actual (10-mm) thickness; however, it is proportional to impulse in the 2.0 mm thickness model. It is expected that there is an effect between the thick plate (vertical H-beam) and the thin plate (blast wall plate).

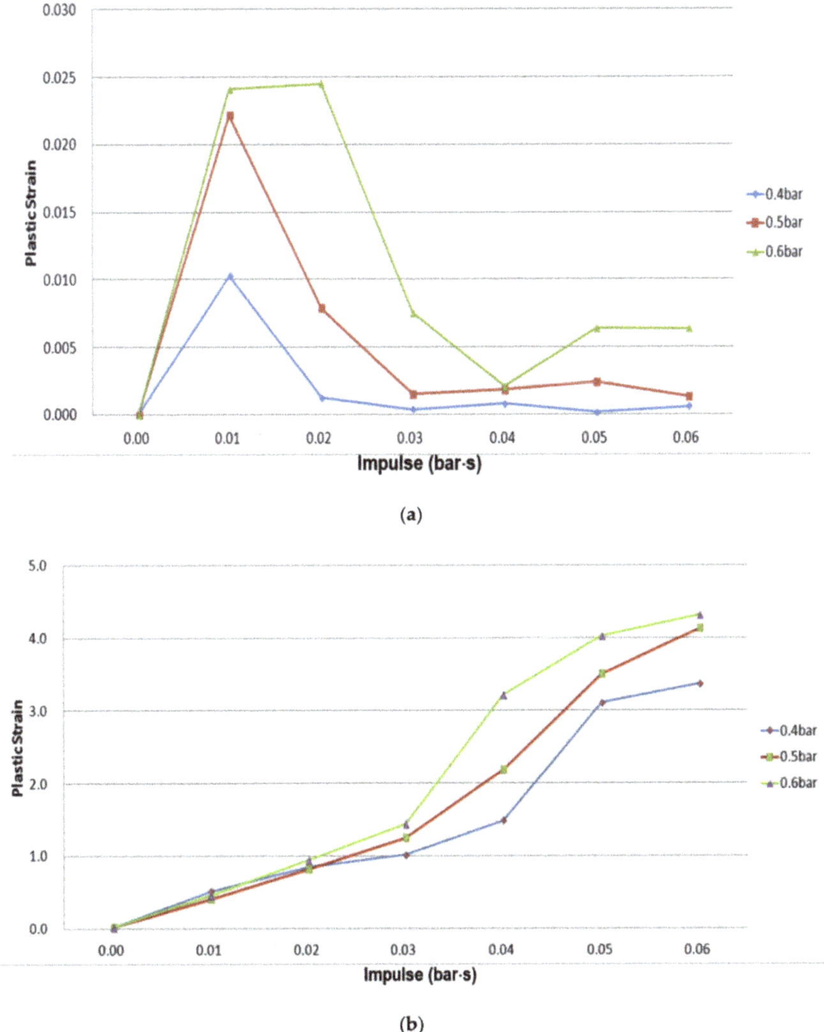

Figure 11. Plastic strain comparison. Actual plate thickness and thin plate. (**a**) Blast wall with actual plate thickness (**b**) Blast wall with 2.0.mm of plate thickness.

The maximum deflection is slightly higher than the permanent deflection, and the deflection with a peak pressure of greater than 1.0 bar is proportional to the peak pressure and impulse. However, it also has an unusual range whose deflection is not proportional to the peak pressure and impulse. Figure 12 shows that the P-I curve for a previous study [10] and that for this study are very similar. This means that the elastic and plastic responses are nearly the same for the corrugated blast wall and the stiffened blast wall in this study when the same peak pressure and impulse are applied.

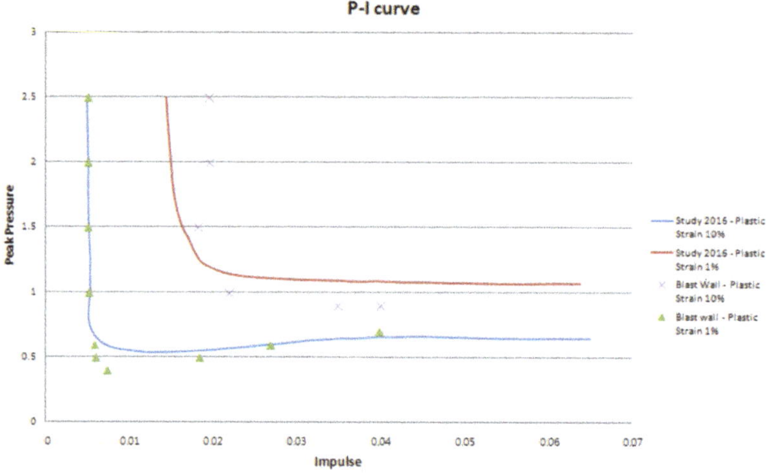

Figure 12. P-I curve comparison.

Table 3 shows the suggested critical strain [26] for various materials. In this study, the material of the blast wall was AH36, with a yield strength of 355 MPa and a plastic strain limit assumed to be 4%. Table 4 summarizes the resultant plastic strain according to the peak pressure and impulse. The under-bar values in the table are more than 4% of the plastic strain. Up to 0.6 bar, plastic strain did not occur at any location. With the peak pressure at 1.0 bar and above, the plastic strain was greater than 4% in most cases in which the impulse was higher than 0.1 bar and the higher peak pressure had a higher plastic strain.

Table 3. Maximum plastic strain, DNV-RP-C203 [26].

	Maximum Plastic Strain			
Critical Strain	S235 0.05	S355 0.04	S420 0.03	S460 0.03

Table 4. Resultant plastic strain for each peak pressure and impulse (at maximum point).

Impulse (bar·s)	0.2 bar	0.3 bar	0.6 bar	1.0 bar	1.5 bar	2.0 bar	2.5 bar	3.0 bar
0.0010	0.0000	0.0000	0.0000	0.0000	0.0000	0.0000	0.0000	0.0000
0.0100	0.0000	0.0014	0.0241	0.0350	*0.0438*	*0.0466*	*0.0468*	*0.0461*
0.0200	0.0000	0.0000	0.0245	*0.0901*	*0.1085*	*0.1040*	*0.1043*	*0.1068*
0.0300	0.0000	0.0000	0.0076	*0.1298*	*0.1720*	*0.1850*	*0.2000*	*0.2230*
0.0400	0.0000	0.0000	0.0021	*0.1495*	*0.2430*	*0.3090*	*0.3450*	*0.3920*
0.0500	0.0000	0.0000	0.0064	*0.1481*	*0.3200*	*0.4820*	*0.5820*	*0.6230*
0.0600	0.0000	0.0000	0.0064	*0.1381*	*0.4110*	*0.6810*	*0.7830*	*0.8280*

In the actual condition, as with the load profile [18], the negative phase is considered. To evaluate the effects of the negative phase of the explosion load, the load profile in Figure 13 has been simplified and applied to the case with 0.6 bar in Table 4. In this case, 25% of the maximum pressure is applied for the minimum pressure, and the duration of the negative phase is twice that of the positive phase.

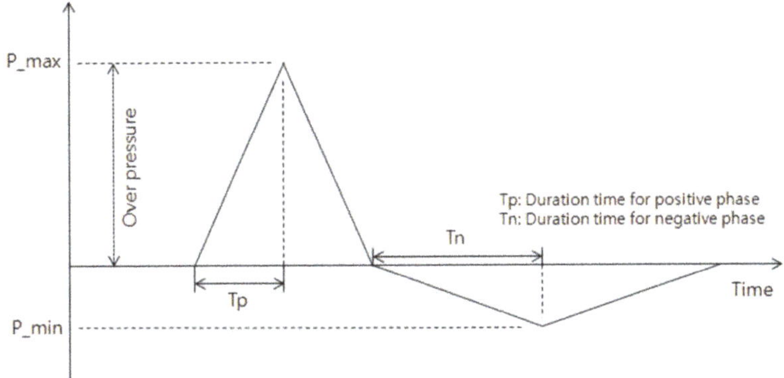

Figure 13. Idealized load profile with negative phase.

Figure 14 shows the plastic strain results: the plastic strain in the case with negative pressure is slightly greater than that in the case without negative pressure, and it is still below 5%. With these results, it can be concluded that the current blast wall structure has enough strength to endure the explosion pressure during the well test. The applied blast pressure in the calculation of the current design is 0.2 bar, but this study showed that the plastic strain begins to occur at 1.0 bar, which indicates a very low possibility that the blast wall will be deformed permanently because the explosion pressure 1.0 bar is not likely to occur. The high stresses in the blast wall structure are concentrated at the connection of the vertical H-beam and the boundary (deck plate), and the other locations experience very low stress relative to the end connection.

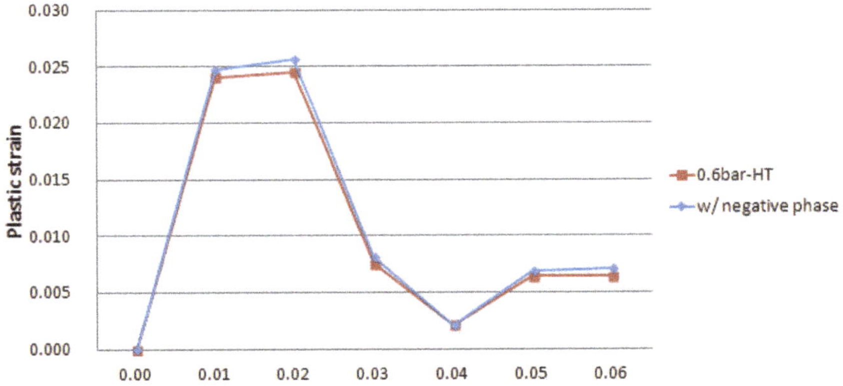

Figure 14. Plastic strain comparison with negative phase.

It can be proposed that this difference is because the current structure is evaluated following the linear beam theory, so the structure has a large vertical H-beam as a "primary" member, and the wall plates are connected to the H-beam as a "secondary" member. Therefore, the deformation of the blast wall is insufficient, and the explosion energy cannot be absorbed by the blast wall structure. A typical blast wall is thin corrugated steel panel because the blast pressure energy can be absorbed by the deformed corrugated panel. The deflection and plastic strain results for the stiffened panel blast wall within the area with permanent deformation are very similar to those of previous studies with a corrugated blast wall. This finding indicates that the stiffened panel blast wall and the corrugated blast wall have similar structural dynamic responses.

3. Application to the Design of Blast Walls

The FE results from previous investigations show that the maximum plastic strain occurs at the bottom connection between the vertical girder (24-mm flange plate) and the blast wall plate (10-mm plate). Therefore, several alternative design applications are suggested to reduce the fabrication cost of a blast wall for application in current industrial practices. In this section, an applicable alternative design for existing blast wall is determined to reduce the plastic strain with following alternatives.

- Alt #1: Weld the blast wall plate and deck.
- Alt #2: Replace the blast wall with a thicker plate.
- Alt #3: Replace the blast wall and the support with mild steel.

With the three alternatives, the maximum von Mises stress and plastic strain are compared with the results of the current wall structure. Alt #1 and Alt #2 are compared with the result in Figures 15 and 16 (peak pressure: 1.5 bar; impulse: 0.02 bar), and Alt #3 is compared with the result in Figure 17 (peak pressure: 0.6 bar; impulse: 0.02 bar).

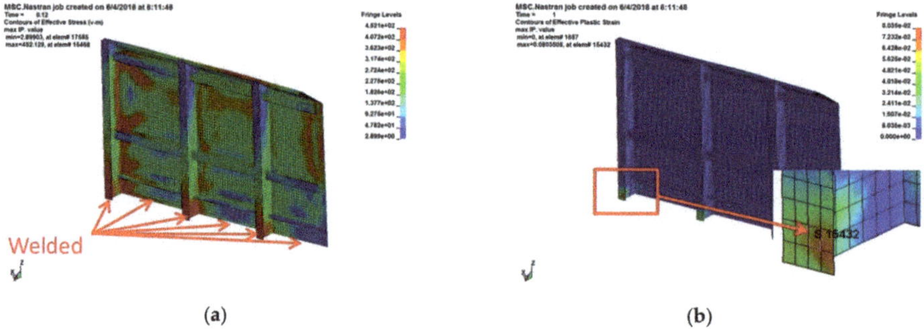

Figure 15. Results of alternative method Alt #1. (**a**) Updated boundary condition and stress distribution. (**b**) Strain distribution.

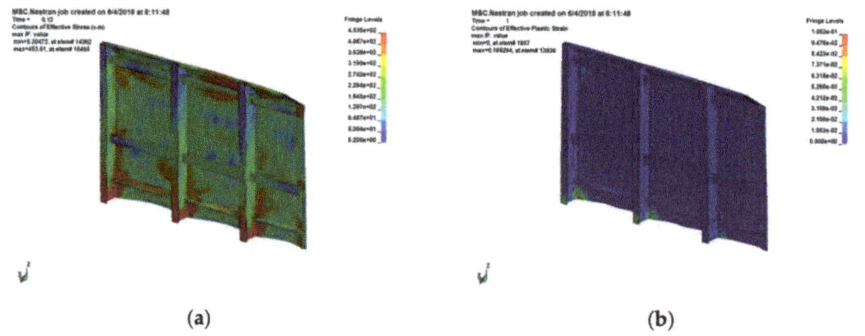

Figure 16. Updated blast wall plate thickness for wall plate thickness of 4 mm of Alt #2. (**a**) Von Mises stress distribution. (**b**) Strain distribution.

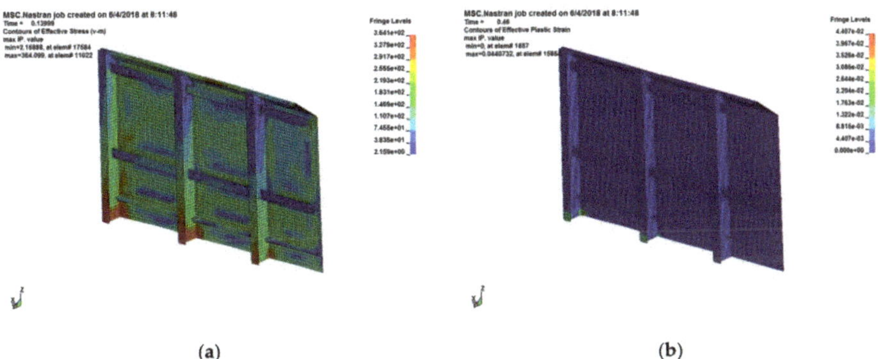

Figure 17. Maximum von Mises stress and plastic strain with wall mild steel of Alt #3. (**a**) Von Mises stress distribution. (**b**) Strain distribution.

The first alternative method is to change the welding conditions to weld the bottom of the blast wall plate and the existing welding stiffeners, as shown in Figure 15. Therefore, the location of the maximum plastic strain is changed from the plate to the stiffener. When the bottom of the blast wall plate is welded, the von Mises stress of the blast wall plate is greatly increased, but the maximum von Mises stress does not change. However, the plastic strain is more than 20% lower than in the original condition.

The second alternative method is to reduce the blast wall plate thickness (4.0 and 6.0 mm) to increase the absorbed energy. The maximum von Mises stress and plastic strain are shown with the different blast wall thicknesses. When the blast wall thickness is reduced, the von Mises stress and the plastic strain on the wall plate are increased.

The last alternative method is to use mild steel, which will reduce the cost of the current industrial project. Therefore, mild steel is used for the blast wall structure, and the results are compared with the current structure. To evaluate the plastic strain with the rule requirement, a case with a peak pressure of 0.6 bar and impulse of 0.02 bar is considered.

Figure 18 shows the plastic strain of the maximum point for the three cases such as original, Alt #1 (Full welding) and Alt #2 (P06, P15, P20, P04). The plastic strain is reduced in the case with a full welding boundary (Alt. #1), but the use of a thinner (or thicker) plate (Alt. #2) for the blast wall causes little change in plastic strain. This means that the blast wall thickness does not affect the plastic strain because the vertical supports are very rigid: in other words, the blast wall thickness can be reduced to the thinner plate.

Figure 19 shows the effects of different materials in the case of Alt. #3. The use of mild steel nearly doubled the plastic strain with high-tensile steel. Even though the plastic strain is increased in the case of mild steel, it is still below the rule (DNV-RP-C203) [26] requirement (5% in mild steel), which means that mild steel can be used for the blast wall. Therefore, it can be determined whether the current blast wall is over-designed due to blast loads.

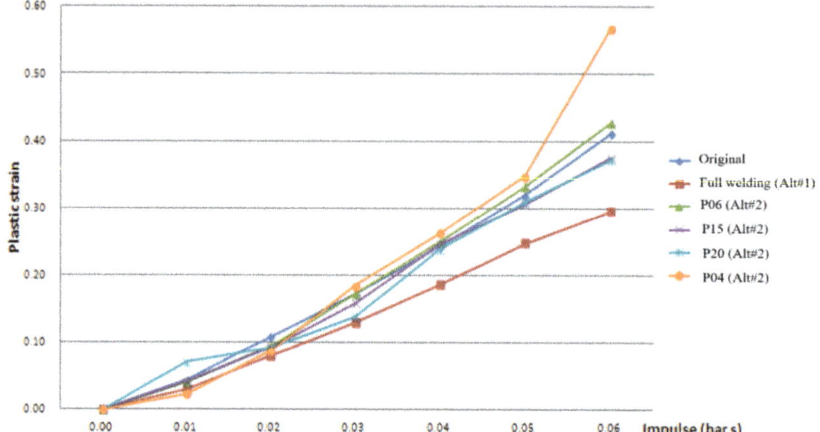

Figure 18. Plastic strain vs. impulse curves for Alt #1 and Alt #2 (at maximum strain point).

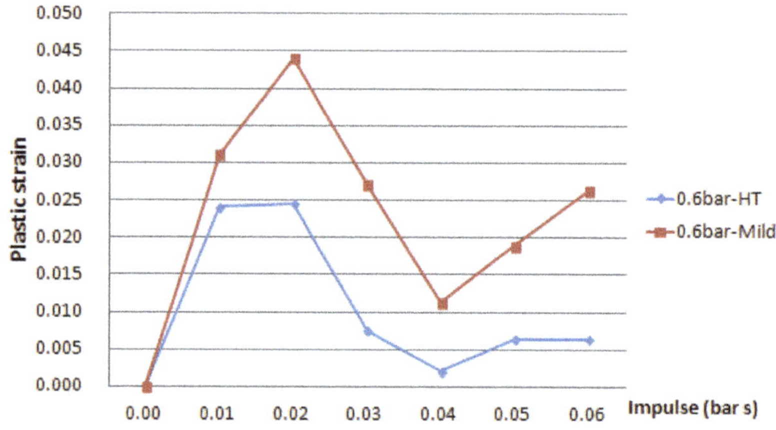

Figure 19. Plastic strain vs. impulse curve for Alt. #3.

4. Conclusions

The focus of this study is to determine whether the current scantling of the blast wall in a well-test area is adequate for the blast pressure. Three alternatives are also suggested to improve the strength characteristic of the blast wall and to reduce the cost of construction. The results of the analysis in this study lead us to the following conclusions.

The study result with the stiffened panel blast wall and the existing study results for a corrugated blast wall were very similar in deflection, plastic strain, and P-I curve. This means that the stiffened blast wall and the corrugated blast wall had very similar structural characteristics in the elastic and plastic regions and that developments can be made in the stiffened blast wall according to the corrugated blast wall study results.

The alternative with the additional weld between the blast wall and the deck has a great effect on reducing and delaying plastic strain. The blast wall thickness did not affect the plastic strain result because the vertical supports are very rigid, and the loads are all concentrated on the supports. This means that the thickness could be reduced if the vertical supports are retained in the blast wall structure. The blast wall could be changed from high-tensile steel to mild steel. The plastic strain increases significantly with a mild steel structure, but it is still below the rule requirement. The cost of construction can be reduced with the use of mild steel. The results of the non-linear analysis in this study will be a good reference for future studies to apply a higher peak pressure than the current industrial practices.

Author Contributions: Conceptualization, B.J. and J.K.S.; Methodology, B.J. and J.K.S; Formal analysis, B.J. and J.H.K.; Investigation, B.J.; Data curation, B.J. and J.K.S.; Writing-original draft preparation, B.J.; Writing-review and editing, J.K.S.; Visualization, B.J., J.H.K., and J.K.S.; Supervision, J.K.S.; project administration, J.K.S.; funding acquisition, J.K.S. All authors have read and agreed to the published version of the manuscript.

Funding: This research was a part of the project titled "Development of guidance for prevent of leak and mitigation of consequence in hydrogen ships", funded by the Ministry of Oceans and Fisheries, Korea.

Conflicts of Interest: The authors declare no conflict of interest.

Appendix A Load Scenario (Peak Pressure, 0.3 to 3.0 Bar)

Table A1. Load scenario (0.3 bar).

Peak Pressure (bar)	Case Number	Duration Time (s)	Peak Time (s)	Impulse (bar)
	P03M01	0.0067	0.0033	0.001
	P03M10	0.0667	0.0333	0.010
	P03M20	0.1333	0.0667	0.020
0.3	P03M30	0.2000	0.1000	0.030
	P03M40	0.2667	0.1333	0.040
	P03M50	0.3333	0.1667	0.050
	P03M60	0.4000	0.2000	0.060

Table A2. Load scenario (0.6 bar).

Peak Pressure (bar)	Case Number	Duration Time (s)	Peak Time (s)	Impulse (bar)
	P06M01	0.0033	0.0017	0.001
	P06M10	0.0333	0.0167	0.010
	P06M20	0.0667	0.0333	0.020
0.6	P06M30	0.1000	0.0500	0.030
	P06M40	0.1333	0.0667	0.040
	P06M50	0.1667	0.0833	0.050
	P06M60	0.2000	0.1000	0.060

Table A3. Load scenario (1.0 bar).

Peak Pressure (bar)	Case Number	Duration Time (s)	Peak Time (s)	Impulse (bar)
	P10M01	0.0020	0.0010	0.001
	P10M10	0.0200	0.0100	0.010
	P10M20	0.0400	0.0200	0.020
1.0	P10M30	0.0600	0.0300	0.030
	P10M40	0.0800	0.0400	0.040
	P10M50	0.1000	0.0500	0.050
	P10M60	0.1200	0.0600	0.060

Table A4. Load scenario (1.5 bar).

Peak Pressure (bar)	Case Number	Duration Time (s)	Peak Time (s)	Impulse (bar)
	P15M01	0.0013	0.0007	0.001
	P15M10	0.0133	0.0067	0.010
	P15M20	0.0267	0.0133	0.020
1.5	P15M30	0.0400	0.0200	0.030
	P15M40	0.0533	0.0267	0.040
	P15M50	0.0667	0.0333	0.050
	P15M60	0.0800	0.0400	0.060

Table A5. Load scenario (2.0 bar).

Peak Pressure (bar)	Case Number	Duration Time (s)	Peak Time (s)	Impulse (bar)
	P20M01	0.0010	0.0005	0.001
	P20M10	0.0100	0.0050	0.010
	P20M20	0.0200	0.0100	0.020
2.0	P20M30	0.0300	0.0150	0.030
	P20M40	0.0400	0.0200	0.040
	P20M50	0.0500	0.0250	0.050
	P20M60	0.0600	0.0300	0.060

Table A6. Load scenario (2.5 bar).

Peak Pressure (bar)	Case Number	Duration Time (s)	Peak Time (s)	Impulse (bar)
	P25M01	0.0008	0.0004	0.001
	P25M10	0.0080	0.0040	0.010
	P25M20	0.0160	0.0080	0.020
2.5	P25M30	0.0240	0.0120	0.030
	P25M40	0.0320	0.0160	0.040
	P25M50	0.0400	0.0200	0.050
	P25M60	0.0480	0.0240	0.060

Table A7. Load scenario (3.0 bar).

Peak Pressure (bar)	Case Number	Duration Time (s)	Peak Time (s)	Impulse (bar)
	P30M01	0.0007	0.0003	0.001
	P30M10	0.0067	0.0033	0.010
	P30M20	0.0133	0.0067	0.020
3.0	P30M30	0.0200	0.0100	0.030
	P30M40	0.0267	0.0133	0.040
	P30M50	0.0333	0.0167	0.050
	P30M60	0.0400	0.0200	0.060

References

1. Tanaka, S.; Okada, Y.; Ichikawa, Y. Offshore drilling and production equipment. In *Civil Engineering*; Encyclopedia of Life Support Systems: Oxford, UK, 2009.
2. DNV-GL. *Worldwide Offshore Accident Databank (WOAD)*; DNV-GL: Oslo, Norway, 2017.
3. Fattakhova, E.Z.; Barakhnina, V.B. Accident rate analysis on the offshore oil and gas production installation and platforms. *Int. J. Appl. Fundam. Res.* **2015**, *1*, 1–7.
4. Ali, M.; Jarni, H.H.; Aftab, A.; Ismail, A.R.; Saady, N.M.C.; Sahito, M.F.; Keshavarz, A.; Iglauer, S.; Sarmadivaleh, M. Nanomaterial-based drilling fluids for exploitation of unconventional reservoirs: A review. *Energies* **2020**, *13*, 3417. [CrossRef]

5. Aftab, A.; Ali, M.; Sahito, M.F.; Mohanty, U.S.; Jha, N.K.; Akhondzadeh, H.; Azhar, M.R.; Ismail, A.R.; Keshavarz, A.; Iglauer, S. Environmental-friendly and high-performance of multifunctional Tween 80/ZnO nanoparticles added water-based drilling fluid: An experimental approach. *Sustain. Chem. Eng.* **2020**. [CrossRef]

6. Aftab, A.; Arif, M.; Al-khdeawee, E.; Saady, N.M.C.; Ismail, A.R.M.; Sahito, M.F.; Keshavarz, A.; Iglauer, S. Influence of tailored-made TiO2/API bentonite nanocomposite in drilling mud; implications for enhanced drilling operations. *Appl. Clay Sci.* **2020**. in Press.

7. Hifu, Q.; Xueguang, L. Approach to blast wall structure computing in ocean engineering. In Proceedings of the International Conference on Industrial and Information Systems, Haikou, Hainan, 24–25 April 2009; IEEE Computer Society: Washington, DC, USA, 2009. [CrossRef]

8. Vignjevic, R.; Liang, C.; Hughes, K.; Brown, J.C.; Vuyst, T.D.; Djordjevic, N.; Campbell, J. A numerical study on the influence of internal corrugated reinforcements on the biaxial bending collapse of thin-walled beams. *Thin-Walled Struct.* **2019**. [CrossRef]

9. LS-DYNA. *User's Manual for ANSYS/LS-DYNA, Version 14.5*; ANSYS Inc.: New York, NY, USA, 2016.

10. Sohn, J.M.; Kim, S.J.; Seo, J.K.; Kim, B.J.; Paik, J.K. Strength assessment of stiffened blast walls in offshore installations under explosions. *Ships Offshore Struct.* **2016**. [CrossRef]

11. Kang, K.Y.; Choi, K.H.; Choi, J.W.; Ryu, Y.H.; Lee, J.M. Explosion induced dynamic responses of blast wall on FPSO topside: Blast loading application methods. *Int. J. Nav. Archit. Ocean Eng.* **2016**, *9*, 135–148.

12. Syed, Z.I.; Mohamed, O.A.; Rahman, S.A. Non-linear finite element analysis of offshore stainless steel blast wall under high impulsive pressure loads. *Procedia Eng.* **2016**, *145*, 1275–1285. [CrossRef]

13. Sohn, J.M.; Kim, S.J. Numerical investigation of structural response of corrugated blast wall depending on blast load pulse shapes. *Lat. Am. J. Solids Struct.* **2017**, *14*, 1710–1722. [CrossRef]

14. Xiao, W.; Andrae, M.; Ruediger, L.; Gebbeken, N. Numerical prediciton of blast wall effectiveness for structural protection against air blast. *Procedia Eng.* **2017**, *199*, 2519–2524. [CrossRef]

15. Djordjevic, N.; Vignjevic, R.; Kiely, L.; Case, S.; Vuyst, T.D.; Campbell, J.; Hughes, K. Modelling of shock waves in fcc and bcc metals using a combined continuum and dislocation kinetic approach. *Int. J. Plast.* **2018**, *105*, 211–224. [CrossRef]

16. Vignjevic, R.; Hughes, K.; Vuyst, T.D.; Djordjevic, N.; Campbell, J.; Stojkovic, M.; Gulavani, O.; Hiermaier, S. Lagrangian analysis led design of a shock recovery plate impact experiment. *Int. J. Impact Eng.* **2015**, *77*, 16–29. [CrossRef]

17. Aghar, H.; Carie, M.; Elshahawi, H.; Young, J.; Pinguet, B.; Swainson, K.; Takla, E.; Theuveny, B.; Hani, A. The expanding scope of well testing. *Oilfield Rev.* **2007**, *19*, 44–59.

18. Bae, M.H.; Paik, J.K. Effects of structural congestion and surrounding obstacles on the overpressure loads in explosions: Experiment and CFD simulations. *Ships Offshore Struct.* **2018**, *13*, 165–180. [CrossRef]

19. Biggs, J.M. *Introduction to Structural Dynamics*; McGraw-Hill Inc.: New York, NY, USA, 1964.

20. Shi. *Fire and Explosion Risk Analysis: Report 1420023-03 rev0*; Samsung Heavy Industries: Geoje, Korea, 2018.

21. DNV•PHAST. *QRA and Risk Analysis Software–PHAST*; DNV-GL: Oslo, Norway, 2018.

22. MSC/NASTRAN. *Programmers's Manuals*, 3rd ed.; The MacNeal-Schwendler Corporation: Los Angeles, CA, USA, 1986.

23. Paik, J.K. *Explosion and Fire Engineering of FPSOs (Phase III): Nonlinear Structural Consequence Analysis, Report No.EFEF-04*; Pusan National University: Busan, Korea, 2011.

24. Kim, B.J.; Kim, B.H.; Sohn, J.M.; Paik, J.K.; Seo, J.K. A parametric study on explosion impact response factors characteristics of offshore installation's corrugated blast wall. *J. Korean Soc.Ocean Eng.* **2011**, *26*, 46–54.

25. DnV-GL. (DNV-RP-C208). *Determination of Structural Capacity by Non-linear FE Analysis Methods*; Det Norske Veritas: Oslo, Norway, 2013.

26. DnV-GL. (DNV-RP-C203). *Fatigue Design of Offshore Steel Structures*; Det Norske Veritas: Oslo, Norway, 2005.

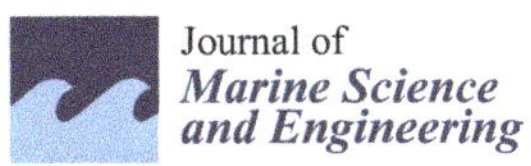
Journal of
Marine Science and Engineering

Article

Anti-Explosion Performance of Composite Blast Wall with an Auxetic Re-Entrant Honeycomb Core for Offshore Platforms

Fang Luo, Shilian Zhang * and Deqing Yang

State Key Laboratory of Ocean Engineering, School of Naval Architecture, Ocean and Civil Engineering, Shanghai Jiao Tong University, 800 Dongchuan Road, Shanghai 200240, China; shaker900@163.com (F.L.); yangdq@sjtu.edu.cn (D.Y.)
* Correspondence: slzhang@sjtu.edu.cn; Tel.: +86-136-2190-3228

Received: 31 January 2020; Accepted: 5 March 2020; Published: 7 March 2020

Abstract: To improve the anti-explosion performance of blast wall in offshore platforms, an auxetic re-entrant blast wall (ARBW) was proposed and designed based on the indentation resistance effect of an auxetic structure. Based on the numerical nonlinear dynamic analysis method verified by the explosion experiment of a conventional steel corrugated blast wall (CBW), the failure mechanisms of ARBW, steel honeycomb sandwich blast wall (HSBW) and CBW were investigated under distributed impulse loads. Computational results demonstrated the excellent anti-explosion performance of the proposed ARBW design. Concerning the minimal deformation at the mid-point of the proposed protective structures, the ARBW performed best. As regards the minimal deformation at the connection, both ARBW and HSBW worked well. The stress distribution of the connection illustrated the different energy absorption and transmission modes of the three blast walls.

Keywords: blast wall; explosion; distributed impulse loads; auxetic; re-entrant honeycomb; corrugated plate; negative Poisson's ratio; offshore platform

---

1. The State-of-the-Art Blast Wall Design in Offshore Platforms

Studies on impulse destruction are an important part of marine engineering protective design. Different protective structures lead to disparate anti-explosion performances under the same impulse loading. At the present time, a popular form of protective structures is corrugated blast wall (CBW), which has been applied to offshore platforms due to its considerable blast-resistant performance and easy availability. A usual design of the CBW in offshore platforms is shown in Figure 1 [1]. Because of the number of explosion accidents on offshore platforms in recent years [2], higher defense requirements have been put forward regarding the impact resistance of the blast wall. Although the analysis method in the design guidance of protective structures has developed from single degree of freedom (SDOF) to nonlinear finite element analysis (NLFEA) [3–5], the typical design scheme in the guidance is a blast wall made by a corrugated plate.

Several methods have been used to strengthen the capability of blast resistance. J.W.Boh et al. [6] presented a passive impact barrier system placed at a certain offset behind the walls. Nwankwo E. et al. [7] presented the development of a rapid assessment tool which provided an understanding of the effect of a composite patch on the blast resistance of profiled blast walls. Christian W. et al. [8] carried out an investigation into the influence of inclined angle of a Vee stiffener used on a blast wall structure. The results showed that the alteration of the inclined angle had a considerable effect on the dynamic response of the blast wall structure. These methods offer a certain reinforcement to the CBW, but the reinforcement is somewhat limited. New shapes of blast wall with high anti-explosion performance need to be studied.

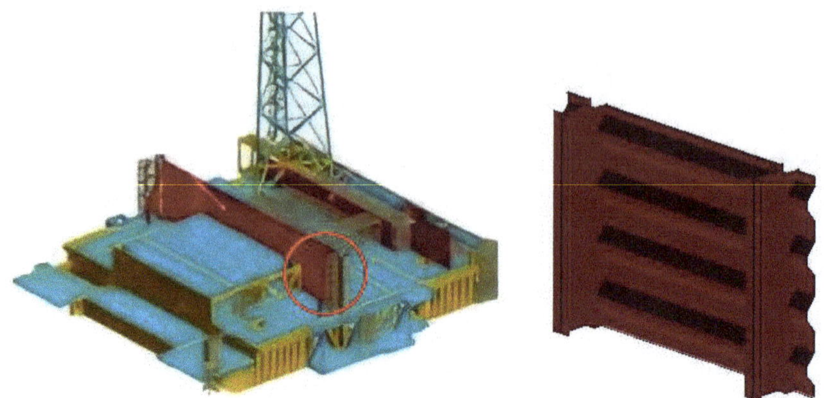

Figure 1. A usual design of the corrugated blast wall (CBW) in offshore platforms.

Honeycomb sandwich plates are another choice for the anti-blast structure. Some research works have shown that sandwich structures possess anti-blast behavior [9]. M.D. Theobald et al. [10] conducted air-blast tests on sandwich panels composed of steel face sheets with unbonded aluminum foam or hexagonal honeycomb cores. The results showed that face sheet thickness has a significant effect on the performance of the panels relative to an equivalent monolithic plate. Zhu F et al. [11] simulated the blast impact of square metallic sandwich panels and compared them with the results of experimental investigation. Pydah A. et al. [12] analyzed transient elasto-plastic deformations of two-core sandwich plates with and without a bumper and subjected them to blast loads with the objective of ascertaining the energy dissipated due to plastic deformations.

Auxetic re-entrant structures are different from traditional sandwich structures. They shrink laterally during compression and expand laterally during stretching, which provides better fracture toughness, shear modulus, energy absorption and anti-fatigue crack growth [13,14]. Junhua Zhang et al. [15] discussed in detail the case of nonlinear transient response and showed that honeycomb sandwich plates with a negative Poisson's ratio perform better than those with a positive Poisson's ratio. Deqing Yang et al. [16] compared the explosion resistance of honeycomb structures with different negative Poisson's ratios and layer number arrangements, which simulated the penetration and failure mode of the structure under underwater explosion shock. Chang Qi et al. [17] investigated experimentally and numerically the responses of auxetic honeycomb-cored sandwich panels as protective systems under impact. Gabriele Imbalzano et al. [18] compared the different fragmentation forms of negative Poisson's ratio honeycomb structures and traditional honeycomb structures under a localized explosion. Hohammad M.H. [19] reinforced the facesheets of auxetic honeycomb plates by carbon nanotubes (CNTs) considering agglomeration effects. At present, there is insufficient research into the honeycomb structure with an auxetic effect under impulse loading. Further applications are needed regarding the design of explosion-proof structures for offshore platforms.

This paper proposes a new design of ARBW, and also a HSBW design for comparison for offshore platforms. The deformation, stress and strain distribution of the ARBW and HSBW under distributed impulse loading are numerically investigated. For the purpose of verification, the responses of the CBW with the same outer dimensions and mass are obtained. The failure mechanism of three kinds of blast wall are revealed. The computational results showed that both the ARBW and HSBW have excellent energy absorption performance.

2. Schematic Design of ARBW, HSBW and CBW

2.1. Design of CBW

The CBW design is based on a non-symmetric trapezoidal deep trough profile, with angle connections at the top and bottom and free sides. The corrugated part is 880 mm wide, 915 mm long and 40.5 mm deep. The whole part is 2 mm thick. The connection is 195 mm deep and 35 mm long. It is composed of 3 mm and 4 mm stainless-steel angles welded to a 12 mm thick angle that forms the primary framework. The mass of the blast plate, made of stainless steel, is 41.5 kg. The model shown in Figure 2 is consistent with the test scheme [20]. For the test results, the 10.27 KPa·s distributed impulse is selected for the simulation, which caused a 22.2 mm permanent deformation of the mid-point [21].

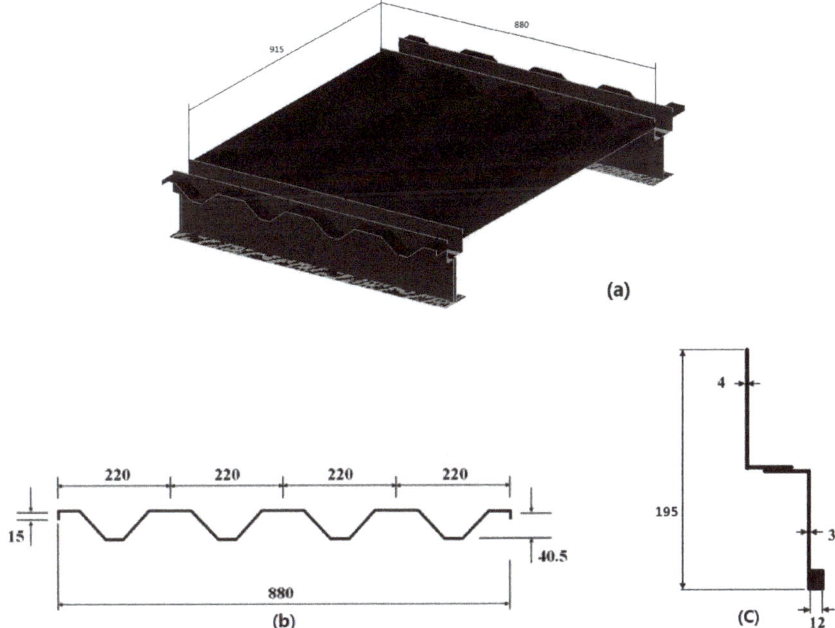

Figure 2. CBW (a) model (b) dimensions of corrugated panel (c) dimensions of connection.

2.2. Design Descriptions of ARBW and HSBW

The ARBW is composed of the auxetic unit cell with a re-entrant shape. According to [16], when the cell angle is 60° and there are layers, the auxetic honeycomb structure presents the smallest plastic deformation. Therefore, the topological cell is selected as a single-cell configuration. The horizontal length h is 52.15 mm, the inclined length l is 26.31 mm, the cell wall thickness is 2.2 mm, and the cell angle θ is 60°. The cell configuration and design of the ARBW are shown in Figure 3. The thickness of the upper and lower faceplates is 0.1 mm, respectively. The negative Poisson's rate is calculated as −3.87 [16]. The relative density $\rho*_{auxetic}$ is 4.74%, which is obtained by dividing the area occupied by the structure and the total area.

The HSBW is composed by the conventional honeycomb unit cell with a typical cellular structure. The cell angle is 120 degrees, the height h is 27 mm, the inclined length l is 15.59 mm, and the cell wall thickness is also 2.2 mm. The cell configuration and design of the HSBW re shown in Figure 4. The thickness of the upper and lower faceplates is 0.1 mm, respectively. The relative density $\rho*_{honeycomb}$ is 2.67%, which is obtained by dividing the area occupied by the structure and the total area.

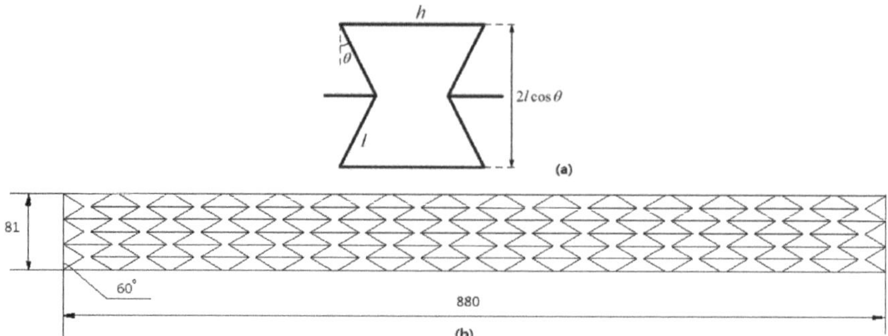

Figure 3. Cell configuration and design of the auxetic re-entrant blast wall (ARBW). (**a**) auxetic unit cell, (**b**) auxetic layout.

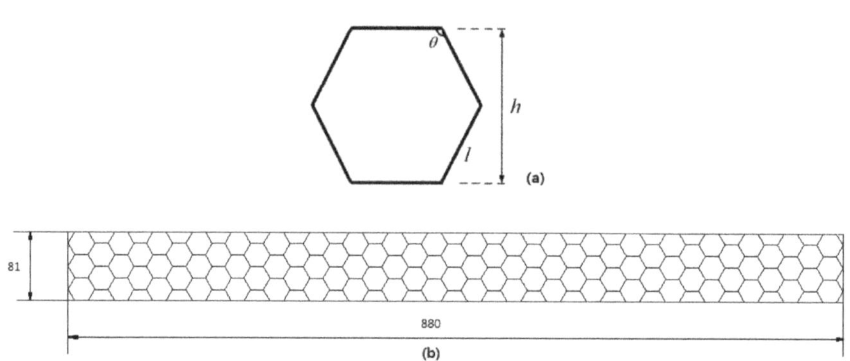

Figure 4. Cell configuration and design of the honeycomb sandwich blast wall (HSBW). (**a**) honeycomb unit cell, (**b**) honeycomb layout.

2.3. Model Diagrams of ARBW and HSBW

The model diagrams of the ARBW and HSBW are presented in Figure 5. These structures are both used to form a three-layer composite blast wall. The length, width, and height are 915 mm, 880 mm, and 81 mm, respectively. The thickness of the upper and lower faceplates is 0.1 mm. The overall mass is under 41.5 kg using stainless steel. Both models are entirely fixed to the framework made by the connection. The connection of the two models is the same as the connection of the CBW model.

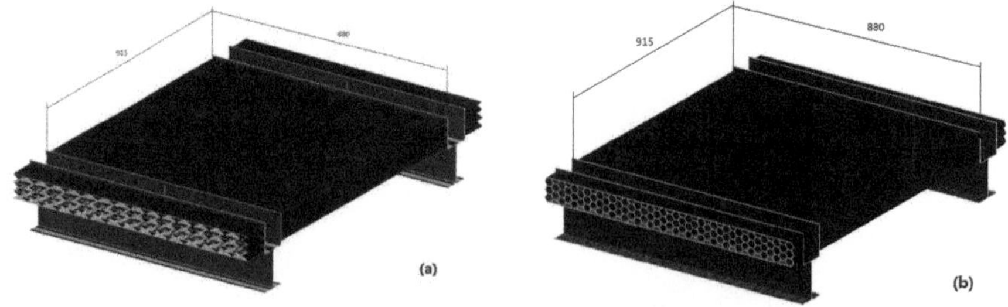

Figure 5. Model diagrams: (**a**) ARBW, (**b**) HSBW.

3. Numerical Models of Anti-Explosion Analysis for the Proposed Blast Wall

3.1. Material Model

Stainless steel is used in the experimental CBW model. The Cowper–Symonds yielding model, as shown in Equation (1), is suitable for describing the properties of stainless-steel structures if the thermal deformation effect is negligible. This model accurately describes the large deformation of the material and the high strain change.

$$\frac{\sigma_y}{\sigma_0} = 1 + (\frac{\dot{\varepsilon}}{D})^{\frac{1}{q}}$$

(1)

where σ_0 is the static yield strength, $\dot{\varepsilon}$ is an equivalent plastic strain, and D and q are material constants. The parameters of the steel are presented in Table 1 [21].

Table 1. Material parameters of stainless steel.

E (GPa)	ν	P (Kg/m3)	σ_0 (MPa)	$\dot{\varepsilon}$	D	q
210	0.3	7850	276	17.76%	2720	5.78

3.2. Finite Element Model

Finite element models of the three kinds of blast wall are shown in Figure 6. All are simulated by plate elements, and all elements are defined by pshell. The element numbers of ARBW, HSBW and CBW are 74,648, 88,730 and 20,252, respectively. Point 1 is the mid-point of the whole panel and point 2 is the junction between the panel and connection. These evaluation points are used to measure the anti-explosion performance after the calculation is finished.

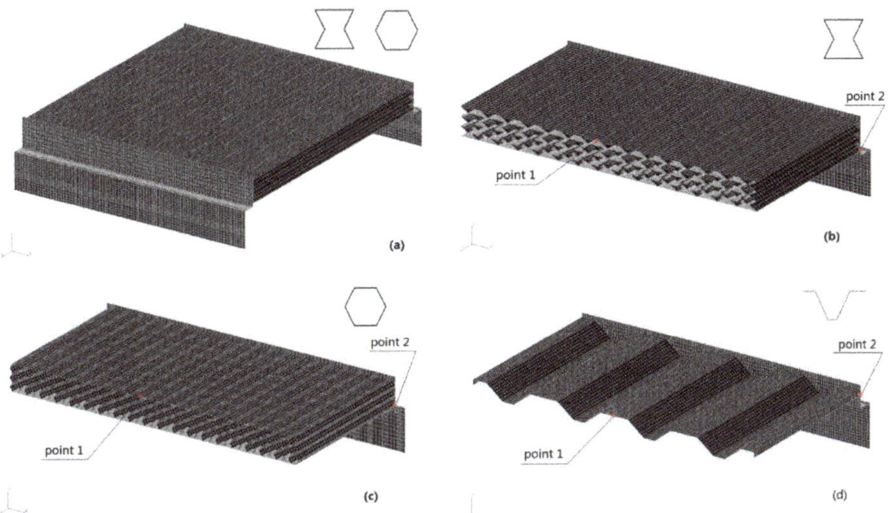

Figure 6. Finite element models: (**a**) the whole model, (**b**) 1/2 symmetry of the ARBW, (**c**) 1/2 symmetry of Table 1. symmetry of the CBW.

3.3. Loadings and Constraints

The bottom of each model is fixed. The distributed impulse loading is applied to the faceplate along the y-direction. The distributed impulse loading is a triangular shock wave, with a peak value of 123 KPa. The applied time is 0.167 seconds, so the unit impulse is 10.27 KPa·s. The distributed impulse loading and constraints are shown in Figure 7.

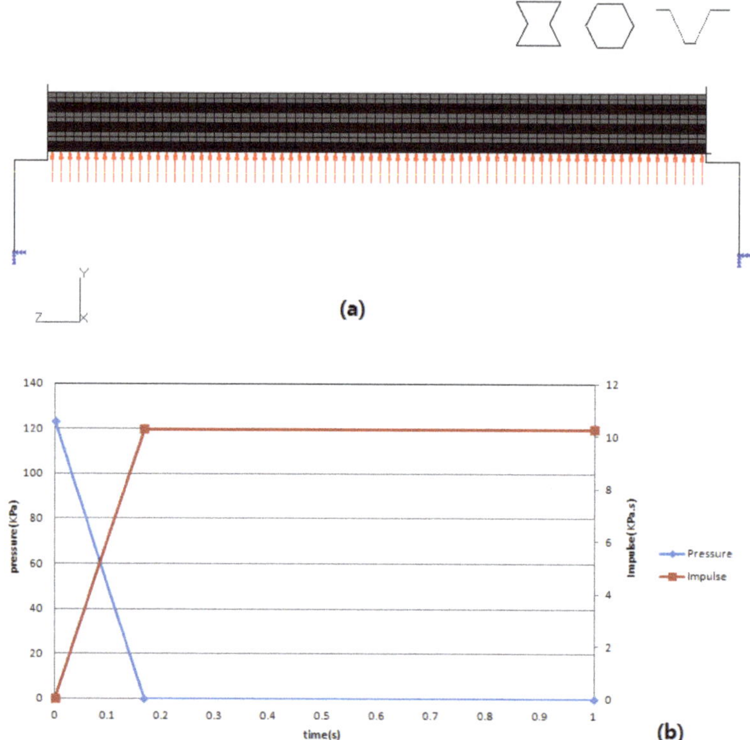

Figure 7. Distributed impulse loading and constraints. (**a**) The distributed impulse loading and constraints. (**b**) The pressure and impulse applied on the models.

4. Comparative Analysis of Anti-Explosion Performance

The computational results of the ARBW and HSBW under distributed impulse loading are compared with the CBW results. Deformation, stress and strain are selected to measure the anti-explosion performance.

4.1. Deformation

The deformation of the ARBW is shown in Figure 8. The bottom faceplate was deformed plastically. The maximum deformation occurred at the free side. The upper faceplate showed a small bulging from the beginning of the test until it reached stability. The auxetic core was crushed, reducing the transmitted pressure. The connection was vertically stretched.

The deformation of the HSBW is shown in Figure 9. Evidently, the honeycomb panel did not present any core densification effect. The whole sandwich panel worked as a bending beam. The upper and bottom faceplates showed the same deformation behavior. No maximum deformation and cell collapse occur at the free side. The connection was stretched vertically and horizontally.

The deformation of the CBW is presented in Figure 10. The whole panel is bent at the beginning of the test. The middle part recovered, but the free side buckled permanently by the end of the test. Evidently, the maximum deformation occured at the free side. The connection was stretched vertically and horizontally.

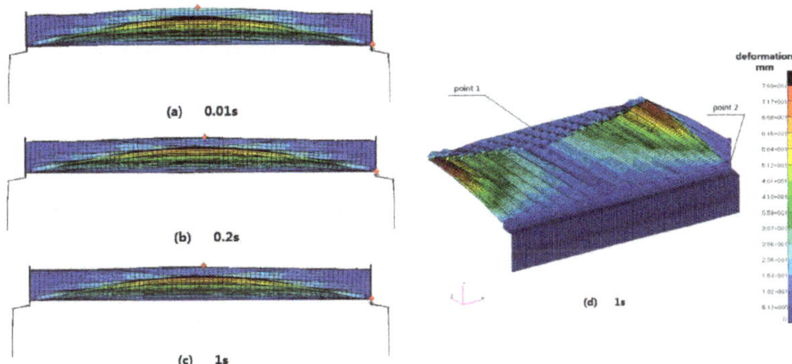

Figure 8. Deformation of the ARBW. (**a**) the moment of 0.01 s, (**b**) the moment of 0.2 s, (**c**) the moment of 1 s, (**d**) the moment of 1 s in isometric view.

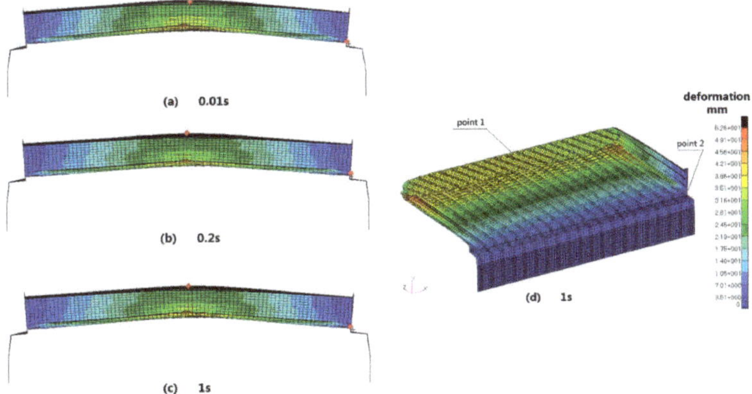

Figure 9. Deformation of the HSBW. (**a**) the moment of 0.01 s, (**b**) the moment of 0.2 s, (**c**) the moment of 1 s, (**d**) the moment of 1 s in isometric view.

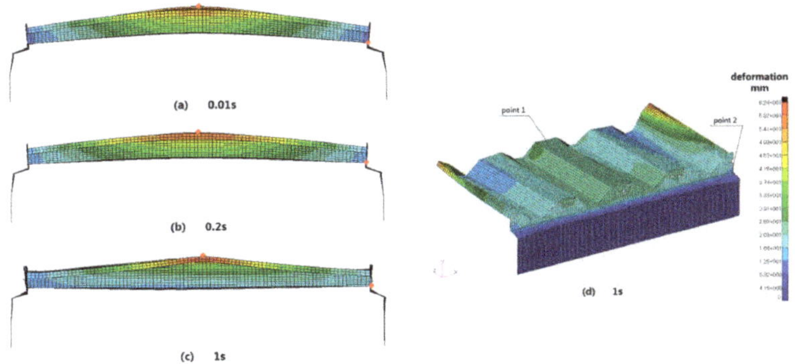

Figure 10. Deformation of the CBW. (**a**) the moment of 0.01 s, (**b**) the moment of 0.2 s, (**c**) the moment of 1 s, (**d**) the moment of 1 s in isometric view.

4.2. Resistance Performance Evaluation

The deformation histories of the evaluation points on the blast walls are shown in Figure 11. Point 1 presents the deformation of the whole blast wall. It can be seen that the peak deformation occurred at the beginning of the test, and the peak value of ARBW was less than the others. The permanent deformation appeared at the end. The permanent value of CBW was 23 mm, which corresponds with the test result mentioned in Section 2.1. The permanent value of ARBW was still the lowest. Both the values of ARBW and HSNW were stable during the shock. Point 2 shows the deformation of the connection. The permanent values of ARBW and HSBW were under 8 mm.

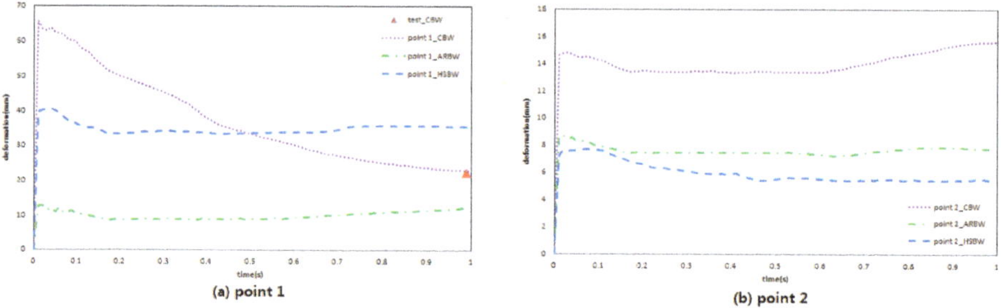

Figure 11. Deformation history of the evaluation points of blast walls: (**a**) point 1, (**b**) point 2.

The deformation difference history shown in Figure 12 is between point 1 and point 2. This is used to measure the energy absorption through large deformation. Obviously, the deformation differences of HSBW and CBW corresponded with the deformations of point 2. The energy absorption of ARBW could hardly be evaluated by the deformation difference, but this could be demonstrated by the stress diagram.

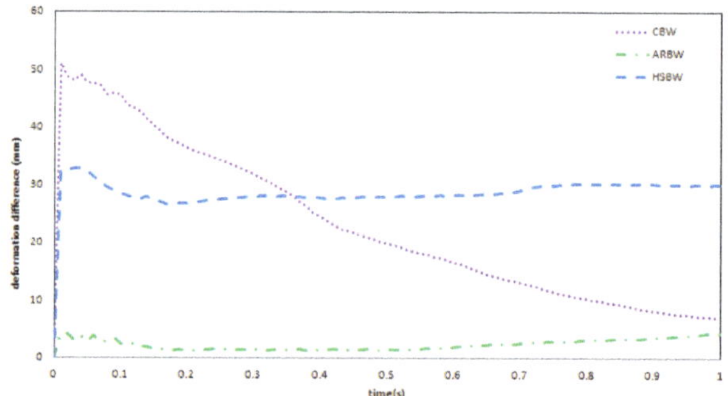

Figure 12. Deformation difference history of blast walls.

4.3. Stress Distribution

The stress distribution of the ARBW is shown in Figure 13. The cells close to the facesheet collapsed immediately at the beginning of the test and redistributed the shock. The neighboring cells were drawn to the center of the panel due to the auxetic behavior. After stress redistribution, the entire ARBW was uniformly damaged, extending from the panel to the connection. The stress distribution of the connection was almost homogeneous except for the 12 mm angle after shock.

The stress distribution of the HSBW is shown in Figure 14. Evidently, the honeycomb panel absorbed the shock energy through bending. The stress of the connection was lower than the stress of the honeycomb panel during the explosion. The maximum stress occurred at cells which were near the free sides.

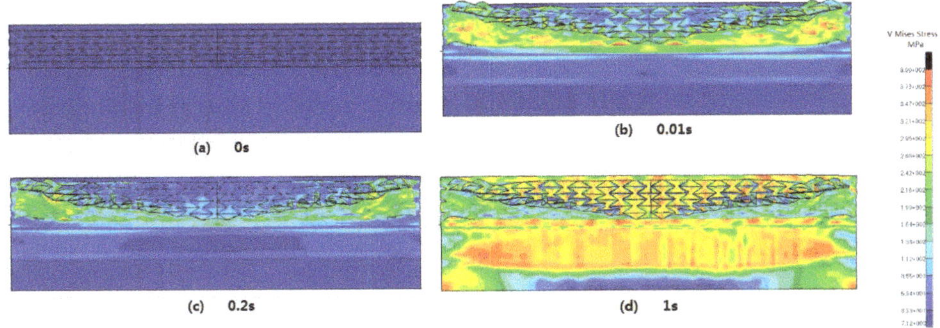

Figure 13. Stress distribution of the ARBW under impulse loads. (**a**) the moment of 0 s, (**b**) the moment of 0.01 s, (**c**) the moment of 0.2 s, (**d**) the moment of 1 s.

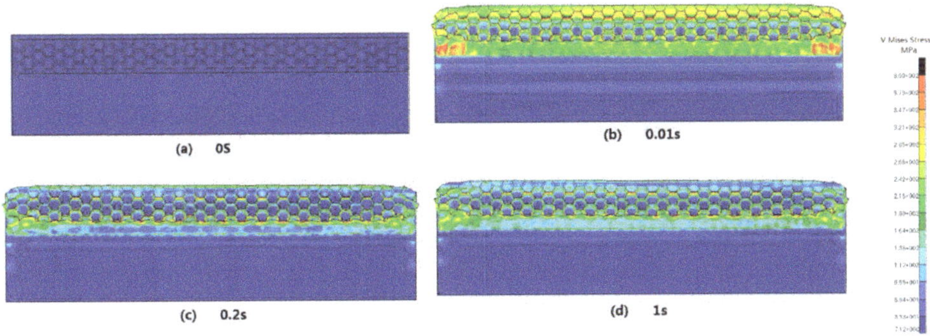

Figure 14. Stress distribution of the HSBW under impulse loads. (**a**) the moment of 0 s, (**b**) the moment of 0.01 s, (**c**) the moment of 0.2 s, (**d**) the moment of 1 s.

The stress distribution of the CBW is shown in Figure 15. The maximum stress occurred on the connection at the beginning of the explosion. After the panel recovered, the stress was transmitted to the connection non-uniformly. The disappearance of some elements showed that a localized failure occurred on the 4 mm angle near the panel. The strain distribution was able to explain the element failure better.

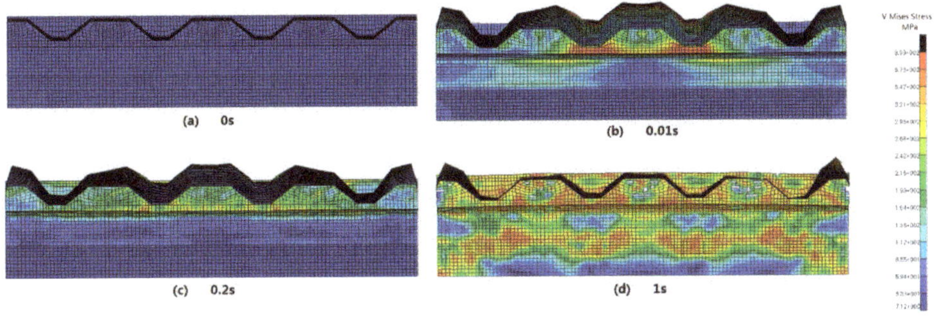

Figure 15. Stress distribution of the CBW under impulse loads. (**a**) the moment of 0 s, (**b**) the moment of 0.01 s, (**c**) the moment of 0.2 s, (**d**) the moment of 1 s.

4.4. Failure Mechanism

The permanent strain distribution of the blast wall is shown in Figure 16. The maximum plastic strain of ARBW and HSBW was under 17.76%. Although the cells at the panel edge collapsed, the most plastic strain of ARBW happened at the panel corner, which shows that the cell-collapsed edge was strengthened due to the densification effect. Besides the panel corner, most of the plastic strain of HSBW occurred symmetrically at the panel center. The maximum plastic strain of CBW happened at the connection, at 17.76%. The connection could not withstand the shock energy and partially failed at the 4 mm angle near the panel, although the 3 mm flexible angle yielded at first.

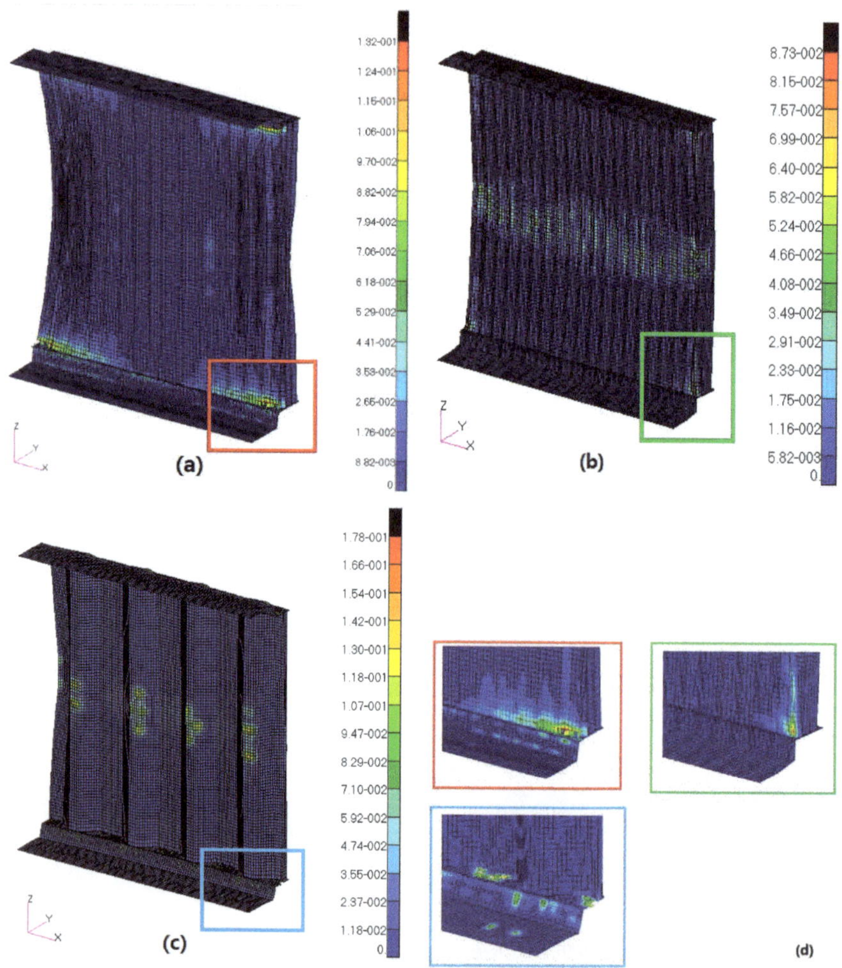

Figure 16. Permanent strain distribution of the blast walls: (**a**) ARBW; (**b**) HSBW; (**c**) CBW; (**d**) corner of blast wall.

5. Conclusions

Numerical investigations of the dynamic responses and energy-absorbing properties of ARBW, HSBW and CBW under distributed impulse loading were conducted. Comparison with experimental results revealed different protective mechanisms of the three kinds of blast walls. This study provided support for the protective structure design of offshore platforms. Key observations are as follows:

- For the ARBW and the HSBW, the distributed impulse loading was absorbed through panel deformation. The ARBW works with crushable cells with auxetic behavior, which the HSBW achieves through whole panel bending.

- To measure the anti-explosion performance, the deformation difference can be used for the HSBW and the CBW, but this hardly works for the ARBW. The stress diagram of the ARBW can demonstrate energy absorption and redistribution because of the auxetic mechanism.

- As regards the deformation of point 2, both the ARBW and the HSBW protected the connection well. With regard to the stress of the connection, the HSBW performed better. Concerning point 1, the ARBW provided the best anti-explosion performance.

- The cell-collapsed edge of ARBW was strengthened due to the densification effect. The plastic failure of CBW happened at the connection first, which meant that the connection absorbed shock energy and partially failed.

Author Contributions: Model analysis and writing—original draft preparation, F.L.; writing—review and editing, D.Y. and S.Z.; All authors have agreed in both contents and form of the final version, being F.L. the responsible of this article. All authors have read and agreed to the published version of the manuscript.

Funding: This research was funded by the National Natural Science Foundation of China (51479115) and Opening Project by the State Key Laboratory of Ocean Engineering (GKZD010071).

Conflicts of Interest: The authors declare no conflict of interest.

References

1. Syed, Z.I.; Mohamed, O.A.; Rahman, S.A. Non-linear Finite Element Analysis of Offshore Stainless Steel Blast Wall under High Impulsive Pressure Loads. *Procedia Eng.* **2016**, *145*, 1275–1282. [CrossRef]
2. Summerhayes, C. Deep Water—The Gulf Oil Disaster and the Future of Offshore Drilling. *Underw. Technol.* **2011**, *30*, 113–115. [CrossRef]
3. Brewerton, R. *Design Guide for Stainless Steel Blast Wall, Technical Note 5*; Fire and Blast Information Group: London, UK, 1999.
4. Walker, S.; Bleach, R.; Carney, S.; Fairlie, G.; Louca, L.A. New guidance on the design of offshore structures to resist the explosion hazard. In Proceedings of the OMAE03, Cancun, Mexico, 8–13 June 2003.
5. Paik, J. Design of offshore facilities to resist gas explosion hazard, engineering handbook. *Ships Offshore Struct.* **2006**, *1*, 165–169. [CrossRef]
6. Boh, J.W.; Louca, L.A.; Choo, Y.S. Energy absorbing passive impact barrier for profiled blastwalls. *Int. J. Impact Eng.* **2005**, *31*, 976–995. [CrossRef]
7. Nwankwo, E.; Fallah, A.S.; Langdon, G.S.; Louca, L.A. Inelastic deformation and failure of partially strengthened profiled blast walls. *Eng. Struct.* **2013**, *46*, 671–686. [CrossRef]
8. Christian, W.; Kim, B.T. Behaviors of a Blast Wall According to the Angle of Vee Stiffener. *Appl. Mech. Mater.* **2012**, *152*, 856–859.
9. Dharmasena, K.P.; Wadley, H.N.G.; Xue, Z.; Hutchinson, J.W. Mechanical response of metallic honeycomb sandwich panel structures to high-intensity dynamic loading. *Int. J. Impact Eng.* **2008**, *35*, 1063–1074. [CrossRef]
10. Theobald, M.D.; Langdon, G.S.; Nurick, G.N.; Pillay, S.; Heyns, A.; Merrett, R.P. Large inelastic response of unbonded metallic foam and honeycomb core sandwich panels to blast loading. *Compos. Struct.* **2010**, *92*, 2465–2475. [CrossRef]
11. Zhu, F.; Zhao, L.M.; Lu, G.X. A numerical simulation of the blast impact of square metallic sandwich panels. *Int. J. Impact Eng.* **2009**, *36*, 687–699. [CrossRef]
12. Pydah, A.; Batra, R.C. Blast loading of bumper shielded hybrid two-core Miura-ori/honeycomb core sandwich plates. *Thin Walled Struct.* **2018**, *129*, 45–57. [CrossRef]
13. Zhang, X.W.; Yang, D.Q. Mechanical Properties of Auxetic Cellular Material Consisting of Re-Entrant Hexagonal Honeycombs. *Materials* **2016**, *9*, 900. [CrossRef] [PubMed]

14. Bezazi, A.; Scarpa, F. Mechanical behaviour of conventional and negative Poisson's ratio thermoplastic polyurethane foams under compressive cyclic loading. *Int. J. Fatigue* **2007**, *29*, 922–930. [CrossRef]

15. Zhang, J.H.; Zhu, X.F.; Yang, X.D.; Zhang, W. Transient nonlinear responses of an auxetic honeycomb sandwich plate under impact loads. *Int. J. Impact Eng.* **2019**. [CrossRef]

16. Yang, D.Q.; Zhang, X.W.; Wu, B.H. The Influence Factors of Explosion and Shock Resistance Performance of Auxetic Sandwich Defensive Structures. *J. Shanghai Jiaotong Univ.* **2018**, *52*, 379–387.

17. Qi, C.; Remennikov, A.; Pei, L.Z.; Yang, S.; Yu, Z.H.; Ngo, T.D. Impact and close-in blast response of auxetic honeycomb-cored sandwich panels: Experimental tests and numerical simulations. *Compos. Struct.* **2017**, *180*, 161–178. [CrossRef]

18. Imbalzano, G.; Linforth, S.; Ngo, T.D.; Lee, P.V.S.; Tran, P. Blast resistance of auxetic and honeycomb sandwich panels: Comparisons and parametric designs. *Compos. Struct.* **2018**, *183*, 242–261. [CrossRef]

19. Hajmohammad, M.H.; Kolahchi, R.; Zarei, M.S.; Nouri, A.H. Dynamic response of auxetic honeycomb plates integrated with agglomerated cnt-reinforced face sheets subjected to blast load based on visco-sinusoidal theory. *Int. J. Mech. Sci.* **2019**. [CrossRef]

20. Langdon, G.S.; Schleyer, G.K. Inelastic deformation and failure of profiled stainless steel blast wall panels. part ii: Analytical modelling considerations. *Int. J. Impact Eng.* **2005**, *31*, 371–399. [CrossRef]

21. Schleyer, G.; Langdon, G. *Pulse Pressure Testing of 1/4 Scale Blast Wall Panels with Connections*; HSE Research Report: Sudbury, UK, 2003.

Journal of
Marine Science and Engineering

Article

Investigations of the Potential Application of k-out-of-n Systems in Oil and Gas Industry Objects

Vladimir V. Rykov [1,2,*], **Mikhail G. Sukharev** [1,*] **and Victor Yu. Itkin** [1,*]

[1] Dep. Applied Mathematics and Computer Modelling, National University of Oil and Gas "Gubkin University", 65 Leninsky Prospekt, 119991 Moscow, Russia

[2] Dep. of Applied Informatics and Probability Theory, Peoples' Friendship University of Russia (RUDN University), 6 Miklukho-Maklaya str., 117198 Moscow, Russia

[*] Correspondence: vladimir_rykov@mail.ru (V.V.R.); mgsukharev@mail.ru (M.G.S.); itkin.v@gubkin.ru (V.Y.I.)

Received: 4 October 2020; Accepted: 12 November 2020; Published: 16 November 2020

Abstract: The purpose of this paper was to demonstrate the possibilities of assessing the reliability of oil and gas industry structures with the help of mathematical models of k-out-of-n systems. We show how the reliability of various structures in the oil and gas complex can be described and investigated using k-out-of-n models. Because the initial information about the life and repair time of components of systems is only usually known on the scale of one and/or two moments, we focus on the problem of the sensitivity analysis of the system reliability indices to the shape of its components repair time distributions. To address this problem, we used the so-called markovization method, based on the introduction of supplementary variables, to model the system behavior with the help of the two-dimensional Markov process with discrete-continuous states. On the basis of the forward Kolmogorov equations for the time-dependent process' state probabilities, relevant balance equations for the process' stationary probabilities are presented. Using these equations, stationary probabilities and some reliability indices for two examples from the oil and gas industry were calculated and their sensitivity to the system component's repair time distributions was analyzed. Calculations show that under "rare" component failures, most system reliability indices become practically insensitive to the shape of the components repair time distributions.

Keywords: k-out-of-n-type systems; reliability; probability of system failure; failure-free time; repair time; stationary mode

1. Introduction and Motivation

The purpose of this work was to highlight the k-out-of-n model for analyzing the reliability of oil and gas facilities. We demonstrate the capabilities of this model using two examples of offshore and onshore structures of the oil and gas complex. Hydrocarbons are successfully produced in the seas and oceans both on the shelf and in deep water. The scale of production is steadily expanding. The extracted raw materials are transported to places of consumption on oil tankers and methane carriers, but the main mode of transport is pipelines. When designing large-scale facilities for oil and gas pipeline transportation systems, and developing plans for their operation and development in the medium and long term, it is always envisaged to reserve production capacities. This is due to the importance of hydrocarbons in modern energetics and economics. According to Russian [1] and international standards, there are various types of redundancy, among which one of the most common is structural redundancy. Structural reservation is the use of redundant structural elements. While structural reservation at pumping and compressor stations is standard practice, structural redundancy at the line section is relatively rare. Reservation of pipelines is provided when the route crosses high-risk water barriers, mainly rivers. Some rivers are characterized by natural phenomena accompanied by increases in water discharge that are difficult to predict. Therefore, crossings over problem rivers should be reserved. To increase the reliability of the pipeline system, an additional, parallel pipeline (it is called siphon barrels) is laid across the river. The diameter of the siphon barrels is less than the diameter of the main pipeline, but the total throughput of the crossing exceeds the throughput of the main pipeline. In addition, a section of the route is often reserved in dangerous mountainous areas prone to landslides, avalanches, or other natural phenomena that lead to ground movement.

Special research is usually carried out to justify reservation decisions in each specific situation. A linear, unique example of redundancy is the multi-line crossing through the Baydaratskaya Bay (Figure 1).

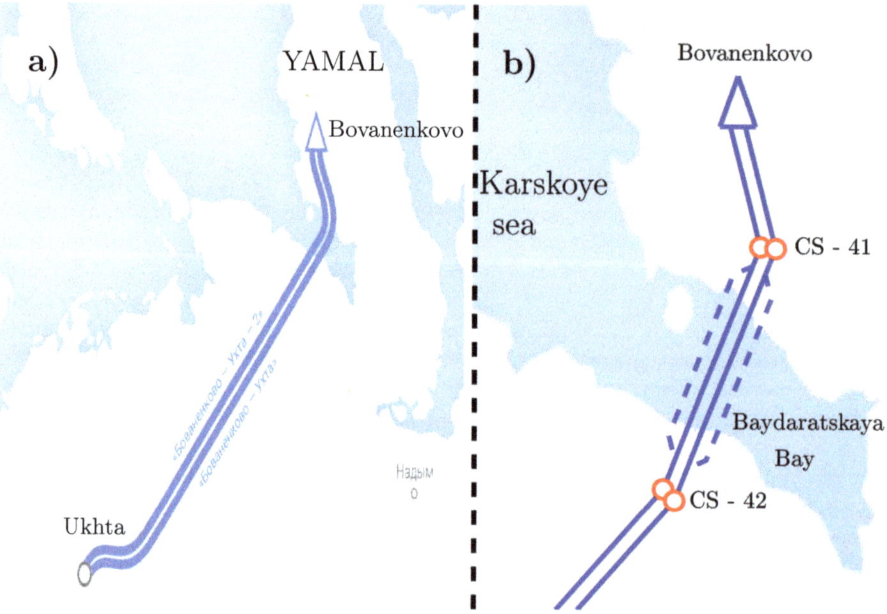

Figure 1. (**a**) The Bovanenkovo–Ukhta gas pipeline. (**b**) The crossing through the Baydaratskaya Bay.

The crossing is part of the technical corridor of the Bovanenkovo–Ukhta trunk gas pipelines, designed to transport gas from the Yamal Peninsula to the Unified Gas Supply System of Russia. Currently, Yamal gas production is carried out at the largest Yamal field—Bovanenkovskoye. The first stage of the corridor was commissioned in 2012, the second, in 2017.

The diameter of the main pipe is 1400 mm, the working pressure is 11.8 MPa (120 at), and the length of the underwater section is 70 km. The crossing is the most difficult section of the corridor and is made of concrete steel pipes with a diameter of 1200 mm. The development of the corridor has been outlined, including the crossing through the Baydaratskaya Bay. The transition lines included in the project, but not yet built, are shown in Figure 1b dotted line.

The creation of the Bovanenkovo–Ukhta gas transmission corridor is one of the largest and most complex projects in the history of pipeline construction. During its implementation, innovative technologies and highly reliable equipment were used. The Baydaratskaya Bay, along the bottom of which the track is laid, is covered with ice for most of the year. The bay is distinguished by special natural and climatic conditions [2]: at a shallow depth, it is characterized by frequent stormy weather, complex bottom sediments, and freezing to the bottom in winter.

When designing the corridor, it was necessary to make a decision on the structural reserve of the crossing, that is, to choose the number of lines n and consider various options for emergency situations associated with the failure of k lines. To justify the decision, it was necessary to consider various situations that could potentially cause a failure and to evaluate the probabilities of such situations.

Failure (here, a *failure* should be understood as a rupture to a line or damage forcing operations to halt) can occur, for example, as a result of damage caused by icebergs, which are carried into Baydaratskaya Bay by one of the Gulf Stream jets, which weaken at these latitudes.

Note that the Boolean reliability model for a bay crossing system, when the system and each of its elements are only in states 0 and 1, is not directly applicable. The gas pipeline corridor will partially fulfil its functions until all the lines of the crossing are out of order. The degree of its function performance is determined by the throughput of the entire Bovanenkovo—Ukhta gas pipeline (Figure 1a). However, it is possible to calculate this indicator by considering all possible k-out-of-n situations.

According to the authors, the possibility of using k-out-of-n models is determined by the technological specifics of the object. Therefore, the permissible number and connection schemes of the compressor shop units are often uniquely determined by the load on the pipeline (*compressor shop* is a group of compressors working in common mode; on a two-line gas pipeline, the compressor station includes two compressor shops). If the number of gas compressors is insufficient, the shop (all units of the shop) is turned off (the so-called "on-pass" mode).

In other cases, for example, when investigating the reliability of crossing a river-line obstacle, it is necessary to calculate the possibility of several k-out-of-n situations in order to find the distribution of the corridor's capacity, depending on the number of operational lines.

The problem of choosing a structural reserve was further complicated by the multi-stage construction of the corridor and crossing the bay. The life cycle of these objects is determined not only by the predicted durability of the pipelines, but also by the rational rates of gas extraction from the fields of the Yamal Peninsula in the long term. The rate of production should be tailored to the gas needs of the constituent entities of the Russian Federation, export contracts, and the schedules of possible withdrawals from the fields of the Nadym-Purtazovsky region—the main gas supplier in modern times.

The problems of the reliability of large pipeline systems for oil and gas supply was given exceptional attention during the period of their intensive development, when several thousand kilometers of pipelines were laid annually in the USSR. Various mathematical models were developed [3]. A scientific seminar "Methodological issues of researching the reliability of large energy systems" was organized that is still in effect: this year, the 92nd meeting was held. Within the framework of the seminar, a four-volume desk book on the reliability of energy systems was prepared and published. The publication reflected the state of the subject at that time in detail. A separate volume of the desk book (in two books) is devoted exclusively to the reliability of gas and oil supply systems [4,5]. Since then, special studies concerning reliability have been carried out when designing each new main pipeline, in particular for the Yamal–West corridor (1st option) [6]. Various options of the route were compared according to reliability criteria: through the Baydaratskaya Bay and bypassing the bay along the Subpolar Urals. In addition, ref. [7] is devoted to the rationale of the decisions for crossing the bay. Naturally, in applied calculations, the methodological apparatus developed by that time was used. In this paper, a model is proposed, which is particularly applicable to the study of the reliability of the crossing through the Baydaratskaya Bay. The application of this model allows one, at the design stage, to assess the longevity of an object with greater confidence.

Another example, in which it is advisable to use k-out-of-n models, is the compressor stations of the Bovanenkovo–Ukhta gas pipeline, which are located on two shores of the bay. A structural reserve of gas-compressor units is provided in each shop of the compression stations. Going through various options for k operable units out of n installed units, it is possible to fully characterize the impact of the shop on the reliability indicators of the gas pipeline corridor.

The above examples of structural redundancy in oil and gas complex facilities can be expanded by considering the production enterprises (reservation of oil wells, electric motors in the field transport system, etc.), preparation for oil and gas transportation (separators, sedimentation tanks), processing (number of production lines), etc. The calculation of the reliability indicators of engineering systems with a structural reserve, as a rule, can be carried out using standard statements of k-out-of-n models, a brief overview of which is given below.

The redundancy technique is widely used to improve system reliability. A typical form of redundancy is a k-out-of-n configuration. A k-out-of-n ($k \leq n$) system is a repairable redundancy system that consists of n components in parallel, each of which can be in one of two states: operable or not operable.

The repair of the system's components is realized by a single repair unit. A k-out-of-n system may be described in two ways, depending on the definition of the parameter k, as follows: the parameter k may represent the number of components in the system that must work in order for the entire system to work, referred to as a k-out-of-n: G system; or k may represent the number of components in the system that must fail before the entire system fails, referred to as a k-out-of-n: F system [8].

As a result of the wide range of practical applications, a lot of papers are devoted to the study of k-out-of-n systems. The earlier investigations deal with homogeneous binary models, where each component can take only two states "UP" or "DOWN". The probabilities of the state "UP" for each component are equal. There are vast amounts of literature on such studies (see for example Trivedi [9], Chakravarthy et al. [10], and the bibliographies therein).

For heterogeneous system investigation, the method of Universal Generating Functions can be used. The basic ideas of the method were introduced by I. Ushakov in the mid 1980s [11,12]. Since then, this method has been considerably expanded and is currently very popular (see, for example, Levitin [13] and the bibliography therein).

Further investigations of such systems have been directed at the study of systems with non-exponential life and/or repair time distributions. In [14], M.S. Moustafa considered a k-out-of-n system with exponential life and an arbitrarily distributed repair time of its components with the help of the embedded Markov chains method. In addition, ref. [15] contains a detailed analysis of 2-out-of-n and 3-out-of-n systems with general repair time distributions and an evaluation of different reliability characteristics during a system's life cycle with the use of the Laplace transform to stochastic relations. In a series of papers a review to which can be found in [16], these investigations were developed with the help of the supplementary variables method firstly proposed by D. Cox [17].

These works allow to set up and develop investigations in one of the principal directions systems' reliability study—insensitivity or low sensitivity of their characteristics to life and repair times distributions of their components. These investigations are very important, because it is very difficult (or even impossible) to find sufficient reliable information about these distributions (see Section 4.1). In [18–21], using analytical and simulation methods, the sensitivity of different system reliability characteristics to the shapes of their components' life and repair time distributions was studied.

These studies show that (at least for "rare" component failures) system reliability indicators are practically insensitive to the shapes of their components life and repair time distributions. Essential dependents take place only on their two first moments. For these investigations, Markov processes with a discrete–continuous space state are used. At that, the system's time-dependent state probabilities satisfy to the Kolmogorov forward partial differential equations. In [22], this approach was used to calculate the reliability function of a k-out-of-n heterogeneous system. In [23] the general approach to solution of special kind of partial differential equations was proposed, which was used for Kolmogorov forward system of equations for time-dependent state probabilities.

The novelty of the paper and its difference from another studies on reliability of oil and gas equipments consists in:

- A study of the k-out-of-n mathematical model with arbitrary repair time distributions;
- A calculation of the steady state probabilities of this system;
- A demonstration of this model potential applications to study of reliability problems for objects in the oil and gas industry;
- A study of system reliability indicators sensitivity to the shape of system components repair time distributions.

The paper is organized as follows. In the next section, the mathematical model and some notations are presented. Section 3 is devoted to the research regarding the stationary regime of k-out-of-n systems. In Section 4, an analysis of the input information needed for the problem considered in the examples is proposed. Sections 4.2 and 4.3 consider two potential applications ("3-out-of-4" and "2-out-of-4") of these models in engineering systems of the oil and gas industry. Finally, the last section concludes the paper.

2. Mathematical Model

For the considering objects modelling, we use the k-out-of-n: F system, which consists of n components and fails if k components fail. Failed components and the whole system are repaired by a single facility. Suppose that

- The components fail according to a Poisson flow with intensity α;
- The random repair times of components are independent identically distributed (i.i.d.) random variables (r.v.) and their common cumulative distribution function (c.d.f.) $B(t)$ is absolutely continuous with probability density function (p.d.f.) $b(t) = B'(t)$;
- After full system failure (when the system occurs in state k), it is repaired during a random time with absolutely continuous c.d.f. $\Gamma(t)$.

Denote by

- j—the number of components in the "DOWN" state;
- $E = \{j : j = \{0, 1, \ldots k\}\}$—the system's state space, where j denotes the number of failed components, and the state k is the system's "DOWN" state;
- $\lambda_j = (n - j)\alpha$—the system failure intensity in its j-th state;
- $\beta(x) = \frac{B'(x)}{1 - B(x)}$, $\gamma(x) = \frac{\Gamma'(x)}{1 - \Gamma(x)}$ are conditional repair densities of components and the whole system, given that elapsed repair time is x;
- $b = \int\limits_0^\infty (1 - B(x))\, dx$—mean repair time of any of the components;
- $g = \int\limits_0^\infty (1 - \Gamma(x))\, dx$—mean repair time of the system after its full failure.

For the study of the above system, we use the supplementary variables method as firstly proposed by D. Cox [17], which is intensively used in series of works (see, for example, [22] and the bibliography therein). To construct the appropriate Markov process, the elapsed repair time of the component under repair is introduced as a supplementary variable $X(t)$, and consider a two-dimensional random process

$$Z = \{Z(t) = (J(t), X(t)), \ t \geq 0\},$$

where for $j > 0$, value $J(t)$ represents the number of failed components at time t,

$$J(t) = j, \ \text{if at time } t, \text{ the system is in state } j \in E,$$

and the supplementary variable $X(t)$ denotes elapsed time, i.e., the time spent by the repair facility to repair the component. The state space of the process Z is

$$\mathcal{E} = \{0, (j, \mathbb{R}^+) : j = \overline{1, k}\}.$$

As a result of the supplementary variable, the process Z is a Markov one. Denote its state probability (for $j = 0$) and p.d.f.'s by

$$\pi_0(t) = \mathbf{P}\{J(t) = 0\}, \quad \pi_j(t; x) = \mathbf{P}\{J(t) = j, \ X(t) = x\} \quad (j = \overline{1, k}), \tag{1}$$

and appropriate the macro-state probabilities by

$$\pi_j(t) = \mathbf{P}\{J(t) = j\} = \int_0^t \pi_j(t;x)\,dx.$$

The system's lifetime is denoted by T, $T = \inf\{t : J(t) = k\}$, and its reliability function is denoted by $R(t) = \mathbf{P}\{T > t\}$.

The time-dependent probabilities $\pi_j(t;x)$ satisfied to the Kolmogorov forward partial differential equations. This system can be obtained from appropriate system difference equations. The last one can be construct by the usual method of comparison of the system state probabilities in a small interval between two closed time points. Further, by passing to the limit when length of the interval tends to zero, the Kolmogorov system is obtained. This system was proposed in [22]. With the help of the operator

$$D = \left(\frac{\partial}{\partial t} + \frac{\partial}{\partial x}\right),$$

these equations can be written in the form

$$
\begin{aligned}
\frac{d}{dt}\pi_0(t) &= -n\alpha\pi_0(t) + \int_0^t \pi_1(t,x)\beta(x)\,dx + \int_0^t \pi_k(t,x)\gamma(x)\,dx,\\
D\pi_1(t;x) &= -((n-1)\alpha + \beta(x))\pi_1(t;x),\\
D\pi_i(t;x) &= -((n-i)\alpha + \beta(x))\pi_i(t;x) + (n-i+1)\alpha\pi_{i-1}(t;x)\\
&\quad (i = \overline{2,k-1}),\\
D\pi_k(t;x) &= -((n-k)\alpha + \gamma(x))\pi_k(t;x) + (n-k+1)\alpha\pi_{k-1}(t;x)
\end{aligned}
\tag{2}
$$

with the initial $\pi_0(0) = 1$ and boundary conditions for $k > 2$

$$
\begin{aligned}
\pi_1(t,0) &= n\alpha_1\pi_0(t) + \int_0^t \pi_2(t;x)\beta(x)\,dx,\\
\pi_i(t,0) &= \int_0^t \pi_{i+1}(t;x)\beta(x)\,dx \quad (i = \overline{2,k-2}),\\
\pi_{k-1}(t,0) &= \pi_k(t,0) = 0.
\end{aligned}
\tag{3}
$$

For the solution to this kind of partial differential equations, as shown in [23], an algorithm was proposed. The algorithm allows one to find the analytic expressions for the system time dependent state probabilities in terms of their Laplace transforms. These expressions can be used for further analytic and numerical investigations of the objects considered in the paper.

For the system reliability function calculation, the same process Z with absorption in the state k can be used. In [22], the reliability function of the k-out-of-$n : F$ system in terms of its Laplace transform was found and some examples were proposed. This approach can also be used for further investigations of the appropriate objects in the oil and gas industry. Following the main idea of the paper, we focus on the calculation of the stationary system state probabilities and their applications to engineering systems in the oil and gas industry.

3. Stationary Regime Study

In this section, the stationary regime of the above system is considered. Note that the state 0 is a positive atom of the process Z with respect to its invariant measure. Therefore, it is positively recurrent (the Harris process) and thus its limiting probabilities

$$\pi_0 = \lim_{t\to\infty} \pi_0(t), \quad \pi_j(x) = \lim_{t\to\infty} \pi_j(t,x)$$

exist and are the system steady state probabilities.

There are least two different possible ways to calculate the steady state probabilities: (a) by passing the limits in the time-dependent state probabilities from Section 2; or (b) by balancing the equations for the steady state probabilities solution. The second option is preferable because the partial differential equations solution is not a simple problem as one can see in [23].

The balanced equations also follow from equations for the time-dependent probabilities (2) by applying the derivatives with respect to the time variable to zero. As a result, they take the form

$$
\begin{aligned}
n\alpha\pi_0 &= \int_0^\infty \pi_1(x)\beta(x)\,dx + \int_0^\infty \pi_k(x)\gamma(x)\,dx, \\
\frac{d}{dx}\pi_1(x) &= -((n-1)\alpha + \beta(x))\pi_1(x), \\
\frac{d}{dx}\pi_i(x) &= -((n-i)\alpha + \beta(x))\pi_i(x) + (n-i+1)\alpha\pi_{i-1}(x) \\
&\quad (i = \overline{2,k-1}), \\
\frac{d}{dx}\pi_k(x) &= (n-k+1)\alpha\pi_{k-1}(x) - \gamma(x)\pi_k(x),
\end{aligned}
\tag{4}
$$

jointly with the boundary conditions, which, for $k = 2$, are

$$
\begin{aligned}
\pi_1(0) &= n\alpha\pi_0, \\
\pi_2(0) &= 0,
\end{aligned}
\tag{5}
$$

and, for $k > 2$, they are

$$
\begin{aligned}
\pi_1(0) &= n\alpha_1\pi_0 + \int_0^\infty \pi_2(x)\beta(x)\,dx, \\
\pi_i(0) &= \int_0^\infty \pi_{i+1}(x)\beta(x)\,dx \ (i = \overline{2,k-2}), \\
\pi_{k-1}(0) &= \pi_k(0) = 0.
\end{aligned}
\tag{6}
$$

The analytical solution of this equation, in terms of functions $\beta(x)$ and $\gamma(x)$, can be obtained with the constant variation method. The special case 3-out-of-6 was presented in [22]. However, this is too bulky and numerically demanding for real investigations. Therefore, for practical applications, a direct numerical solution is proposed.

To represent the system (4)–(6) as a linear boundary value problem, introduce the functions

$$\pi_{k+i}(x) = \int_0^x \pi_i(x)\,dx \ (i = \overline{1,k}).$$

Thus, the additional equations

$$\frac{d\pi_{k+i}(x)}{dx} = \pi_i(x)\ (i = \overline{1,k}),$$

should be added to system (4)
and the additional relations

$$\pi_{k+i}(0) = 0\ (i = \overline{1,k}).$$

should be added to the boundary conditions (6).

In order to eliminate integrals of unknown functions from the system (4), let us integrate all equations with derivatives over the interval from 0 to ∞.

The left sides of the transformed equations are $\pi_i = \lim\limits_{x\to\infty} \pi_{k+i}(x)$, and the right sides contain integrals of the form $\int_0^\infty \pi_i(x)\beta(x)\,dx$ and $\int_0^\infty \pi_k(x)\gamma(x)\,dx$, which are present in the first equation and in the boundary conditions. Represent these expressions from the transformed system and substitute them into the boundary conditions (6). Represent the obtained system in matrix form:

$$\frac{d\vec{\pi}}{dx}(x) = U(x)\vec{\pi}(x), \tag{7}$$

$$\vec{\pi}(0) = \alpha V \vec{\pi}(\infty) + \vec{h}, \tag{8}$$

where

- $\vec{\pi}(x) = [\pi_1(x), ..., \pi_{2k}(x)]'$ is a vector of unknown functions;
- $\vec{\pi}(\infty) = \lim\limits_{x\to\infty} \vec{\pi}(x)$ is a vector of their limiting values;
- matrix $U(x) \in \mathbb{R}^{2k\times 2k}$ consists of the following non-zero components

$$\begin{aligned}
U_{i,i}(x) &= -(n-i)\alpha - \beta(x), \\
U_{i+1,i}(x) &= (n-i)\alpha, \\
U_{k+i,i}(x) &= 1,\ i = \overline{1,k-1}, \\
U_{k,k}(x) &= -\gamma(x), \\
U_{k+k,k}(x) &= 1.
\end{aligned}$$

The matrix $V \in \mathbb{R}^{2k\times 2k}$ consists of the following non-zero components for $k > 2$:

$$\begin{aligned}
V_{1,k+i+1} &= -n, \\
V_{i+1,k+k-1} &= -(n-k+1), \\
V_{i+1,k+i+1} &= n-i-1,\ i = \overline{1,k-3}, \\
V_{1,k+1} &= -1, \\
V_{1,k+k-1} &= -(2n-k+1), \\
V_{1,k+k} &= -n,
\end{aligned}$$

and for $k = 2$:

$$\begin{aligned}
V_{1,k+k-1} &= -n, \\
V_{1,k+k} &= -n.
\end{aligned}$$

Vector $\vec{h} \in \mathbb{R}^{2k}$ only has one non-zero component $h_1 = n\alpha$.

For example, for the 3-out-of-4 system, these matrices and vectors appear as follows:

$$U(x) = \begin{bmatrix} -3\alpha - \beta(x) & 0 & 0 & 0 & 0 & 0 \\ 3\alpha & -2\alpha - \beta(x) & 0 & 0 & 0 & 0 \\ 0 & 2\alpha & -\gamma(x) & 0 & 0 & 0 \\ 1 & 0 & 0 & 0 & 0 & 0 \\ 0 & 1 & 0 & 0 & 0 & 0 \\ 0 & 0 & 1 & 0 & 0 & 0 \end{bmatrix};$$

$$V = \begin{bmatrix} 0 & 0 & 0 & -1 & -6 & -4 \\ 0 & 0 & 0 & 0 & 0 & 0 \\ 0 & 0 & 0 & 0 & 0 & 0 \\ 0 & 0 & 0 & 0 & 0 & 0 \\ 0 & 0 & 0 & 0 & 0 & 0 \\ 0 & 0 & 0 & 0 & 0 & 0 \end{bmatrix}, \quad \vec{h} = \begin{bmatrix} 4\alpha \\ 0 \\ 0 \\ 0 \\ 0 \\ 0 \end{bmatrix}.$$

For the numerical solution, it is necessary to change the infinite interval of the repair time to the finite one. Furthermore, for some distributions, the conditional probability density function is undefined if $x = 0$. Thus, it should be solved the system for $x > 0$ only. To do this, we need to choose some small parameters p (for example $p = 0.001$) and solve the considered system of equations in interval $[x_p, x_{1-p}]$, where x_p is the minimum of p-quantiles for distributions $B(x)$ and $\Gamma(x)$, and x_{1-p} is the maximum of $(1 - p)$-quantiles for distributions $B(x)$ and $\Gamma(x)$.

The obtained systems (7) and (8) is a linear boundary value problem with non-separated boundary conditions. Such a problem can be solved using the standard *bvp4c* or *bvp5c* solvers in MATLAB. They implement an iterative method with three-point and four-point Lobatto formulas, respectively, which are partial cases of the implicit Runge–Kutta scheme [24]. Procedures *bvp4c* and *bvp5c* use a linear equation solver for general sparse matrices, because it is possible to solve the problem with non-separated boundary conditions (8). The solutions (7) and (8) for the special cases are given in the next section. They provide calculation any reliability indicators such as failure probability π_k and availability $K_{av} = 1 - \pi_k$ etc.

4. Examples

In this section, two examples of applications of the k-out-of-n system for study of the reliability systems in the oil and gas industry are considered.

4.1. Input Information for the Reliability of Engineering Systems

One of the most important and difficult problems in analyzing the reliability of complex systems in general and objects of the oil and gas industry in particular is to obtain reliable information about the reliability of their components. Complete information about the reliability of any technical object contains the distributions of two random variables: the life and repair time of all structural units of the suitable model. However, these distributions are often unknown, and one is usually limited to knowing one or two moments of these distributions: the mean and variance. In this regard, the problem of analyzing the sensitivity of the system's reliability indicators to the shape of their initial information distributions and the variability of their first moments becomes urgent. Estimates of the means and variances of life and repair time components of systems can be obtained from the statistics of the operating equipment. The main indicator of the pipeline system reliability as a whole is the rate of failures (pipe ruptures) [25]. In the 1970s, this indicator for the Unified Gas Supply System of the USSR was equal to one failure/1000 km per year. In the last decade, the failure statistics, according to the Unified Gas Supply System of Russia, indicate the stability of this parameter at the level of about 0.2 failures/1000 km per year [25]. For a 70 km section, it would be 0.014 failures/year.

Technological progress is a characteristic feature of our civilization. The technologies for the production of pipes are being improved comparatively quickly on a historical scale, and the indicators of their reliability (failure-free operation and longevity [1]) are improving. The construction and manufacturing quality of power equipment are being improved. It is methodologically incorrect to transfer the estimators of the reliability indicators of equipment in operation to the designed facilities without making corrections for scientific and technological progress. We have to rely, to one degree or another, on expert knowledge and forecasts. For example, when assessing the reliability indicators of a unique object—the Baydaratskaya Bay crossing—it was necessary to analyze its design features and the conditions of its functioning, the natural features of the territory, and the possible causes of accidents that occur in the practice of global pipeline construction. Furthermore, it was necessary to assess the risks of accidents from non-standard situations according to the increased stability of pipes and reliability of equipment. We believe that the failure flow intensity for the lines crossing through the Baydaratskaya Bay is several times lower than the average. However, examining the sensitivity of the model, we consider options that are quite close to average intensity.

The specificity of the problems concerning the whole pipeline reliability indicator estimation consists in the need for a comparative assessment of the indicators of pipes and power equipment. There is a significant difference between failures of power equipment and pipes. The failure rate of the power equipment units is much higher than the failure rate of pipes. This can be caused by wear and tear on high-speed centrifugal blowers, the gas turbines that drive the blower, and power interruptions. According to the data from long-term operation, the operating time of gas pumping units can be considered to be obeying an exponential distribution; the average overhaul time ranges from 2000 to 5000 h, which corresponds to about 2–5 failures/year. For modern equipment, this figure should be significantly lower.

Repairing a guillotine rupture in the Baydaratskaya Bay crossing would take a very long time, which would include, in particular, moving a specialized offshore pipe layer to the polar region. The uncertainty of the repair time is aggravated by extreme climatic conditions, due to which, for example, work to restore the crossing is in principle possible only for a few months of the year. The main time spent on the repair of the line is the transportation of special equipment. Consequently, the repair would take at least a year, and in difficult cases, even more. Time expenditure depends little on the number of failed lines, i.e., it can be assumed that the distribution of the repair time of one line and the total repair time of all lines are similar.

Power equipment fails more often, but the consequences and time to eliminate the accident are usually shorter than for the linear part. In the oil and gas complex of the Russian Federation, block repairs have taken strong positions in relation to the power equipment of pipelines, i.e., the replacement of the pump, compressor, or driving engine with subsequent repair at a specialized plant. In connection with the development of new methods of repairs and repair equipment, the reliability indicators are improving. The repair of one object can take from 5 to 20 days, and the time of general repair is proportional to the number of objects.

Thus, on the basis of the analysis performed, we can conclude that:

- The failure rate of a line of Baydaratskaya Bay crossing varies from 0.004 to 0.01 failures/year;
- The failure rate of power equipment varies from 0.5 to 2 failures/year;
- The mean repair time for a line varies from 1 to 2 years;
- The mean repair time for power equipment varies from 5 to 20 days.

In conditions of a lack information, the uncertainty of estimates of reliability indicators for unique objects increases the value of robust mathematical models. In this regard, the sensitivity analysis of system reliability indicators to the distribution of its component's life and repair times is a very important. Thus, in our examples, we focus on the investigation of the sensitivity system reliability characteristics to the shapes of their components life and repair time distributions.

4.2. Gas Pipeline through Baydaratskaya Bay

Let us apply the mathematical apparatus described above for modeling the reliability of the pipeline system (Figure 1b) and studying the sensitivity of the model to the parameters of the distributions of time between failures and renewal time.

For the line crossing through Baydaratskaya Bay, we consider the "3-out-of-4" model, i.e., the failure of the system is considered as being the failure of three pipelines. As the main indicator of reliability, we take the stationary probability π_k of falling into the system failure state ($\pi_k = 1 - K_{av}$).

As a result of the uncertainty of the initial information about the reliability of the elements, which was mentioned above, we consider two possible distribution laws of the repair time (the Γ-distribution and the Gnedenko–Weibull distribution) and carry out calculations for various combinations of distribution parameters.

The results are shown in Figures 2 and 3. The abscissa is the variation V, varying in the range from 0.1 to 2; the ordinate is the probability of system failure in the stationary mode π_k.

The graphs are plotted for different values of the failure flow intensity α: in Figure 2, the average repair times are $b = g = 1$ year, and in Figure 3, they are $b = g = 2$ years.

For $V = 1$, both the Γ-distribution and the Gnedenko–Weibull distributions are reduced to an exponential distribution, so we can compare our calculations with the classical solution for the k-out-of-n Markov system.

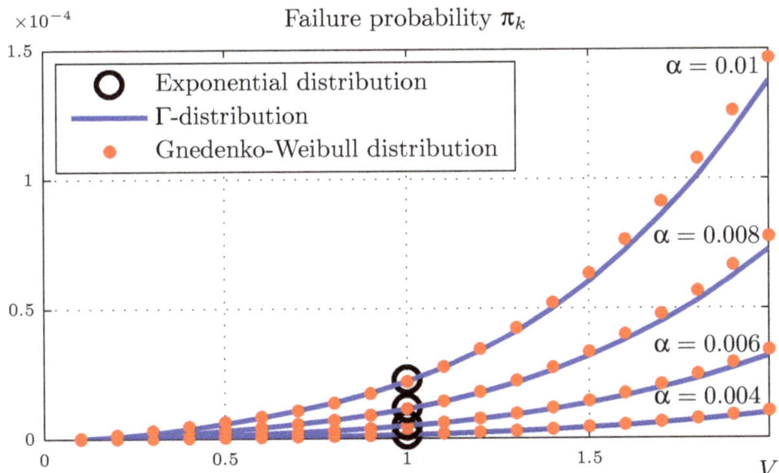

Figure 2. Probabilities of failure of the pipeline sections ("3-out-of-4") for $b = g = 1$ year.

Figures 2 and 3 show that there is a monotonic, but nonlinear increase in the probability of system failure with an increase in both the variation of the repair time and the failure stream intensity. The results obtained for the Γ-distribution and the Gnedenko–Weibull distribution are almost identical, which allows us to hope that the probability of failure of a line of the pipeline system is not sensitive to the type of distribution of the repair time of a line. However, the influence of the variation is quite significant. Therefore, when calculating the stationary probability of an operational state of the system (availability), the second moment of repair time of its components must be taken into account. On the other hand, the insensitivity of the system availability to the type of distribution of the repair time of its components makes it possible to use the Gnedenko–Weibull law to reduce the calculation time.

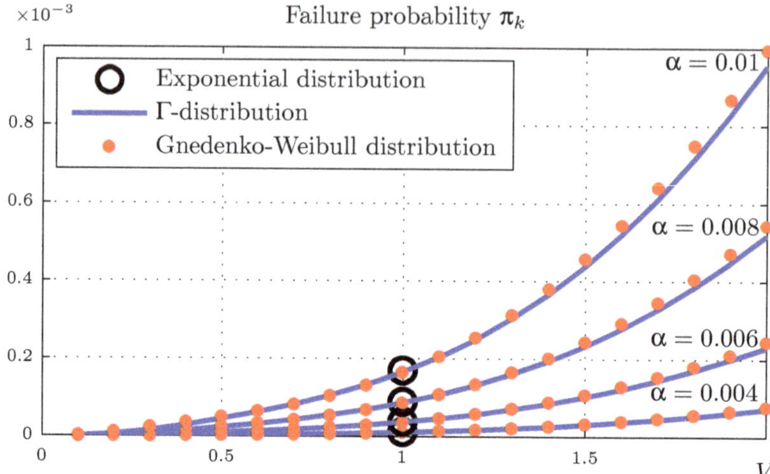

Figure 3. Probabilities of failure of the pipeline sections ("3-out-of-4") for $b = g = 2$ years.

4.3. Compressor Shop Model

Consider a model of a compressor shop equipped with n gas compressors that is in operable state if fit for work are k of them. We believe that the piping manifold provides moving redundancy (standby redundancy with general ratio [1]). Consider the compressor shop as a "2-out-of-4" system and carry out calculations similar to Section 4.2. The calculation results are presented in the Figures 4–6. The abscissa is the variation V, as before, varying in the range from 0.1 to 2; the ordinate is the probability of system failure in the stationary mode π_k. The graphs are dotted for different values of failure stream intensity α and mean repair times b.

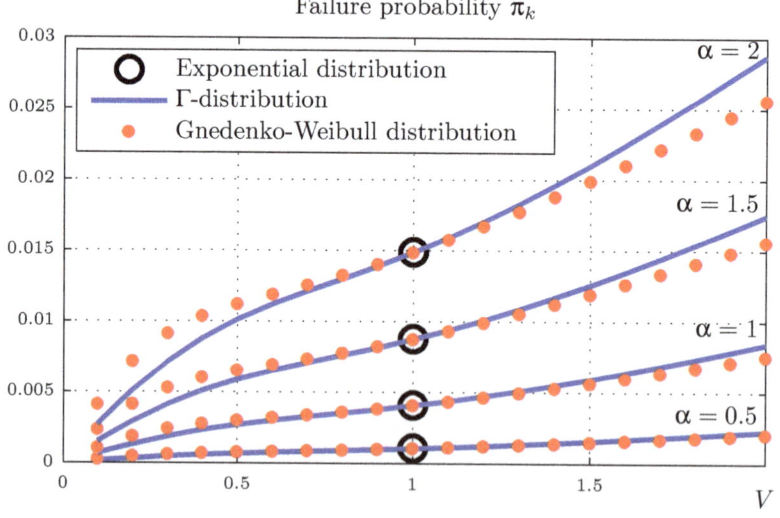

Figure 4. Probabilities of the compressor shop failure ("2-out-of-4") for $b = 5$ days, $g = 10$ days.

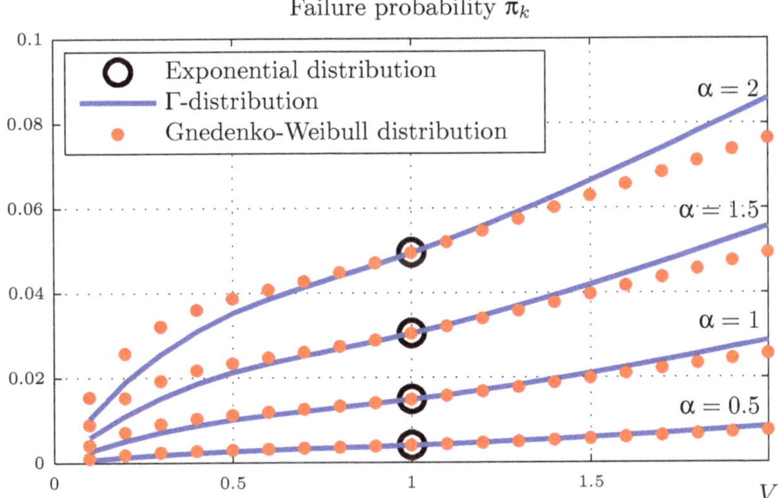

Figure 5. Probabilities of the compressor shop failure ("2-out-of-4") for $b = 10$ days, $g = 40$ days.

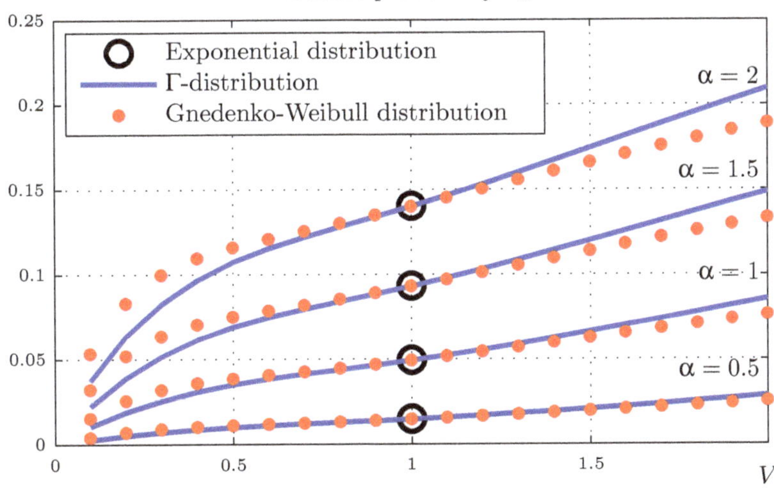

Figure 6. Probabilities of the compressor shop failure ("2-out-of-4") for $b = 20$ days, $g = 40$ days.

It can be seen from Figures 4–6 that in this case, the distribution law has a stronger effect on the probability of system failure, which is explained by the smaller ratio of the mean operating and repair times.

The calculations show that all the parameters of the model—mean operation time to failure and mean repair time: the repair time variation coefficient—significantly affect the reliability of the system. The shape of the repair time distribution is less important, but one should not draw far-reaching conclusions based on this study, in which only two closely related distribution laws were used.

5. Conclusions

The paper shows the possibility of applying the k-out-of-n model to calculating the reliability characteristics of oil and gas equipment. The practical significance of the results obtained in this work lies, first of all, in the fact that the proposed model makes it possible to reflect the conditions of the functioning of the designed or operated object more accurately. The engineer must know how the resulting reliability indicators change depending on the initial information. This can be found out by performing a series of calculations. However, it is impossible to get an idea of the effect on the result of the assumption about the distribution law, for example, the distribution of repair time, if there are not theoretical studies. Decision makers will consider certain measures to improve the reliability justified only if there is confidence in the adequacy of the models taken into account.

Thus, the main results of the paper are as follows:

- On the basis of two examples of engineering systems in the oil and gas industry, we firstly show the possibility to calculate the reliability characteristics of corresponding systems with the help of k-out-of-n models;
- We demonstrated how the non-stationary reliability indicators of the k-out-of-n system with a Poisson flow of failures and an arbitrary distribution of repair time can be calculated by solving a system of partial differential equations for a non-stationary mode;
- An appropriate system of ordinary differential equations for a stationary regime was obtained;
- The system for stationary regime was reduced to a two-point boundary value problem, and its numerical solution was obtained with the help of suitable procedures on the MATLAB platform;
- Taking into account that complete information about a system's component's life and their repair time distributions are usually unknown, we focus on the problems of the systems reliability characteristic's sensitivity to the shape of distributions of its component's life and repair times.
- The calculations show negligible sensitivity of the system availability to their components life and repair times distributions when the ratio mean repair time to mean lifetime of each component is small. However, a system's characteristics essentially depend on the component's repair time variation.

Further investigations will focus in the proposed direction.

Author Contributions: Conceptualization, methodology and formal analysis, V.V.R.; setting tasks for engineering applications, compliance of terms on reliability with technical standards, M.G.S.; computer calculations, visualization and writing—original draft preparation, V.Y.I. All authors have read and agreed to the published version of the manuscript.

Funding: This research received no external funding.

Acknowledgments: The authors thanks the Publisher MDPI and the Guest Editor Pavel Praks for the permission to publish the paper for free price.

Conflicts of Interest: The authors declare no conflict of interest.

References

1. Voropy, N.I. (Ed.) *Reliability of Energy Systems (Recommended Terms)*; IAC "Energiya": Moscow, Russia, 2007; p. 191.
2. Remizov, V.V.; Odisharia, G.E.; Kalatskiy, V.I.; Ryabinin, V.E.; Tsvetintsky, A.S.; Dubikov, G.I.; Sovershaev, V.A.; Vozzhinskaya, V.B.; Danilov, A.I.; Zilbershtein O.I.; et al. *Natural Conditions of the Baydaratskaya Bay. The Main Research Results for the Construction of the Underwater Crossing of the Yamal-Center Gas Trunk Line System*; Geos: Moscow, Russia, 1997; p. 432.
3. Gritsenko, A.I.; Sukharev, M.G.; Stavrovskiy, E.R. The Problem of Ensuring the Reliability of Gas-Supply System and Approaches to their Solution. In *Energy Reviews. Scientific and Engineering Problems of Energy System Reliability*; Section A: Energy Reviews; Harwood Academic Pub.: New York, NY, USA, 1987; Volume 3.
4. Sukharev, M.G. (Ed.) Reliability of gas and oil supply systems. In *Reliability of Energy Systems and Their Equipment. Reference Book in 4 Volumes*; Book 1; Publishing House "Nedra": Moscow, Russia, 1994; Volume 3, p. 414.
5. Sukharev, M.G. (Ed.) Reliability of gas and oil supply systems. In *Reliability of Energy Systems and Their Equipment. Reference Book in 4 Volumes*; Book 2; Publishing House "Nedra": Moscow, Russia, 1994; Volume 3, p. 297.

6. Galiullin, Z.T.; Sukharev, M.G.; Karasevich, A.M.; Merenkov, A.P. Problems of safety and reliability of large projected gas transmission systems (on the example of the Yamal-West GTS). In *Methodological Issues of Research of Reliability of Large Power Systems*; Moscow Institute of Oil and Gas: Moscow, Russia, 1989; pp. 42–49.

7. Sukharev, M.G.; Karasevich, A.M. Transition of the Yamal—West gas transmission system through the Baydaratskaya Bay: Methods for assessing reliability, choosing backup means. In *Methodological Issues of Research of Reliability of Large Power Systems*; SEI SB AS USSR: Irkutsk, Russia, 1990; Volume 1, pp. 66–74.

8. Shepherd, D.K. *k*-out-of-*n* Systems. In *Encyclopedia of Statistics in Quality and Reliability*; American Cancer Society: Atlanta, GA, USA, 2008. Available online: https://onlinelibrary.wiley.com/doi/pdf/10.1002/9780470061572.eqr342 (accessed on 26 February 2019).

9. Trivedi, K.S. *Probability and Statistics with Reliability, Queuing and Computer Science Applications*; Wiley: New York, NY, USA, 2002; p. 830.

10. Chakravarthy, S.R.; Krishnamoorthy, A.; Ushakumari, P.V. A $(k - out - of - n)$ reliability system with an unreliable server and Phase type repairs and services: The (N, T) policy. *J. Appl. Math. Stoch. Anal.* **2001**, *14*, 361–380. Available online: https://cyberleninka.ru/article/n/reliability-of-a-k-out-of-n-system-with-repair-by-a-single-server-extending-service-to-external-customers-with-pre-emption (accessed on 26 February 2019).

11. Ushakov, I. An universal generating function. *Sov. J. Comput. Syst. Sci.* **1986**, *24*, 37–49.

12. Ushakov, I. Optimal standby problem and an universal generating function. *Sov. J. Comput. Syst. Sci.* **1987**, *25*, 61–73.

13. Levitin, G. *The Universal Generating Function in Reliability Analysis and Optimization*; Springer: London, UK, 2005; p. 442. [CrossRef]

14. Moustafa, M.S. Availability of *K*-out-of-*N*: *G* Systems with Exponential Failure and General Repairs. *Econ. Qual. Control* **2001**, *16*, 75–82. [CrossRef]

15. Linton, D.G.; Saw, J.G. Reliability analysis of the *k*-out-of-*n*: *F* system. *IEEE Trans. Reliab.* **1974**, *23*, 97–103. [CrossRef]

16. Rykov, V. On Reliability of Renewable Systems. In *Reliability Engineering. Theory and Applications*; Vonta, I., Ram, M., Eds.; CRC Press: Boca Raton, FL, USA, 2018; pp. 173–196.

17. Cox, D. The analysis of non-Markovian stochastic processes by the inclusion of supplementary variables. *Math. Proc. Camb. Philos. Soc.* **1955**, *51*, 433–441. [CrossRef]

18. Efrosinin, D.; Rykov, V. Sensitivity Analysis of Reliability Characteristics to the Shape of the Life and Repair Time Distributions. *Commun. Comput. Inf. Sci.* **2014**, *487*, 101–112._13. [CrossRef]

19. Rykov, V.; Itkin, V. On Sensitivity of Reliability Systems Operating in Random Environment to Shape of their Input Distributions. *Reliab. Theory Appl.* **2015**, *10*, 71–80. Available online: https://gnedenko.net/Journal/2015/042015/RTA_4_2015-07.pdf (accessed on 26 February 2019).

20. Zaripova, E.R.; Kozyrev, D.V.; Rykov, V.V. Modeling and simulation of reliability function of a homogeneous hot double redundant repairable system. In Proceedings of the 31st European Conference on Modeling and Simulation, Budapest, Hungary, 23–27 May 2017, ; pp. 701–705.

21. Rykov, V.; Zaripova, E.; Ivanova, N.; Shorgin, S. On Sensitivity Analysis of Steady State Probabilities of Double Redundant Renewable System with Marshal-Olkin Failure Model. In *Distributed Computer and Communication Networks*; Springer: Berlin/Heidelberg, Germany, 2018; pp. 234–245.

22. Rykov, V.; Kozyrev, D.; Filimonov, A.; Ivanova, N. On reliability function of a *k*-out-of-*n* system with general repair time distribution. *Probab. Eng. Inf. Sci.* **2020**, 1–18. [CrossRef]

23. Rykov, V.V.; Filimonov, A.M. Hyperbolic systems with multiple characteristic and applications. *UBS* **2020**, *85*, 72–86. [CrossRef]

24. Kierzenka, J.; Shampine, L.F. A BVP solver that controls residual and error. *Eur. Soc. Comput. Methods Sci. Eng. ESCMSE J. Numer. Anal. Ind. Appl. Math.* **2008**, *3*, 27–41.

25. Savonin, S.; Arsentieva, Z.; Moskalenko, A.; Chugunov, A.; Tunder, A. Analysis of the main causes of accidents that occurred on the main gas pipelines. *Eng. Prot.* **2015**, *11*, 52–57.

J. Mar. Sci. Eng. **2020**, *8*, 928

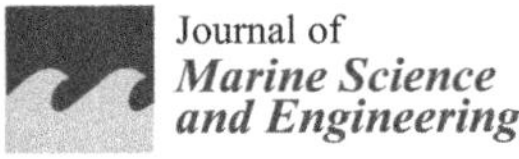

Journal of
Marine Science and Engineering

Article

Preventive Maintenance of a k-out-of-n System with Applications in Subsea Pipeline Monitoring

Vladimir Rykov [1], Olga Kochueva [1,*] and Mais Farkhadov [2,*]

[1] Department Applied Mathematics and Computer Modelling, National University of Oil and Gas "Gubkin University", 65, Leninsky Prospekt, 119991 Moscow, Russia; vladimir_rykov@mail.ru
[2] V. A. Trapeznikov Institute of Control Sciences of Russian Academy of Sciences, Profsoyuiznaya 65, 117997 Moscow, Russia
* Correspondence: olgakoch@mail.ru (O.K.); mais@ipu.ru (M.F.)

Abstract: Environmental safety issues are of particular importance when we design and operate underwater transport systems. To ensure the transport systems function safely, special systems to monitor their condition are being created. Underwater pipeline monitoring systems should continuously operate to detect and prevent emergency and pre-emergency situations in a timely manner. The purpose of this article is to demonstrate the possibility of using a mathematical model of a k-out-of-n system to support decision-making in the preventive maintenance of an unmanned underwater vehicle to monitor the condition of a subsea pipeline. The novelty and feature of this study are that we investigate a strategy of preventive maintenance for a model of a k-out-of-n system, where failures depend not only on the number but also on the location of the failed components in the system. The method to solve this problem, based on the distribution of the members of the variational series of the failing components, is also new. Since the distributions of the system component lifetimes are usually known with an accuracy of only one or two moments, we paid special attention to how sensitive the decision making about preventive maintenance is to the shape of the distributions. Numerical examples are conducted in order to support the theoretical investigations of the paper. The results of the study are applied to specific equipment to monitor the state of the outer surface of the pipeline.

Keywords: subsea pipeline monitoring; unmanned underwater vehicle; k-out-of-n system; preventive maintenance; reliability function; lifetime distribution

Citation: Rykov, V.; Kochueva, O.; Farkhadov, M. Preventive Maintenance of a k-out-of-n System with Applications in Subsea Pipeline Monitoring. *J. Mar. Sci. Eng.* **2021**, *9*, 85. https://doi.org/10.3390/jmse9010085

Received: 31 December 2020
Accepted: 13 January 2021
Published: 15 January 2021

Publisher's Note: MDPI stays neutral with regard to jurisdictional claims in published maps and institutional affiliations.

1. Introduction, Motivation and an Example

1.1. Introduction

Environmental safety issues are one of the main problems of humanity in recent times. Currently, with industrial development and the new challenges that appear in connection with this, the issues of preventive environmental protection are gaining great importance. Work in the field of underwater transportation of hazardous products (such as oil, gas, etc.) draws special attention to these problems. In this regard, projects in the field of construction and operation of underwater oil or gas transportation systems should be accompanied by continuous monitoring of their condition. For such monitoring, special systems and equipment have been developed. However, this equipment is also susceptible to failures, and special preventive maintenance (PM) procedures must be provided to keep it in proper operational condition.

This work is devoted to the development of a mathematical model to organize such maintenance of pipeline transport underwater monitoring equipment based on the k-out-of-n model. The novelty and features of this study are that the point of failure of the system depends on where the system's failing components are located, as well as in the study of the sensitivity of decision making on the type of distribution of their lifetime. This paper is to some extent related to the paper [1], also published in this issue, both of them focus on mathematical models that can contribute to safe, secure and sustainable pipeline transport offshore. However, this paper deals with another problem that can be solved with a k-out-of-n model—a reliability management problem—that is how to choose the best PM mode. So, we develop and investigate new methodology and provide an example of its real life application to subsea pipeline monitoring systems. Nevertheless, some parts of the introduction about k-out-of-n systems and the notations coincide with those in [1].

A k-out-of-n system is a system that contains n components in parallel and may be described in two ways, depending on the definition of the parameter k. The parameter k may represent the number of components in the system that must function for the entire system to work, referred to as a k-out-of-$n : G$ system. On the other hand, parameter k may represent the number of components in the system such that when the components fail, the entire system fails, referred to as a k-out-of-$n : F$ system [2]. Of course, these descriptions are closely connected and each of them is dual to another. Since for our aims it is more convenient to consider the subset of failed components of the system, in this paper we use the second type of system description.

Study of k-out-of-n systems reliability is interesting both from theoretical and practical points of view. From a theoretical point of view, it gives the wide possibility for new mathematical methods and applications. From a practical point of view, there are many investigations devoted to the reliability-centric analysis of k-out-of-n systems. The application of such models can be seen in many real-world phenomena, including telecommunication, transmission, transportation, manufacturing, and service applications. A probabilistic study of a real-world k-out-of-n system often helps to develop an optimal strategy to maintain high-level of system reliability. Thus, the theory of the k-out-of-n repairable systems is quite developed, but it does not cease to attract attention of researchers. Models are being developed taking into account various features of the systems. An important issue is to investigate the practical aspects of the application of these models. One of the applications of such kind of systems is described in the next subsection, where an automated system for remote monitoring of underwater sections of the "Dzhubga-Lazarevskoye-Sochi" gas pipeline is considered.

The paper is organized as follows. In the next subsection, an example of a real monitoring system, where we have to justify our choice of PM strategy, is presented. Further, this example will be used to demonstrate our theoretical research and proposed algorithms with the results of numerical calculations. A short literature review is given in Section 1.3. Section 2 contains the state of the problem and some notations. Section 3 presents general procedure of the PM quality calculation and gives an algorithm to solve the problem. Further, in Section 4, we consider the conditions to make PM efficient for the homogeneous system, where system failure does not depend on the failed components location. A model of PM organization for a system, where system failure depends on the location of its failed components is presented in Section 5. In the conclusion, further directions for research are proposed.

1.2. An Automated System for Remote Monitoring of Underwater Pipeline as an Example of k-out-of-n:F System

As an example of such kind of systems, we consider an automated system for remote monitoring of underwater sections of the "Dzhubga-Lazarevskoye -Sochi" gas pipeline [3]. The annual capacity of the pipeline is up to 3.78 billion cubic meters of gas. Estimated service life is 50 years. Total length is 171.6 km, where the offshore part of the gas pipeline accounts for some 90 per cent of the whole route length. The route runs some 4.5 km off the coast where the sea depth reaches 80 m. Linepipe diameter is 530 mm, wall thickness—15 mm for the offshore and 11.3 mm for the onshore part of the gas pipeline, material—high strength steel.

Figure 1 shows the general concept of an automated system to remotely inspect of the offshore section of the gas pipeline using an unmanned underwater vehicle (UUV).

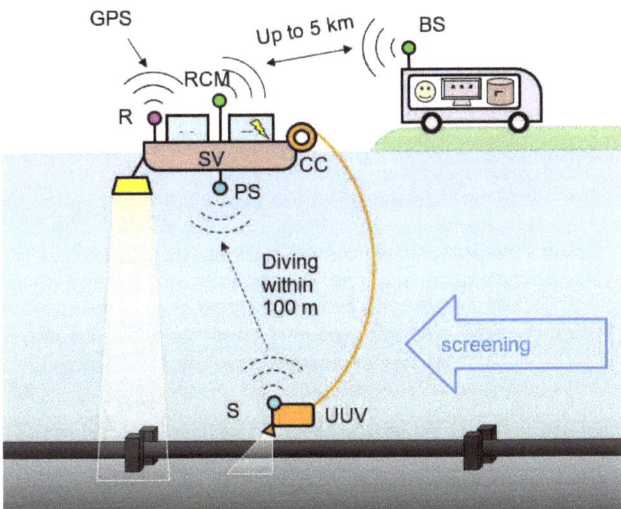

Figure 1. "Dzhubga-Lazarevskoye-Sochi" gas pipeline.

The purpose of the survey is to remotely conduct an automated set of measures to collect diagnostic data on the external state and the surroundings of an offshore gas pipeline. The monitoring systems can detect defects (such as damage) to the external coating, indentations and cracks on the pipe joints, erosion of the seabed, the presence of suspicious objects near the pipeline, and more. The functional diagram of the proposed automated remote survey system assumes there is an accompanying surface vessel (SV) [4], floating on the surface along the gas pipeline; the vessel also carries the UUV. The remotely controlled unmanned underwater vehicle "Vodyanoy-1" (our custom development) with the following functional modules (see Figure 2) is used as the UUV.

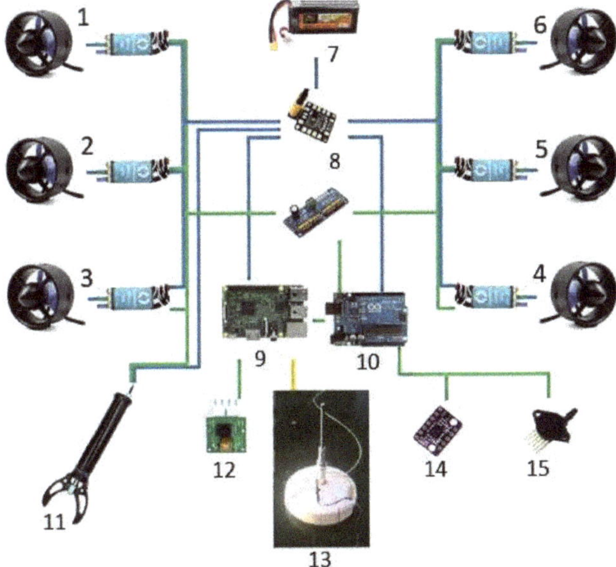

Figure 2. An unmanned multi-functional underwater vehicle.

In Figure 2 the numbers indicate parts of the UUV as given below: 1–6 are the Motor drivers and T200 Thruster, 7 is the LiPo Battery, 8 is the Matek PDB-xt60, 9 is the Raspberry Pi, 10 is the Atmega328P, 11 is the Gripper, 12 is the Front NoIR Camera, for example, Basler camera acA1920-40gc (GigE interface, Sony IMX249 CMOS sensor, 42 fps @ 2.3 Mpix), 13 is the Wi-Fi Beacon, 14 is the IMU 9dof, 15 is the MPX5700. In addition, the UUV has the following attachable functional modules: overview sonar BlueView (to reconstruct 3D underwater scenes, avoid collisions, improve navigation accuracy), miniSVP sound velocity profile meter (to correct calculations). Various technical means and equipment for underwater robotic systems are given on site [5].

The paper [6] introduces a remotely operated underwater smart vehicle. The article demonstrates that smart features can be added to a dumb analogue remotely operated underwater drone by a small team of engineers on tight budget. This UUV maintains compass and depth headings, records video to an onshore terminal. In addition, vehicle has a remotely operated arm with 3 degrees of freedom. UUV "Malakhit" won the first prize on AquaRoboTech 2018 and remotely operated UUV "Vodyanoy-1" won the first AquaRoboTech 2020 competition and has a potential to be upgraded with advanced machine vision algorithms [7].

The article [8] examines the main stages of visual odometry in order to identify the factors that affect the quality of motion assessment, and establish the degree of their importance.

The main functionality of the software simulator developed of the visual odometry system are described. The paper presents the results of experiments conducted on the simulator, and infer certain conclusions out of them.

The functions between the SV (unmanned catamaran OceanAlpha M40 [7]) and the UUV are distributed as follows.

The SV provides:

- power supply for all equipment;
- scanning the bottom topography using the hydroecholocation system;
- global positioning receiver GNSS-H;
- local underwater positioning system (PS);
- wire communication via cable (CC) on an electric winch;
- wireless communication of the module via radio channel with the base station (BS) with Directional antenna, Network hardware;
- receiving and processing control commands.

The underwater vehicle can film underwater objects. Based on the data from the transponder (S), the positioning system calculates the vehicle's coordinates and transmits them via a cable to the surface vehicle to provide autonomous underwater navigation. To improve the navigation accuracy of the underwater vehicle and to perform work on the 3D reconstruction of underwater scenes, in addition to a video camera, it is necessary to install an overview sonar.

The mobile operator station consists of a radio system, an operator's workstation, and a server for recording and processing database data.

The inspection procedure is as follows.

The system receives control commands from the operator to launch certain scenarios. The scenarios are followed automatically, while the operator controls the process and only intervenes when anomalies are detected in the inspected object, or in the event of emergency situations in the system.

Let us highlight the following two basic scenarios to conduct surveys.

1. Continuous. The underwater vehicle dives at the starting point. The SV begins to continuously move along the survey vector, scanning the bottom relief and the gas pipeline, while the UUV is simultaneously sailing behind it and filming the situation. Having reached the endpoint, the UUV ascends.

2. Localized. The SV stops at a given fix. The UUV begins to sequentially dive, survey the surroundings, and ascend. Then the SV continues to move to the next checkpoint.

The operator's mobile station can be additionally equipped with a software analytic system to automatically process the received data in real time to promptly adjust the survey control process.

The UUV carries out a set of measures to externally inspect the offshore section of the gas pipeline to determine its technical condition to detect defects and provide data to subsequently analyze the causes of defects and assess the technical condition of the gas pipeline and its surroundings. This procedure places high demands on the quality, reliability and uninterrupted operation of all components of the integrated automated system.

These requirements are especially stringent to one of the vulnerable components of the complex technology, namely for a remotely operated UUV. Therefore, an important and urgent problem is to assess the reliability characteristics of the underwater vehicle using advanced mathematical models.

The UUV can perform its functions as long as at least two engines located on opposite sides or any three engines are operational. Therefore, based on our agreement, the UUV can be considered as a k-out-of-$n : F$ system. The system consists of $n = 6$ components, its failure depends on the position of its failed components, thus, it could be considered as a combination of $3 + 1$-out-of-6 : F and 5-out-of-6 : F systems. For such a system we will use special notation such as $(5, 3 + 1)$-out-of-6 : F system.

1.3. Literature Review

Due to the wide practical application area, a lot of papers are devoted to the study of k-out-of-$n-$ systems. The literature on such studies is vast and has been reviewed in [1] that is published in this issue. Thus, we will not repeat the review here, and because the paper is devoted to the problems of PM organization for the k-out-of-n models, we focus only on some works devoted to PM problems.

The idea to increase the system reliability by organizing the PM has a long story. A fairly detailed review of PM methods one can find in the monograph of Gertsbakh [9]. Some investigations of the k-out-of-n repairable systems with different strategies of repair and additional services have been considered in a series of work of Dudin, Krishnamurthy, and all [10–15]. Some recent developments on optimal maintenance policies can be found in [16–20].

Since detailed initial information about the reliability of system components is usually not available, it is fundamentally important to study the sensitivity of the system reliability indicators to the shape of system components lifetime distributions. Some research in this direction one can find in [21], in chapter 9 of [22], as well as in [23,24].

In this paper, we study and compare the effectiveness of different PM strategies for k-out-of-$n : F$ systems based on observation for their states. A feature of the model under consideration is the dependence of the system failure on the location of its failing components.

2. The Problem Set and Notations

2.1. The Notations and Assumptions

Consider a heterogeneous k-out-of-$n{:}F$ system that is described in the Introduction. Denote by $A_i : i = 1, 2, \ldots$ the sequence of the system components random lifetimes. Suppose that they are independent identically distribute (i.i.d.) random variables (r.v.'s) with the cumulative distribution function (c.d.f) $A(t) = \mathbf{P}\{A_i \le t\}$ the same for all of them. After any system failure it is repaired with a single facility and the repair times are i.i.d. r.v.'s $B_i^{(0)} : i = 1, 2, \ldots$ with common c.d.f. $B_0(t) = \mathbf{P}\{B_i^{(0)} \le t\}$ and mean value

$$b_0 = \mathbf{E}[B_0] = \int_0^\infty (1 - B_0(t))dt.$$

To increase the reliability of the system, the possibility of PM, based on the system states observation, is assumed. Let $\mathcal{L} = \{0, 1, \ldots, L\}$ be a set of possible PM strategies including running to the system failure for $l = 0$. For the l-th strategy denote by E_l the system "pre-failure" subset of states, where l-th type of maintenance begins. The times of PM are i.i.d. r.v.'s $B_i^{(l)}$ with c.d.f. $B_l(t) = \mathbf{P}\{B_i^{(l)} \le t\}$ and mean value

$$b_l = \mathbf{E}[B_l] = \int_0^\infty (1 - B_l(t))dt.$$

The mean PM time b_l is supposed to be less than the mean repair time b_0, $b_l \le b_0$, but may depend or not on the type of maintenance.

For investigation of reliability of the complex system, where failure depends not only on the number of its failed components but also on their position in the system, let us denote the system state by $\mathbf{j} = (j_1, j_2, \ldots j_n)$, where $j_i = 1$, if the i-th component is in "DOWN" state, and $j_i = 0$, if the i-th component is in "UP" state. Thus, $j = j_1 + \cdots + j_n$ means number of failed components of the system. Let's denote also

$$E = \{\mathbf{j} = (j_1, j_2, \ldots j_n) : (j_i \in (0, 1))\}$$

to be the system set of states and by E_0 and $\bar{E}_0$ subsets of its "DOWN"'and 'UP" states accordingly. Note that the description of these sets is a special problem of concrete applications and should be considered for any special case.

It is supposed that

- in the very beginning the system is absolutely reliable, i.e. it is in zero state $\mathbf{j} = (0, \ldots, 0)$;
- all sequences of r.v.'s (components lifetimes, repair, and PM times) are i.i.d. for each type of r.v.'s;
- after any repair and PM completion the system becomes "as a new one", i.e., goes to the zero state[1].

2.2. The Problem Set

The paper's aim is to compare different PM strategies $l \in \mathcal{L}$ (including running to the system failure for $l = 0$) with respect to some criterion. As a criterion of the PM quality the system availability $K_{av.,l}$ for different PM strategies $l \in \mathcal{L}$ is considered[2],

$$K_{av.,l} = \lim_{t \to \infty} \frac{1}{t}\{\text{the system working time during time } t \text{ under strategy } l \}.$$

For the stated problem solution, let us define a random process $J = \{\mathbf{J}(t) : t \ge 0\}$ with the set of space E by the relation

$$\mathbf{J}(t) = \mathbf{j}, \ \text{ if at time } t \text{ system is in the state } \mathbf{j} \in E$$

and denote by S_l $(l \in \mathcal{L})$ time to the subset E_l destination,

$$S_l = \inf\{t : \mathbf{J}(t) \in E_l\}.$$

Thus, the value S_0 represents lifetime of the system and S_l $(l = \overline{1, L})$—the time till the l-th type maintenance beginning.

In the paper, we are interested to calculate different characteristics of the system, such as

---

[1] The assumption that the system returns to its original state is simplifying, it does not fully correspond to the real situation, however, most studies of real systems are based on this assumption.

[2] Another quality criteria also possible, such as productivity of the system and/or system service cost under different maintenance strategies etc.

- the system reliability function

$$R(t) = \mathbf{P}\{S_0 > t\} \quad \text{and its mean value } M_0 = \int\limits_0^\infty R(t)dt. \tag{1}$$

- distributions of time before starting different maintenance and their mean values

$$F_l(t) = \mathbf{P}\{S_l \le t\}, \quad M_l = \int\limits_0^\infty (1 - F_l(t))dt. \tag{2}$$

- the system availability $K_{\mathrm{av.},l}$ for different PM strategies $l \in \mathcal{L}$.

Because the initial information about system components lifetime is usually very limited and available only up to one or two moments, we focus on the study of how sensitive is a decision on the PM quality to the shape of their distributions.

3. Process J and the General Procedure of the PM Quality Calculation

3.1. Process J

Note first of all that due to our assumptions under any PM strategy, including running to the system failure, the process **J** is a regenerative one, where regenerative epochs are the times of maintenance or repair ends. Denote by Π_0 and Π_l the process regeneration periods for the cases of the system working up to failure (for $l = 0$) or under the l-th type $l = \overline{1, L}$ of maintenance. Thus, the system availability is

$$K_{\mathrm{av.},0} = \frac{\mathbf{E}[S_0]}{\mathbf{E}[\Pi_0]}, \quad K_{\mathrm{av.},l} = \frac{\mathbf{E}[S_l]}{\mathbf{E}[\Pi_l]}. \tag{3}$$

Therefore due to the properties of regenerative processes for availability $K_{\mathrm{av.},l}$ calculation, we need only the mean value $\mathbf{E}[\Pi_l]$ of the regeneration period Π_l and the mean value $M_l = \mathbf{E}[S_l]$ of the working time S_l in it. Since for any PM strategy $l \in \mathcal{L}$, the regeneration period equals to

$$\Pi_l = S_l + B_l,$$

and the mean repair and PM times $b_l = \mathbf{E}[B_l]$ are supposed to be known and measured in the same scale, for the problem solution we need only to calculate the distributions (or only the mean values) of the system working times S_l for the case when it works to failure (for $l = 0$) and for a system that operates under the l-th maintenance strategy.

3.2. The General Procedure of the PM Quality Calculation

For the solution of the declared problem, we need to calculate system availability $K_{\mathrm{av.},l}$ for different preventive maintenance strategies $l \in \mathcal{L}$, including running to the system failure (for $l = 0$). To do that we have to calculate c.d.f.'s (1, 2) of the subsets E_l $(l \in \mathcal{L})$ destination times. Note that the time S_l $(l \in \mathcal{L})$ of the subsets E_l destinations coincides with the corresponding member of the variation series of the failure epochs of the system components. Therefore, for the solution of the stated problem, the following general algorithm should be used.

Remark 1. *The Algorithm 1 can also be used to solve other different problems, for example, to analyze if the preference of one strategy over another is sensitive to the shape of the system components lifetime distributions.*

Further, the Algorithm 1 will be applied to several examples.

4. Homogeneous System Preventive Maintenance

4.1. Preliminary

Consider firstly a homogeneous *k*-out-of-*n:F* system, where failure does not depend on the configuration of its failed components. The subsets of UP and DOWN states of the system in this case are:

$$\bar{E}_0 = \{0, 1, 2, \ldots, k-1\} \quad E_0 = \{k, k+1, \ldots, n\}.$$

Let us define the subset E_l for the PM $l \in \mathcal{L}$ beginning as a single state $E_l = \{l\}$ with $l \le k - 1$. Thus, in this case we can investigate k strategies $l = \{0, 1, \ldots, k-1\}$, where 0-strategy means allow the system to operate up to its failure.

In this case, the general Algorithm 1 gets essentially simpler because the time to the subset E_l destination coincides with respective member $A_{(l)}$ of the variation series of the times to the system components failures $A_i : (i = \overline{1, n})$.

The analytical expressions for mean values M_l are not always accessible. However, their numerical calculation in accordance with the Algorithm 1 is not too difficult and it will be proposed in the next subsection for the case of a k-out-of-$n : F$ system for $k = 4$, $n = 6$.

4.2. Numerical Results

To obtain the concrete results apply our Algorithm 1 to the 4-out-of-6 : *F*-system that can be considered as a model for the example of the Section 1.2. In this case, only four strategies of system control are possible.

- Strategy 0 is that the system operates up to its failure.
- Strategy l ($l = 1, 2, 3$) is to begin the PM when the system occurs in the state l.

In order to compare Strategy l with the Strategy 0 (to work without any PM up to the system failure), we need to know the ratio $\frac{b_l}{b_0}$. It is supposed that the values of mean repair and PM times b_0, b_l as well as their ratios are known to a DM. Therefore, to make a decision about the preference of one strategy over the other, one only needs to know the ratios of the mean working time of the system $\frac{M_l}{M_0}$ for them. Let's consider the respective ratios for an exponential distribution and compare different strategies of PM $l = \overline{1, L}$ with the strategy to work up to full system failure $l = 0$ for different distributions and variations.

To conduct a numerical experiment, the program code on the MATLAB platform is generated. This computing environment is chosen due to a wide range of built-in functions, covering cdf functions and numerically evaluation of the integrals, including improper integrals. MATLAB additional advantage are simple plotting functions and friendly interface. The figures and the tables shown below in the article are the output of the developed program.

In order to investigate the sensitivity of the PM quality to the shape of the distribution of system components lifetime in numerical experiments, four types of distributions: exponential with parameter α, $Exp(\alpha)$, Gamma distribution, $\Gamma(\Theta, k)$, Gnedenko-Weibull distribution, $GW(\lambda, k)$ and log-normal distribution $LN(\mu, \sigma^2)$ are considered.

The parameters of all distributions in experiments are chosen such that their expectations coincide for different distributions and equal to 1 (it means that we scaled it with respect to mean components lifetime), while the coefficient of variation $c = \frac{\sigma}{\mu}$ is varied in the interval $c \in [0.3, 5.0]$.

The results of the experiments are presented in Figure 3 and Tables 1 and 2. In Figure 3 the ratios of mean working times M_l under different PM strategies $l = 1, 2, 3$ to mean working time M_0 for the system operating up to its failure ($l = 0$) for different distributions of components lifetime versus the coefficient of variation are shown. Bold dashed horizontal lines correspond to the ratios of the mean PM time b_l for any strategy l to mean repair time b_0. Intersections of these lines with the curves M_l / M_0 for different distributions determine the boundary values c^* of the coefficient of variation, where the preference of appropriate strategy l is changed to the preference for "the system working up to the failure". If the coefficient of variation exceeds the boundary value $c > c^*$ "the system working up to the failure" strategy ($l = 0$) is preferable for given distribution, otherwise the appropriate PM strategy should be used.

Algorithm 1: General algorithm to choose a PM strategy

Start. Determine: Integers n, k, distribution $A(t)$ of components lifetime, subsets E_l ($l \in \mathcal{L}$) for the PM beginnings or of the system failure for $l = 0$, mean PM and system repair times b_l ($l \in \mathcal{L}$).

Step 1. Describe the duration of the subsets E_l ($l \in \mathcal{L}$) destinations in terms of members of the variation series,

$$A_{(1)}, \ldots, A_{(j)}, \ldots, A_{(n)}$$

of the system components failure times (i.i.d r.v.) $A_1, \ldots, A_j, \ldots, A_n$ that bring the system to the subset E_l.

Step 2. Calculate distributions of the respective members of the variation series

$$A_{(j)}(t) = \mathbf{P}\{A_{(j)} \leq t\} = \sum_{j \leq i \leq n} \binom{n}{i} A^i(t)(1 - A(t))^{n-i}. \tag{4}$$

Step 3. Calculate the distributions $F_l(t)$ of the subsets E_l destination times in terms of distributions of respective series members and their expectations,

$$M_l = \mathbf{E}[S_l] = \int_0^\infty (1 - F_{(l)}(t))dt.$$

Step 4. Compare different PM strategies with respect to maximizing the system availability given by (3). We know that in terms of mean times to the destination of the respective subsets, the system availability can be represented as

$$K_{\mathrm{av.},l} = \frac{M_l}{M_l + b_l}$$

and because the l-th PM strategy is preferred over the j-th one if $K_{\mathrm{av.},l} > K_{\mathrm{av.},j}$ from the inequality

$$\frac{M_l}{M_l + b_l} > \frac{M_j}{M_j + b_j}$$

it follows that the l-th strategy is preferred over the j-th one ($l \succeq j$) if and only if

$$M_l b_j > M_j b_l, \quad \text{or in terms of dimensionless indexes} \quad \frac{b_l}{b_j} < \frac{M_l}{M_j}. \tag{5}$$

Step 5. Print results in terms of mean operational times M_l ($l \in \mathcal{L}$) and their ratio $\frac{M_j}{M_l}$ as advice to a Decision Maker (DM) in order to choose the best strategy accordingly to inequality (5).

Stop.

The boundary values c^* of the coefficient of variation when the PM strategy $l = 1, 2, 3$ is preferable to strategy $l = 0$ (running to the system failure) for two values of the ratios mean PM duration b_l to mean repair time $b_l/b_0 = 0.5$ (upper horizontal line in Figure 3) and $b_l/b_0 = 0.2$ (lower horizontal line in Figure 3) for different distributions of system components lifetime are represented in the Table 1.

The Figure 3 demonstrates an almost evident fact that the highest value of the ratios $\frac{M_l}{M_0}$ is achieved for $l = 3$, which means that the strategy $l = 3$ is preferable over other strategies in the case when all mean PM times are equal, $b_l = b$ for $l = 1, 2, 3$. Moreover this strategy will be better than the strategy $l = 0$ "the system runs to failure" until the coefficient of variation is less than the boundary value c^* for any specified ratio $\frac{b}{b_0}$. If the coefficient of variation $c > c^*$ strategy $l = 0$ will be preferable to all others.

Table 1. Boundary values for the coefficient of variation c^* for 4-out-of-6 : F system.

Distribution	$b_l/b_0 = 0.5$			$b_l/b_0 = 0.2$		
	$l = 1$	$l = 2$	$l = 3$	$l = 1$	$l = 2$	$l = 3$
Γ- distribution	0.44	0.76	1.51	0.93	1.57	3.30
GW-distribution	0.38	0.71	1.76	0.91	1.96	> 5
Log-normal distribution	0.51	0.99	4.08	1.66	> 5	> 5

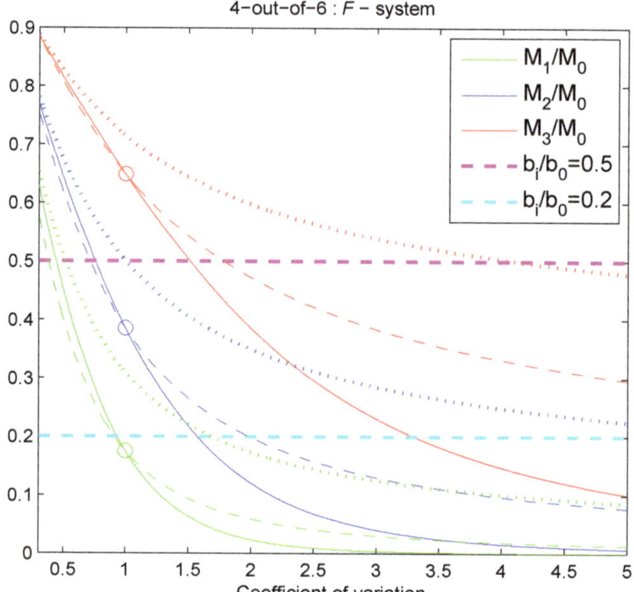

Figure 3. The dependence of the ratios M_l / M_0 for different distributions of system components lifetime versus their coefficient of variation for 4-out-of-6 : F system. Solid lines—Γ-distribution, dashed lines—GW-distribution, dotted lines—log-normal distribution, circles—exponential distribution.

However, depending on the coefficient of variation, the decision about the choice of the PM is sensitive to the distribution of system components lifetime. With an increase in the coefficient of variation, the ratio M_l / M_0 decreases, but the difference between distributions grows.

Suppose the repair time is twice longer than PM time (the violet line in Figure 3). Assuming an exponential distribution of components lifetime, $l = 3$ is preferable for 4-out-of-6 : F system. The decision for other distributions of components lifetime depends on the coefficient of variation. If $c > 1.51$ for Γ-distribution or $c > 1.76$ for GW-distribution, the strategy $l = 0$ should be chosen. In case components lifetime follows a log-normal distribution, the strategy $l = 0$ should be chosen if $c > 4.08$.

If the repair time is five times longer than PM time (the cyan line in Figure 3) and the coefficient of variation is no greater than 3.3, the strategy $l = 3$ will be the best regardless of the type of components lifetime distribution. However, as it is possible to see from Figure 3, for $c > 3.3$ the choice of the strategy significantly depends on the components lifetime distribution.

In case of different mean PM times b_l, strategies $l = 1, 2, 3$ can be compared one with another. We compare the strategies $l = 2$ and $l = 3$ for the studied distributions, and present the results in Table 2. The table provides the boundary values for the coefficient of variation c^{**}, where the preference for the strategy $l = 2$ is changed to the preference for the strategy $l = 3$ for $b_2/b_3 = 0.5$ or $b_2/b_3 = 0.2$. If the coefficient of variation exceeds the boundary value $c > c^{**}$ the strategy $l = 3$ "to start PM after 3 components failure" for given distribution is preferable. Otherwise the strategy $l = 2$ "to start PM after 2 components failure" should be used. So according to the Table 2 if the PM time for $l = 3$ is twice as much as the PM time for $l = 2$ and Γ-distribution with the coefficient of variation $c > 1.275$ is taken, the strategy $l = 3$ is better than the strategy $l = 2$. The same conclusion can be made if components lifetime follows GW-distribution and $c > 1.427$, and for log-normal distribution if $c > 3.771$. If the PM time for $l = 3$ is five times longer than the PM time for $l = 2$ assuming GW- or log-normal distribution, the strategy $l = 2$ is preferable for the whole interval $c \in [0.3, 5.0]$, and for a Γ- distribution it will be the best choice if $c < 2.749$.

Table 2. Boundary values for the coefficient of variation c^{**} to compare strategies $l = 2$ and $l = 3$ for 4-out-of-6 : F system.

Distribution	$b_l/b_0 = 0.5$	$b_l/b_0 = 0.2$
Γ-distribution	1.275	2.749
GW-distribution	1.427	>5
Log-normal distribution	3.771	>5

Thus, the conducted study allows us to draw the following conclusions regarding the sensitivity of the decision on the choice of a PM strategy to the type of distributions of the system components lifetime and the coefficient of variation.

At low values of the coefficient of variation, strategies with PM gain an advantage over the strategy of working until complete system failure and repair (in terms of a higher value of the availability factor). It is possible to distinguish intervals of values of the coefficient of variation where the conclusion about the use of the certain strategy will be the same for any of the considered distributions.

Γ-distribution and GW-distribution have insignificant differences in M_l/M_0 when the coefficient of variation is in the interval $c \in [0.3, 1.1]$, the log-normal distribution tends to widen the range of the coefficient of variation values when the strategy $l = 3$ can be adopted as the preferred strategy in comparison to $l = 0$.

Calculated with Matlab, the system mean working times under different PM strategies and the mean system operational time until its failure (for $l = 0$) for the special case of the exponential distribution of components lifetime with parameter $\alpha = 1$ are: $M_1 = 0.1667$, $M_2 = 0.3667$, $M_3 = 0.6167$, $M_0 = M_4 = 0.95$. Corresponding ratios are

$$\frac{M_1}{M_0} = 0.1754, \quad \frac{M_2}{M_0} = 0.386, \quad \frac{M_3}{M_0} = 0.649.$$

These values, calculated for the exponential distribution, are marked with circles in Figure 3. Consequently, the strategy $l = 2$ is better than $l = 0$ if the system repair time b_0 is $\frac{1}{0.386} \approx 2.6$ or more times longer than the mean PM time b_2. The ratios to compare strategies $l = 1$ and $l = 2$ or $l = 2$ and $l = 3$ are also shown below

$$\frac{M_1}{M_2} = 0.454, \quad \frac{M_2}{M_3} = 0.595.$$

The strategy $l = 2$ is better than $l = 1$ if the mean PM time b_1 is not less than 45.4% of the mean PM time b_2.

However, due to the properties of an exponential distribution, this case can be studied analytically, as it is shown in the next subsection.

4.3. Special Case: Exponential Distribution of Components Lifetime

Under the assumption about an exponential distribution of the lifetimes of the system components A_i,

$$A(t) = \mathbf{P}\{A_i \leq x\} = 1 - e^{-\alpha t},$$

due to the independence of the residual lifetimes of all other components on the failure time of any one of them, another approach is possible. In this case, the intervals T_i between the $i-1$-th and the i-th failures are

$$T_i = \min\{A_1, A_2, \ldots A_{6-(i-1)}\},$$

and therefore

$$F_i(t) = \mathbf{P}\{T_i \leq t\} = 1 - \mathbf{P}\{T_i > t\} = 1 - (1 - A(x))^{6-i} = 1 - e^{(6-(i-1))\alpha t}.$$

Thus, it holds $m_i = \mathbf{E}[T_i] = [(6 - (i-1))\alpha]^{-1}$ and therefore

$$\frac{M_1}{M_2} = \frac{m_1}{m_1 + m_2} = \frac{\frac{1}{6\alpha}}{\frac{1}{6\alpha} + \frac{1}{5\alpha}} = \frac{5}{11} \approx 0.4545.$$

From here it follows that the necessary and sufficient condition (5) for the first PM to be preferable over the second one is

$$\frac{b_1}{b_2} > \frac{M_1}{M_2} \approx 0.4545,$$

or the mean time of the second strategy PM should be more than twice longer than the relevant value for the first one. Analogously the inequality

$$\frac{b_2}{b_0} > \frac{M_2}{M_4} = \frac{m_1 + m_2}{m_1 + m_2 + m_3 + m_4} = \frac{\frac{1}{6\alpha} + \frac{1}{5\alpha}}{\frac{1}{6\alpha} + \frac{1}{5\alpha} + \frac{1}{4\alpha} + \frac{1}{3\alpha}} = \frac{22}{57} \approx 0.386$$

shows that the second PM strategy is preferable to the system running to its failure with the following repair only if its mean PM time is less than 38% of the mean repair time. The results of this section are in complete agreement with the numerical calculations given for exponential distribution in Section 4.2.

5. Preventive Maintenance of a System, Where Failures Depend on the Location of the Failed Components

5.1. Preliminary

If the system failure depends on the location of the failed components, the comparison of strategies, including "running to the system failure", and the decision about the choice of PM are system-specific and depend on the exploitation conditions. Thus, it is impossible to solve these problems in general settings. Therefore, in this section we consider this problem for the concrete $(3+1,5)$-out-of-6 : F system with specific conditions of its exploitation.

5.2. Example: Model $(3+1,5)$-out-of-6

Turn back to the investigation of the model that has been proposed in Section 1.2 under the condition that the system fails, when four (moreover three from one side and one from the other side) or five motors fail. In other words, it means that the system operates if any three or at least one from one side and one from the other side of its motors operate. Thus, this system could be considered as a combination of $3+1$-out-of-6 : F and 5-out-of-6 : F systems. For simplicity in Section 1.2 for such kind of systems a special notation $(3+1,5)$-out-of-6 : F system is proposed.

For the convenience, a binary code is used to indicate system states, namely the number of the state $\mathbf{j} = (j_1, j_2, \ldots j_6)$ is given in accordance with the formula

$$j = |\mathbf{j}| = \sum_{0 \leq i \leq 6} j_i 2^{6-i}.$$

Then the subset of failure states E_0 includes the states with the numbers

$$E_0 = \{\mathbf{15}, \mathbf{23}, 31, \mathbf{39}, 47, 55, \mathbf{57}, \mathbf{58}, 59, \mathbf{60}, 61, 62, 63\},$$

where the states with 3 failures on the same side and 1 on the other are highlighted in bold. By analogy with how it is defined in Section 4.2 consider four strategies:

- Strategy 0 is to run to the system failure (do not use any PM). It means that the repair begins when 4 failures occur at that 3 of them on one side and one on the other or 5 failures occur. The subset of the states for the repair beginning is E_0.
- Strategy l $(l = 1, 2, 3)$ is to begin the PM after the failure of any l components.

In this case the ordinal statistics do not determine uniquely the distribution time to the corresponding subset of states destination. Thus, the Algorithm 1 takes the following form.

5.3. Numerical Analysis

The results of the numerical experiments performed in accordance with Algorithm 2 are presented in Figure 4 and Table 3. As in the previous case (Section 4.2) in Figure 4 the ratios $\frac{M_l}{M_0}$ of mean system working times M_l under different strategies $l = 1, 2, 3$ to mean system working time M_0 up to its failure ($l = 0$) for different distributions of system components lifetime versus the coefficient of variation for $(3 + 1, 5)$-out-of-6 : F system are given. Four failure distributions: Γ-distribution, GW-distribution, log-normal distribution, and exponential distribution are examined. Bold dashed horizontal lines correspond to the ratios of the mean PM time b_l for any PM strategy l to the mean repair time b_0. The intersections of these lines with curves M_l / M_0 for different distributions determine the boundary values c^* of the coefficient of variation, where the preference of appropriate strategy l is changed to the preference for the strategy "the system running to the failure" ($l = 0$). If the coefficient of variation exceeds the boundary value $c > c^*$, the strategy $l = 0$ for given distribution is the preferable one. Otherwise appropriate PM strategy l should be chosen. The boundary values c^* of the coefficient of variation when PM strategy $l = 1, 2, 3$ is preferable to strategy $l = 0$ for $b_l / b_0 = 0.5$ (upper horizontal line in Figure 4) and $b_l / b_0 = 0.2$ (lower horizontal line in Figure 4) for the two values $b_l / b_0 = 0.5$ and $b_l / b_0 = 0.2$ numerically calculated for different distributions of system components lifetime are represented in Table 3.

Table 3. Boundary values for the coefficient of variation c^* for $(3 + 1, 5)$-out-of-6 : F system.

Distribution	$b_l/b_0 = 0.5$			$b_l/b_0 = 0.2$		
	$l = 1$	$l = 2$	$l = 3$	$l = 1$	$l = 2$	$l = 3$
Γ-distribution	0.37	0.6	0.98	0.82	1.26	2.05
GW-distribution	0.34	0.56	0.98	0.78	1.36	2.85
Log-normal distribution	0.42	0.68	1.27	1.2	2.49	> 5

Figure 4 shows that the difference between distributions grows as the coefficient of variation increases. So, depending on the coefficient of variation, the decision about the choice of the PM is sensitive to the distribution of system components lifetime.

Since collection of data on real equipment failures takes considerable time and the confidence intervals for the mathematical expectation and standard deviation of the investigated random variables can be wide enough, it makes sense when choosing the best strategy to focus not on a specific value of the coefficient of variation, but on the range $c \in [c_{min}, c_{max}]$, and make a decision on the choice of a strategy on the assumption that the coefficient of variation can be in the specified range. For example, with the ratio of PM time to repair time $b_l/b_0 = 0.2$ if the variation coefficient is supposed to be in the interval $c \in [0.3, 2.05]$, the strategy $l = 3$ will be the best for any of the considered distributions. For $c \in [2.05, 2.85]$ for a Γ-distribution of the system components lifetime, one should prefer the strategy of operation until complete failure $l = 0$, for a GW- and a log-normal distributions strategy $l = 3$ will remain the best. For $c \in [2.85, 5]$ the strategy $l = 0$ should be chosen for a Γ- and a GW- distributions. For a log-normal distribution the choice is the strategy $l = 3$ regardless the coefficient of variation. So the choice of the strategy significantly depends on the distribution of components lifetime.

Algorithm 2: The choice of a PM strategy for a heterogeneous system

---

Start is repeated from the Algorithm 1.

Step 1. Determine the times to set of states E_l destination in terms of ordinal statistics $A_{(j)}$. Because the system failure occurs when 4 (3+1) or 5 motors fail, so the time S_0 to the subset E_0 destination has the following form

$$S_0 = \begin{cases} A_{(4)}, \text{ if four motors fail, at that three from one side and one from} \\ \text{the other side (3+1),} \\ A_{(5)}, \text{ if four motors fail, at that two from one side and two from the} \\ \text{other side (2+2), and the system failure occurs after the fifth failure.} \end{cases}$$

The times S_l to the subsets E_l $(l = 1, 2, 3)$ destination coincide with the relevant variation series members, namely:

$$S_l = A_{(l)}, \text{ the PM under strategy } l \text{ begins after the failure of } l \text{ motors.}$$

Step 2. Calculate the distributions of times to the corresponding subsets destination in terms of the ordinal statistics distributions. Since the system failure occurs when (3+1) or 5 motors fail, and (3+1) failures state contains 6 of the 15 states from the complete subset E_4 of states with 4 failures, taking into account that the probabilities of any component failures are equal, so the probability of time to the destination of subset E_0 has a form

$$F_0(t) = \frac{2}{5} A_{(4)}(t) + \frac{3}{5} A_{(5)}(t).$$

The distributions of the subset of states E_l $(l = 1, 2, 3)$ destination are

$$F_1(t) = A_{(1)}(t), \quad F_2(t) = A_{(2)}(t), \quad F_3(t) = A_{(3)}(t),$$

where accordingly to (4) distributions $A_{(j)}(t)$ are

$$A_{(j)}(x) = P\{X_{(j)} \leq x\} = \sum_{j \leq i \leq n} \binom{n}{i} A^i(x)(1 - A(x))^{n-i}.$$

Step 3. Calculate the expectations times to destinations of the subsets E_l.

$$M_0 = \frac{2}{5} E[A_{(4)}] + \frac{3}{5} E[A_{(5)}], \quad M_1 = E[A_{(1)}], \quad M_2 = E[A_{(2)}], \quad M_3 = E[A_{(3)}].$$

Step 4. With the help of obtained values compare different PM strategies using the necessary and sufficient condition to prefer the j-th strategy over the l-th one in the form of inequality (5).

Step 5. Print results in terms of mean operational times M_l $(l \in \mathcal{L})$ and their ratio $\frac{M_j}{M_l}$ as advice to a DM in order to choose the best strategy.

Stop.

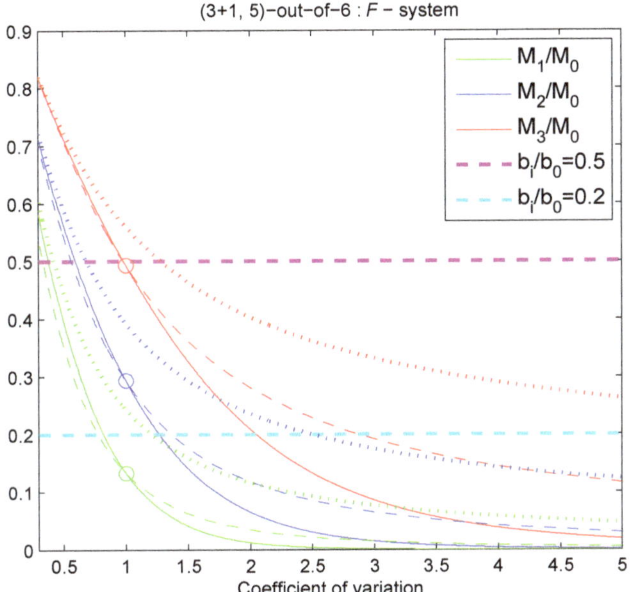

Figure 4. The dependence of the ratios M_l / M_0 for different distributions of system components lifetime versus their coefficient of variation for $(3 + 1, 5)$-out-of-6 : F system. Solid lines—Γ-distribution, dashed lines—GW-distribution, dotted lines—log-normal distribution, circles—exponential distribution.

5.4. Special Case

In the special case of exponential system components lifetime distribution, it is possible to use the same approach as before and the problem can be solved analytically.

In this case the time S_2 of the subset E_2 destination coincides with the distribution of the second ordinal statistics $A_{(2)}$, but it also equals the sum $S_2 = T_1 + T_2$ of the time to the first component failure T_1 and the time interval between the first and the second component failure T_2, where the distributions are exponential with parameters 6α and 5α. Thus, the mean time to the second component failure is

$$M_2 = m_1 + m_2 = \frac{1}{6\alpha} + \frac{1}{5\alpha} = \frac{11}{30\alpha}$$

Analogously the mean time to the subset E_3 destination is

$$M_3 = m_1 + m_2 + m_3 = \frac{1}{6\alpha} + \frac{1}{5\alpha} + \frac{1}{4\alpha} = \frac{37}{60\alpha}$$

Thus, the condition (5) of preference of the strategy $l = 2$ to the strategy $l = 3$ $K_{\text{av.},2} > K_{\text{av.},3}$ takes the form

$$\frac{b_2}{b_3} > \frac{m_1 + m_2}{m_1 + m_2 + m_3} = \frac{22}{37} \approx 0.59.$$

This means that the strategy $l = 2$ will be preferable to the strategy $l = 3$ if the mean PM time b_2 is less than 59% of b_3. This result coincides with the one we obtained for the corresponding strategies for the 4-out-of-6 : F system examined in Section 4.3 as the states with 2 or 3 failures are the same for both systems, the difference will be only in calculating the mean time before system failure with the following repair.

Compare now each of PM strategy with the regime of the system operating up to its failure with the following repair. To do that we should compare the mean times M_l ($l = 2, 3$) of the subsets E_l ($l = 2, 3$) destination with the mean time M_0 of the system operation up to its failure without any PM.

$$M_0 = \frac{2}{5}\mathbf{E}[A_{(4)}] + \frac{3}{5}\mathbf{E}[A_{5)}] = \frac{2}{5}M_4 + \frac{3}{5}M_5 = \frac{2}{5}\frac{57}{60\alpha} + \frac{3}{5}\frac{87}{60\alpha} = \frac{5}{4\alpha}$$

The comparison shows that the strategy $l = 2$ is preferable over working without PM $2 \geq 0$ iff

$$\frac{b_2}{b_0} \geq \frac{M_2}{M_0} = \frac{22}{75} \approx 0.29$$

and the strategy $l = 3$ is preferable over the working without PM iff

$$\frac{b_3}{b_0} \geq \frac{M_3}{M_0} = \frac{37}{75} \approx 0.49.$$

In Figure 4 the blue circle corresponds to the ratio $\frac{b_2}{b_0} = 0.29$ and the red circle corresponds to the ratio $\frac{b_3}{b_0} = 0.49$ which demonstrates the coincidence of analytical and numerical calculations for exponential distribution.

6. Conclusions

In this paper we investigate different PM strategies for a k-out-of-n system based on its states. The novelty and the feature of the paper are that the system failure depends on the position of its failed components. We propose a new method to solve this problem, based on the distribution of the members of the variational series of the failing components. Also, we investigated how sensitive the decision about PM beginning is to the shape of the system components' lifetime distributions. This investigation is very actual because the initial information about system components lifetime is usually very limited and available only up to one or two moments.

We propose an algorithm to compare different PM strategies and to choose the best among them with respect to the system availability maximization. The algorithm can be used for any k-out-of-n : F system with any system failure set. The algorithm is applied to analyze PM strategies of the UUV to monitor the condition of a subsea pipeline. A series of numerical experiments made it possible to draw conclusions about how sensitive the choice of PM strategy to the shape of system components lifetime distribution. It is also possible to determine the intervals for the coefficient of variation, where the decision to choose a preferable strategy does not depend on the type of distribution.

The proposed approach can be expanded with other quality criteria, such as productivity of the system or system service cost under different PM strategies. Multi-criteria assessment of the PM effectiveness is also possible.

Author Contributions: Conceptualization, methodology, formal analysis, and writing—original draft preparation, V.R.; investigation, software and computer calculations, visualization, and writing—original draft preparation, O.K.; setting tasks for engineering applications, data curation, writing—original draft preparation M.F. All authors have read and agreed to the published version of the manuscript.

Funding: This research received no external funding.

Acknowledgments: The authors thanks the MDPI Publishing House and the Issue Guest Editor Pavel Pracs for the possibility to publish this paper with 50% discount. The authors are grateful to the reviewers for their valuable comments that allowed us to improve the paper.

Conflicts of Interest: The authors declare no conflict of interest.The funders had no role in the design of the study; in the collection, analyses, or interpretation of data; in the writing of the manuscript, or in the decision to publish the results.

Abbreviations

The following abbreviations are used in this manuscript:

PM	Preventive Maintenance
UUV	Unmanned Underwater Vehicle
SV	Surface Vessel
i.i.d.	independent and identically distributed
r.v.	random value

References

1. Rykov, V.V.; Sukharev, M.G.; Itkin, V.Y. Investigations of k-out-of-n systems application possibilities to objects of oil and gas industry. *Mar. Sci. Eng.* **2020**, *8*, 928. [CrossRef]
2. Shepherd, D.K. k-out-of-n Systems. In *Encyclopedia of Statistics in Quality and Reliability*; American Cancer Society: Atlanta, GA, USA, 2008. Available online: https://onlinelibrary.wiley.com/doi/pdf/10.1002/9780470061572.eqr342 (accessed on 26 February 2019).
3. Gazprom. Dzhubga—Lazarevskoye—Sochi. The First Russia-Based Offshore Gas Pipeline. Available online: https://www.gazprom.com/projects/dls/ (accessed on 19 November 2020).
4. OceanAlpha Group Ltd. Autonomous Hydrographic Survey Boat. Available online: https://www.oceanalpha.com/product-item/m40/ (accessed on 19 November 2020).
5. Blue Robotics. BlueROV2. Available online: https://bluerobotics.com/product-category/rov/bluerov2/ (accessed on 19 November 2020).
6. Farkhadov, M.; Abramenkov, A.; Abdulov, A.; Eliseev, A. Portable Remotely Operated Underwater Smart Vehicle with a Camera and an Arm. In Proceedings of the 2019 10th International Power Electronics Drive Systems and Technologies Conference (PEDSTC) Shiraz, Iran, 12–14 February 2019; pp. 178–183. [CrossRef]
7. Advanced Research Foundation. AquaRoboTech 2020 Competition. Available online: http://aquarobotech.ru/#about (accessed on 19 November 2020).
8. Abdulov, A.; Abramenkov, A. Visual odometry system simulator, In Proceedings of the 2017 International Siberian Conference on Control and Communications (SIBCON), Astana, Kazakhstan, 29–30 June 2017; pp. 1–5.
9. Gertsbakh, I. Preventive Maintenance Models Based on the Lifetime Distribution. In *Reliability Theory with Applications to Preventive Maintenance*; Springer: Berlin/Heidelberg, Germany, 2000; pp. 67–106.
10. Dudin, A.N.; Krishnamoorthy, A.; Narayanan, V.C. Idle time utilization through service to customers in a retrial queue maintaining high system reliability. *J. Math. Sci.* **2013**, *191*, 506–517. [CrossRef]
11. Krishnamoorthy, A.; Viswanath, C.N.; Deepak, T.G. Reliability of a k-out-of-n-system with Repair by a Service Station Attending a Queue with Postponed Work. *Int. J. Reliab. Qual. Saf. Eng.* **2007**, *14*, 379–398. [CrossRef]
12. Krishnamoorthy, A.; Viswanath, C.N.; Deepak, T.G. Maximizing of Reliability of a k-out-of-n-system with Repair by a facility attending external customers in a Retrial Queue. *Reliab. Theory Appl.* **2007**, *2*, 21–33.
13. Krishnamoorthy, A.; Sathian, M.K.; Viswanath C.N. Reliability of a k-out-of-n system with repair by a single server extending service external customers with pre-emption. *Reliab. Theory Appl.* **2016**, *11*, 61–93.
14. Krishnamoorthy, A.; Sathian, M.K.; Viswanath C.N. Reliability of a k-out-of-n system with a single server extending non-preemptive service to external customers- Part I. *Reliab. Theory Appl.* **2016**, *11*, 62–75.
15. Krishnamoorthy, A.; Sathian, M.K.; Viswanath C.N. Reliability of a k-out-of-n system with a single server extending non-preemptive service to external customers-Part II. *Reliab. Theory Appl.* **2016**, *11*, 76–88.
16. Finkelstein, M.; Levitin, G. Preventive maintenance for homogeneous and heterogeneous systems. *Appl. Stoch. Model. Bus. Ind.* **2019**, *35*, 908–920. [CrossRef]
17. Finkelstein, M.; Cha, J.H.; Levitin, G. On a new age-replacement policy for items with observed stochastic degradation. *Qual. Reliab. Eng. Int.* **2020**, *36*, 1132–1146. [CrossRef]
18. Park, M.; Lee, J.; Kim S. An optimal maintenance policy for k-out-of-n system without monitoring component. *Qual. Technol. Qual. Manag.* **2019**, *16*, 140–153. [CrossRef]
19. Endharth, A.J.; Yun, W.Y.; Yamomoto H. Preventive maintenance policy for linear-consequtive -k-out-of-n : F system. *J. Oper. Res. Soc. Jpn.* **2016**, *59*, 334–346.
20. Hamdan, K.; Tavangar, M.; Asadi M. Optimal preventive maintenance for repairable weighted k-out-of-n systems. *Reliab. Eng. Syst. Safe* **2021**, *205*, 107267. [CrossRef]
21. Kozyrev, D.; Kolev, N.; Rykov, V. Reliability function of renewable system under marshall-lkin Failure Model. *Reliab. Theory Appl.* **2018**, *13*, 39–46.
22. Rykov, V. On Reliability of Renewable Systems. In *Reliability Engineering. Theory and Applications*; Vonta, I., Ram, M., Eds.; CRC Press: Boca Raton, FL, USA, 2018; pp. 173–196.

23. Rykov, V.; Kozyrev, D. On the reliability function of a double redundant system with general repair time distribution. *Appl. Stoch. Model. Bus. Ind.* **2019**, *35*, 191–197. [CrossRef]
24. Rykov, V.; Kozyrev, D.; Filimonov, A.; Ivanova, N. On reliability function of a *k*-out-of-*n* system with general repair time distribution. *Probab. Eng. Inf. Sci.* **2020**, *5*, 1–18. [CrossRef]

Journal of
Marine Science and Engineering

MDPI

Review

Kalman Filters for Leak Diagnosis in Pipelines: Brief History and Future Research

Lizeth Torres [1,*], Javier Jiménez-Cabas [2], Omar González [3], Lázaro Molina [4] and Francisco-Ronay López-Estrada [5,*]

1   Instituto de Ingeniería, Universidad Nacional Autónoma de México, Mexico City 04510, Mexico
2   Departamento de Ciencia de la Computación y Electrónica, Universidad de la Costa, 080002 Barranquilla, Colombia; jjimenez41@cuc.edu.co
3   Universidad Tecnológica de México Campus Atizapán, Mexico City 04510, Mexico; omar180282@hotmail.com
4   Universidad Autónoma de Nuevo León, Monterrey 64460, Mexico; lazaro.molinasp@uanl.edu.mx
5   Tecnológico Nacional de México/IT Tuxtla Gutiérrez, TURIX-DYNAMICS Diagnosis and Control Group, Carretera Panamericana km 1080, Tuxtla Gutierrez C.P. 29050, Mexico
*   Correspondence: ftorreso@iingen.unam.mx (L.T.); frlopez@ittg.edu.mx (F.-R.L.-E.)

Received: 13 February 2020; Accepted: 1 March 2020; Published: 5 March 2020

Abstract: The purpose of this paper is to provide a structural review of the progress made on the detection and localization of leaks in pipelines by using approaches based on the Kalman filter. To the best of the author's knowledge, this is the first review on the topic. In particular, it is the first to try to draw the attention of the leak detection community to the important contributions that use the Kalman filter as the core of a computational pipeline monitoring system. Without being exhaustive, the paper gathers the results from different research groups such that these are presented in a unified fashion. For this reason, a classification of the current approaches based on the Kalman filter is proposed. For each of the existing approaches within this classification, the basic concepts, theoretical results, and relations with the other procedures are discussed in detail. The review starts with a short summary of essential ideas about state observers. Then, a brief history of the use of the Kalman filter for diagnosing leaks is described by mentioning the most outstanding approaches. At last, brief discussions of some emerging research problems, such as the leak detection in pipelines transporting heavy oils; the main challenges; and some open issues are addressed.

Keywords: leak detection; Kalman filter; pipelines

---

1. Introduction

Because of the operation conditions, onshore and offshore pipelines are subjected to environmental loads (wind, waves, current, seabed movements, and earthquakes) that can provoke undesirable vibrations, stress, fatigue problems, and the propagation of cracks [1,2]. In particular, this last issue is an important source of leaks together with the aging of the pipelines, failures in the installation, illegal extractions, and terrorist sabotage. For this reason, to avoid environmental disasters, the oil and gas sectors invest generous resources for the development of robust and reliable leak detection systems, which according to the API RP 1130 standard, can be classified as external or internal systems [3]. External systems use local sensors (e.g., acoustic microphones or fiber optic cables) to send an alarm when a leak occurs, and they do not perform the computation for diagnosing a leak. Internal systems utilize field instrumentation outputs, which monitor internal pipeline parameters (e.g., pressure, temperature, flow rate, and viscosity), and algorithmic monitoring tools. Therefore, internal systems are also known as computational pipeline monitoring systems (CPM systems) [4].

Among the algorithmic tools that have been extensively used for dealing with the fault diagnosis of pipelines, the state observers have proved to be a powerful tool for the estimation in real-time of the internal state variables of a given system (e.g., a pipeline). These estimations are based on the knowledge of available measurements (inputs and outputs of such a system) and other known parameters. Concretely, a state estimator is a model of a system with an online adaptation (correction) based on available measurements for reconstructing unknown information; see Figure 1 [5]. Usually, the model is given in a state-space representation, which can be, in general, continuous-time or discrete-time, deterministic or stochastic, and finite-dimensional or infinite-dimensional.

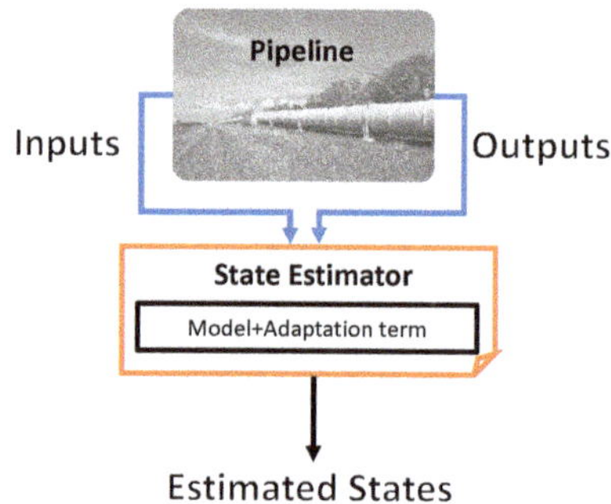

Figure 1. Architecture of a computational pipeline monitoring (CPM) system based on a state estimator.

Several types of state estimators, such as Kalman filters, Luenberger-type observers, high gain observers, and sliding mode observers, have been used for leak detection and localization. Three good reviews that summarize the research and development of state estimator-based leak detection systems for liquid pipelines are given in [6,7]. The literature, however, still lacks a more in-depth review of state of the art in leak detection using Kalman filters, which are the most commonly used estimators for detecting and localizing leaks, and according to D. Simon, are "the best linear estimators" [8]. For this reason, this work aims to fill this gap, since the approaches based on these filters deserve a separate study.

There are several versions of the Kalman filter for dealing with the diversity of physical systems and their associated problems. For example, the ensemble Kalman filter (EnKF), which is suitable for problems with a large number of variables, such as those described by partial differential equations [9]. For estimating the states of nonlinear systems, there are ad hoc versions, such as the extended Kalman filter (EKF), the unscented Kalman filter (UKF), and the particle filter (PF). This paper focuses only on the versions that are used for leak detection, such as The Kalman filter, the discrete Kalman filter, and the extended Kalman filter, among others (see Appendix A for mathematical details). It is important to note that in recent years, several Kalman-based pipeline leak diagnosis methodologies have been proposed, which demonstrates their applicability and support from the scientific community.

This paper is organized as follows. Section 2 presents a brief review of basic concepts for understanding the functioning of state observers. Section 3 introduces a tentative classification of the Kalman filter-based methods proposed to this day. Section 4 presents the history of the evolution of the Kalman filter-based methods in the area of leak detection. Concretely, this section highlights the contributions that have been a milestone in leak diagnosis. Finally, in Section 5, some recommendations for future research are given. Appendix A presents the mathematical structure of different types of Kalman filters that have been employed in leak detection tasks.

2. State Estimators

In many engineering applications, some variables cannot be directly measured, either because there are no sensors for them or because the cost is prohibitive. An alternative to addressing this problem is to obtain a dynamical estimation of the required variables by using state estimators.

A general definition of a state estimator is as follows: *an algorithmic tool that estimates the variables of a process using (1) a mathematical model represented in state space, (2) the available measurements of the process (inputs and outputs), and (3) an error correction (adaptation) term to ensure the convergence of the algorithm.*

To derive a general structure of a state estimator, let us consider the general structure of the continuous model of a system in a state-space representation given as follows:

$$\dot{x}(t) = f\left(x(t), u(t)\right),$$
$$y(t) = h\left(x(t)\right),$$

(1)

where $x(t) \in \mathbb{R}^n$ is the state vector; $\dot{x}(t) \in \mathbb{R}^n$ is the state derivative vector; $u(t) \in \mathbb{R}^m$ is the external (exogenous) input vector or control signal; $y(t) \in \mathbb{R}^p$ represents the output vector, i.e., the measured states (variables) acquired by sensors; $f \in \mathbb{R}^n$ represents the vector field; and $h \in \mathbb{R}^p$ is the continuous output function.

Since a state estimator is the model of the system plus a correction (adaptation term), this can be expressed as follows:

$$\dot{\hat{x}}(t) = \underbrace{f\left(\hat{x}(t), u(t)\right)}_{\text{Model Copy}} + \underbrace{K\left(\hat{x}(t)\right)\left(y(t) - \hat{y}(t)\right)}_{\text{Correction Term}},$$
$$\hat{y}(t) = h\left(\hat{x}(t)\right),$$

where $\hat{x}(t)$ and $\hat{y}(t)$ are the online estimations of $x(t)$ and $y(t)$, respectively; and $K(\hat{x}(t))$ is the gain of the observer. Thus, the design of the state observer consists of choosing an appropriate gain $K(\hat{x}(t))$ so that the estimation error tends to 0 when $t \rightarrow \infty$ with the desired properties of time convergence and robustness.

If the observation error $e(t)$ is defined as follows,

$$e(t) = x(t) - \hat{x}(t),$$

(2)

the dynamics of the error observation can be derived from (1) and (2), and expressed as

$$\dot{e}(t) = f(\hat{x}(t) + e(t), u(t)) - f(\hat{x}(t), u(t)) - K(\hat{x}(t))(h(\hat{x}(t) + e(t)) - h(\hat{x}(t))).$$

(3)

An observer connected to a pipeline has the structure of the block diagram shown in Figure 2. The inputs in a pipeline can be the flow rate provided by a pump, the level of a tank, the flow rate, or the pressure that results from the opening or closing of a valve. These inputs, or at least a subset of them, must be registered to be injected into the state estimators. The state, which is the smallest possible subset of system variables that can represent the complete state of a system at any time, can be either the pressure or flow rate at any coordinate along the pipeline or the position of a leak. The measured outputs are the measurements provided by *in situ* sensors (flow meters, pressure transducers, or thermocouples).

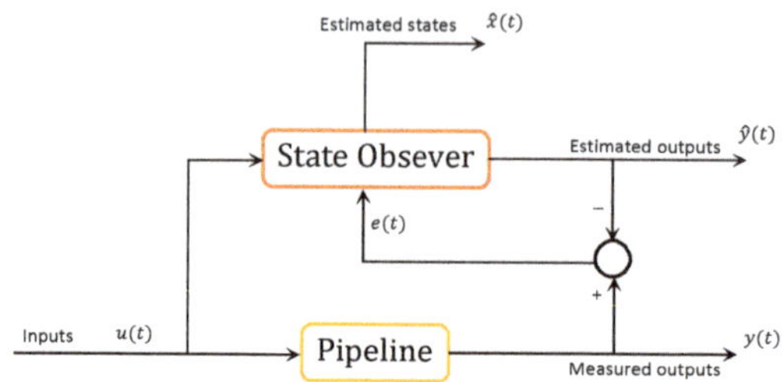

Figure 2. State estimation process for the pipeline.

The design and choice of a state estimator depends on many factors: the application in which the estimates will be used, the nature of the system, the nature of the variables to be estimated, the type of information that will be available for performing the estimation, the nature of such information (e.g., discrete or continuous), and the characteristics of the required estimates. In this spirit, an abbreviated procedure for designing a state estimator for leak diagnosis purposes is proposed in Figure 3.

Steps for designing a state estimator

- **Step 1:** *Identify the available information (observations, data, measurements, and records) for performing the estimation.*

- **Step 2:** *Formulate a model assuming convenient assumptions and constraints.*

- **Step 3:** *Set the model in a state-space representation.*

- **Step 4:** *Set the equations of the state observer.*

- **Step 5:** *Compute the gain of the state observer.*

Figure 3. Procedure to design a state estimator

3. A Proposed Classification for the Kalman Filter-Based Approaches

The CPM systems that have used the Kalman filter as the principal algorithmic tool can be categorized into three approaches, according to the architecture of the leak diagnosis algorithm. (1) The approaches based on the estimations of a bank of filters; (2) the approaches based on the estimation of variables (e.g., pressures and flow rates) at different locations along the pipeline; and (3) the approaches based on the direct estimation of the leak parameters, which are added to the pipeline flow model as if they were states.

This classification is inspired by three influential contributions that were presented in three different years, as shown in the timeline infographic in Figure 4. In 1980, Tørris Digernes proposed the first contribution based on a bank of Kalman filters [10]. In 1988, Benkherouf and Allidina introduced the first contribution based on the estimation of the hydraulic variables at different points of the pipeline [11], and in 2007, Besançon et al. proposed the first approach based on the direct estimation of the leak parameters [12]. The following describes each of these approaches.

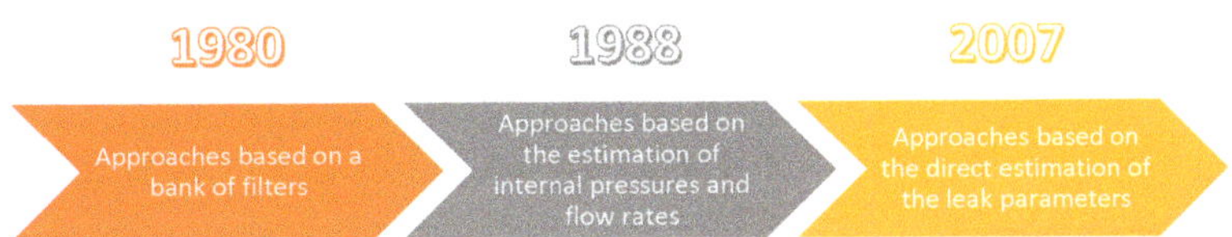

Figure 4. Timeline infographic of the evolution of the Kalman filter-based approaches.

Approaches based on a bank of filters. These approaches were the first proposed for detecting and localizing leaks in pipelines by using Kalman filters. The architecture of these approaches is illustrated in Figure 5, which is a set of Kalman filters that act (or perform an estimate) in parallel. Each filter is different from the other because each filter is constructed from a pipeline model with a leak in a prescribed position that is different from the leak positions involved in the other models that are used to build the other filters. For example, a leak diagnosis algorithm for a 100-meter pipe can be constructed with ten filters. The first filter can be designed to detect a leak in the first 10 m of the pipe, the second in the next 10 m, and so on.

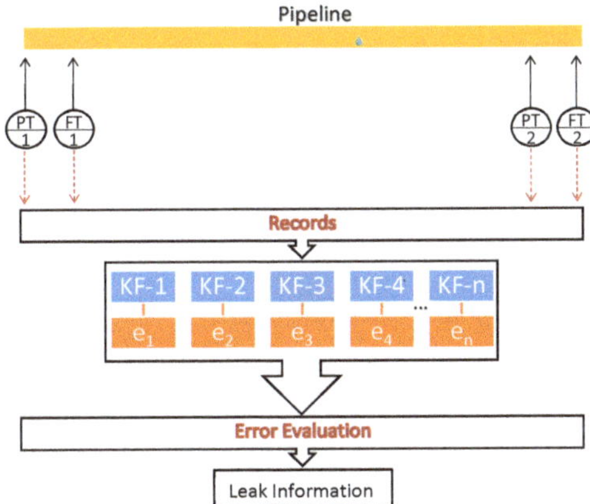

Figure 5. Architecture of the approaches based on a bank of Kalman filters.

On the one hand, the filters forming the bank can be independent of each other, or they can be dependent, that is, interconnected. The interconnection between filters can be cascading or peer-to-peer. On the other hand, the information that each filter receives to make the estimate can be the same (pressures and/or flow rates at the ends) or different (pressures and/or flow rates measured at certain positions of the pipe).

A bank of estimators has been successfully used by commercial leak detection systems, such as PipePatrol software supplied by KROHNE Group [13]. The main disadvantage of these approaches, however, is that in order to have better accuracy regarding the leak position, a bank of many filters must be designed, which implies that a high computational cost is required for finding the numerical solution of each filter. For this reason, another class of approaches was proposed for reducing the computational burden: the approaches based on the estimations of internal variables.

Approaches based on the estimation of internal pressures and flow rates. These approaches use a unique Kalman filter. The states of the model used for designing the Kalman filter are the internal pressures and flow rates at different (positions) coordinates distributed along the pipeline. Once the states are estimated by the Kalman filter, the leak is localized by using the estimations for solving auxiliary algebraic equations (e.g., head loss balances). The accuracy of the leak location is achieved by increasing the number of estimated internal variables. This fact implies that more states must be estimated, and therefore, a greater computational burden is imparted. The architecture of these approaches is illustrated in Figure 6.

Approaches based on the direct estimation of the leak parameters. These kinds of approaches were proposed to avoid using several filters and to discretize the space in many nodes. In this class of approaches, the leak localization is performed by a unique Kalman filter, which is designed from a mathematical model that involves the leak parameters as state variables in order to be estimated. Usually, these approaches involve the Kalman filter described in Appendix A.3. The architecture of these approaches is illustrated in Figure 7.

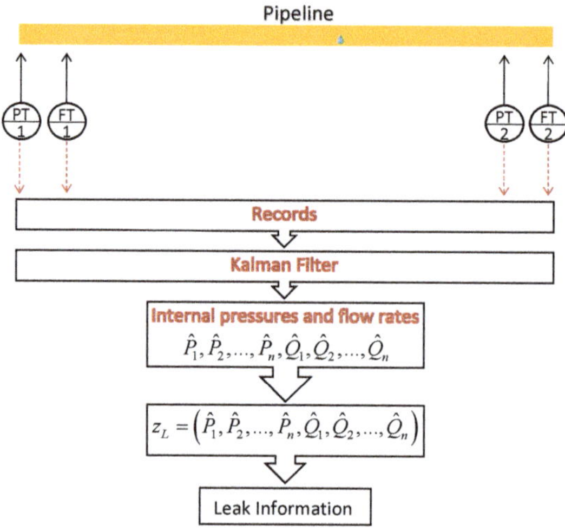

Figure 6. Architecture of approaches based on the estimation of internal pressures and flow rates.

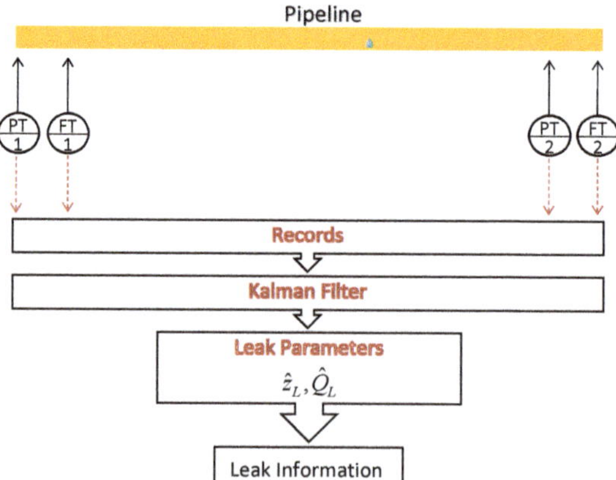

Figure 7. Architecture of approaches based on the direct estimation of the leak parameters.

The following section details different Kalman-based approaches that were presented over time, and highlights the main characteristics, advantages, and disadvantages of the main contributions.

4. Brief History

To the best of our knowledge, the first work presenting an approach based on the Kalman filter to detect and localize faults in pipelines was written by Tørris Digernes [10]. This work, entitled *Real-time failure detection and identification applied to supervision of oil transport in pipelines*, presents a methodology based on multiple parallel filters that are independent of each other: a bank of filters. Each filter was designed from a dynamical model representing a prescribed fault situation. In particular, two fault types were treated by Digernes: single leaks and sensor faults. By applying this methodology, the fault recognition is performed by identifying the filter having the highest probability of representing the plant in the fault situation. The probability is calculated by using the multiple-model hypothesis probability test, which when performed, requires the error estimation between the available measurements from the pipeline and their estimates provided by the filter. To show the potentiality of his methodology, Digernes presented some simulations' results. In such simulations, the features of the oil pipeline between Ekofisk in the North Sea and the terminal in Teesside, UK, were used. The filters were designed from finite models expressed by space-discrete equations that represent the mass and the momentum balance of the fluid in a pipeline. To compute the estimation errors, pressures and flow rates at the inlet and outlet of the pipeline were used, as were pressures at two points along the pipeline. The principal disadvantage of this approach is that in case of a leak, a large number of filters is needed to obtain an acceptable resolution of the leak position. This pioneering work has inspired another important contribution. For example, in [14,15], the same approach was tested in a simulation environment for a long-distance pipeline of water, and only the following aspects were different: the use of a backward time-central space discretization scheme for lumping the continuity and momentum equations together, the use of a modified version of the Kalman filter, and the introduction of a feedforward law for computing the leak magnitude.

Years later, Benkherouf and Allidina (B&A) presented the work entitled *Leak detection and location in gas pipelines*, which proposes a Kalman filter for detecting and finding the position of a single leak [11]. The filter is based on a lumped version of a distributed one-dimensional isothermal model (two partial differential equations describing the continuity and momentum conservation) that describes the gas flow through a single pipeline with fictitious leaks distributed along with it. To obtain the distributed model, both viscous and turbulent effects of the flow were neglected, and it was assumed that both the temperature changes within the gas and the heat exchanges with the surroundings of the pipeline are small. The lumped model was formulated using the method of characteristics (MOC). By using this approach, the position of the leak is determined through the following algebraic equations that relate the fictitious leaks estimated with the Kalman filter to the real one:

$$Q_L(t) = \sum_{i=1}^{N} Q_{L_i}(t), \tag{4}$$

$$Q_L(t)z_L(t) = \sum_{i=1}^{N} Q_{L_i}(t)z_{L_i}(t), \tag{5}$$

where z_L is the real position leak, Q_{L_i} and z_{L_i} are the flows and positions of the fictitious leaks, i is the fictitious leak index, and N is the total number of fictitious leaks.

The methodology of B&A surpasses the Digernes approach in the sense that only one filter has to be designed. For this reason, it has also been the inspiration for a significant number of works. For example, in [16], the authors used the same approach with a slight modification in the covariance formula to locate a leak in a pipeline of water. Moreover, they tested the approach in simulations and in the laboratory. In [17], the same approach was tested together with a technique called the extended boundary approach. In [18], a comparison between B&A's approach and the algorithm proposed in [19] (which does not have the Kalman filter as a core of the diagnosis system) was presented. According to the authors' conclusions, the cycle time of B&A's method is longer, but gains on accuracy.

In 2001, Verde presented the work entitled *Multi-leak detection and isolation in fluid pipelines* [20], which proposes a bank of Kalman filters for localizing leaks in a hydraulic pipeline. Each Kalman filter is designed to diagnose a section of the pipeline, which, in fact, is divided into N sections. Concretely, each KF is designed to estimate the states (pressures and flow rates) at prescribed points (locations) of the pipeline by considering that a leak is occurring in a pipeline section delimited by two prescribed points. If the pipeline is divided into N sections, as small as desired, N Kalman filters must be designed: each one by considering a leak in a different section. If there are many sections, the computational cost is higher. The estimation error of each KF is used to localize the leak. If a leak develops in a given section, the error associated with the section remains around 0, and the rest do not.

Because the methodology was proposed for a hydraulic pipeline, the Kalman filters were designed from a space-discrete version of the water hammer (WH) equations given as follows:

$$\frac{\partial Q(z,t)}{\partial t} = -gA_r \frac{\partial H(z,t)}{\partial z} - J_s(Q(z,t)), \tag{6}$$

$$\frac{\partial H(z,t)}{\partial t} = -\frac{b^2}{gA_r} \frac{\partial Q(z,t)}{\partial z}, \tag{7}$$

which were proposed by Chaudhry in his prestigious work *Applied Hydraulic Transients* [21]. For WH equations, $(z,t) \in [0,L] \times [0,\infty)$ gathers the space (m) and time (s) coordinates, respectively; L is the length of the pipe; $H(z,t)$ is the pressure head (m); $Q(z,t)$ is the flow rate (m^3/s); b is the wave speed in the fluid (m/s); g is the gravitational acceleration (m/s^2); A_r is the cross-sectional area of the pipe (m^2); ϕ is the inside diameter of the pipe (m); and Js is the quasi-steady friction term, which may be expressed by the Darcy-Weisbach relation as

$$J_s(Q(z,t)) = \frac{f(Q(z,t))}{2\phi A_r} Q(z,t)|Q(z,t)|, \tag{8}$$

where f is the Darcy-Weisbach friction factor.

The method used to numerically solve the WH equation was the first-order finite difference method (FDM). By assuming pressures at the ends of the pipeline as the boundary conditions and after applying the FDM, the fluid model can be represented as n sets of coupled nonlinear dynamic equations given in state-space representation.

In 2003, Verde et al. showed that the isolation (localization) of two simultaneous leaks is not feasible only with steady-state data of the fluid in a pipeline [22]. For this reason, Besançon et al. presented an approach based on a single extended Kalman filter and suitable inputs to obtain unsteady data from the pipeline [12]. The filter was constructed from a model deduced from a. The order of this model is the minimal to represent two leaks, so we can say that this model is a minimal-order model for two leaks. In order to estimate two leaks, four states with a constant dynamic that represent both positions and both leak coefficients were joined to the minimal-order model.

Since the pressures at the ends of the pipeline were considered as inputs, in order to excite the pipeline, they were manipulated to be triangular. The estimation of the positions, and the estimation of the coefficients, were both achieved. The estimation results have shown, however, that the estimation is sensitive to the initial conditions. Moreover, experimental results were not presented to validate the approach. Torres et al. presented similar methodologies in [23,24], but a lumped model obtained via the orthogonal collocation method was used. There are two main reasons why this algorithm could not work with experimental data. The first reason: the reduced order of the finite model that resulted from the discretization of the spatial domain into three sections; this would not be a problem if auxiliary inputs were not needed to ensure the convergence of the estimation. Usually, however, these inputs are periodical with fixed or variable frequency. If the frequency of the required input is too high, the finite model is no longer representative of a real pipeline. A solution to this concern may be the increase of the order such that the model becomes representative to high frequencies. The second reason may be that the values of the leak positions may take values between 0 and L. A solution to this concern could be a reduction of this interval.

In 2010, Dos Santos et al. introduced a new approach for detecting gas leaks in high-pressure distribution networks [25]. Each pipeline of the network was modeled as a linear parameter-varying (LPV) system driven by the source node mass flow together with the pressure as the scheduling parameter. The mass flow at the offtake was considered as the system output. The leak position was added as a state of the LPV system, from which a Kalman filter was designed. The effectiveness of the CPM system was illustrated with real and simulated data.

In 2011, Navarro et al. proposed an extended Kalman filter for locating leaks in a plastic pipeline, which was constructed from a discretized model both in time and in space. For the design of the filter, the space discretization was nonuniform and was a function of the unknown leak location; furthermore, the time domain was discretized by using Heun's method. The method was validated in real-time in a laboratory [26].

In 2015, Verde and Rojas presented an iterative scheme for locating sequential leaks, namely, one leak after another. The core of the method is a continuous extended Kalman filter with a prescribed degree of stability, which is constructed from the model of the flow in a pipeline with an equivalent leak; check Appendix A.4. The equivalent leak is a fictitious leak with a position in which a single leak would have to produce specific values of pressure and flow rate along the pipeline at a steady-state, but the values are actually caused by two or more leaks [22]. The equivalent leak must satisfy two criteria: (1) water loss equivalence and (2) energy equivalence [27]. In the case of a pipeline with a single branch and a leak, the equivalent leak is caused by the branch and leak outflows; in addition, it always has a position between both extractions.

In order to address the same concern, in 2016, Delgado et al. presented an approach for localizing sequential leaks by using an extended Kalman filter (Appendix A.2) for estimating parameters (Appendix A.3): the parameters of each sequential leak such as position and size. The filter was designed from a discrete time-space model derived from the WH equations and was modified at each new leak occurrence via an adaptation strategy to augment the size of the model's state vector. The augmentation of the state is done to include the parameters of the actual sequential leak. The approach was validated by using experimental data [28].

In 2016, Verde et al. presented a Kalman-based approach for detecting and localizing single leaks in a pipeline with a branch junction [29]. The approach requires the following information for producing a diagnosis: the flow rate together with the pressure head at the pipeline ends, the position (the spatial coordinate) of the branch junction, and the flow rate through the branch. The approach involves a selector algorithm (a simple algebraic equation that can be deduced from a head loss balance), and two localization algorithms, which are two Kalman filters designed from two different mathematical models, each one representing the flow dynamics of the pipeline before and after the branch, respectively. The goal of the selector algorithm is to indicate whether the leak is to the left (upstream) or to the right (downstream) of the branch. Depending on the indication of the selector, one of the two Kalman filters can be used to estimate the position of the leak. The approach was numerically tested with synthetic and experimental data from a hydraulic test apparatus.

In 2017, Delgado et al. described the successful localization of a leak in a pipeline of the water distribution network in Guadalajara, Mexico [30]. The localization was achieved by using a discrete-time extended Kalman filter (Appendix A.3), which was constructed by a lumped version of the WH equations. Additionally, Navarro et al. presented a two-stage leak isolation methodology based on a fitting loss coefficient calibration. In the first stage, an extended Kalman filter is used to fix the equivalent straight length (ESL) of the pipeline. Once the leak is detected, an algebraic observer allows for estimating the leak position from the ESL fixed by the extended Kalman filter. Since leak isolation is performed in equivalent length coordinates, a transformation to the original coordinates is necessary [31].

In 2018, Santos-Ruiz et al. introduced a methodology for leak detection and localization based on data fusion from two approaches: a steady-state estimation and an extended Kalman filter [32]. The proposed method considers only pressure-head and flows rate measurements at the pipeline end. The approach was tested in real-time by using a USB device for the data acquisition. The novelty of this approach is that the steady-state solution for a nonlinear pipeline model of the pipeline is merged with the dynamic state estimation obtained from the EKF observer, by using Bayesian data fusion in order to refine the leak diagnosis. In the same year, Delgado et al. proposed a method based on two steps for solving the leak diagnosis problem in pipeline networks [33]. In a first step, a faulty system and a nominal model are used to generate residuals with an analysis that allows identifying the region where the leak occurs. As a second step and by using the information generated in the first step, the leak position and magnitude are estimated through extended Kalman filters. The proposed two-step methodology minimizes the problem of observer design, since it avoids the design of an observer for each section of the network. On the other hand, Liu et al. suggested handling multi-leak detection problems in oil pipelines by using unscented Kalman filters [34]. Leaks are detected one at a time with an observer; therefore, the number of observers must be increased when a new leak occurs.

In Table 1, all these approaches are classified according to the decade of their presentation. Additionally, this table contains some works that propose methodologies based on the Kalman filter for addressing different problems associated with the pipeline operation monitoring, which does not concern leak or fault detection.

Table 1. Classification of the approaches by decade.

Period	Contributions
1980–1990	[10,11]
1990–2000	-
2000–2010	[12], [14–16], [18], [20], [23,24], [35–39]
2010-Up to the present	[25,26], [28–34], [40–55],

In Table 2, a taxonomy of CPM systems according mainly to the type of Kalman filter employed in the solution formulation of the leak detection and location is presented. In addition, other parameters are highlighted, such as the fluid involved in the study, the type of leak, either single or multiple leaks, and also the type of validation (in a simulation, in a laboratory, or in the field).

Table 2. Taxonomy of CPM systems.

Year	Reference	Country	Fluid	Fault	Testing	Filter Type
1980	Real-time failure-detection and identification applied to supervision of oil transport in pipelines	Norway	Oil	Single leak	Simulation	Kalman Filter
1988	Leak detection and location in gas pipelines	UK	Gas	Single leak	Simulation	Extended Kalman filter
1988	Robust observer design for a fluid pipeline	China	Water	NA	Simulation, Laboratory	Kalman Filter
1988	State estimation of output-decoupled complex systems with application to fluid pipeline	China	Water	NA	Simulation, Laboratory	Extended Kalman filter
1990	An application of Kalman filter to leak diagnosis of long-distance transport pipelines	China				Generalized Kalman filter
2000	Multi-leak detection and isolation in fluid pipelines	Mexico	Water	Simultaneous leak	Simulation, Laboratory	Kalman filter
2002	A non-linear multiple-model state estimation scheme for pipeline leak detection and isolation	Saudi Arabia	Water	Single leak	Simulation	Modified Extended Kalman filter
2004	Minimal order nonlinear observer for leak detection	Mexico	Water	Simultaneous leak	Simulation, Laboratory	Nonlinear Kalman filter
2004	Identificability of multi-leaks in a pipeline	Mexico	Water	Simultaneous leak	Simulation, Laboratory	Extended Kalman filter
2004	Sub-sea pipelines leak detection and location based on fluid transient and FDI		Oil, Gas		Industrial pipeline	Extended Kalman filter
2005	Application of Kalman filter in pipeline leak detection				Laboratory	Kalman filter
2007	Leak detection in pipelines using the extended Kalman filter and the extended boundary approach	Canada	Water	Simultaneous leak	Simulation, Laboratory	Extended Kalman filter

Table 2. *Cont.*

Year	Reference	Country	Fluid	Fault	Testing	Filter Type
2007	Research on state estimation of oil pipeline considering adaptive extended Kalman filtering	China	Oil	NA	Industrial pipeline	Robust Adaptative Kalman filter
2007	Direct observer design for leak detection and estimation in pipelines	Mexico, France	Water	Simultaneous leak	Simulation	Extended Kalman Filter
2007	Comparison of two detection algorithms for pipeline leaks	Mexico, France	Water	Single leak	Laboratory	Extended Kalman filter
2008	A collocation model for water-hammer dynamics with application to leak detection	Mexico	Water	Single, sequential, simultaneous	Simulation	Extended Kalman filter
2009	A combined Kalman filter-discrete wavelet transform method for leakage detection of crude oil pipelines	China	Oil	Sequential leak	Industrial pipeline	Kalman Filter
2009	Estimation of the temperature field in pipelines by using the Kalman filter	Brazil, USA	Oil-gas-water mixture	NA	Industrial pipeline	Kalman filter
2009	Collocation modeling with experimental validation for pipeline dynamics and application to transient data estimations	France	Water	Single, sequential	Laboratory	Extended Kalman Filter
2010	Kalman filtering of hydraulic measurements for burst detection in water distribution systems	UK	Water	Single leak	Laboratory, Industrial pipeline	Adaptative Kalman filter
2010	Gas pipelines LPV modelling and identification for leakage detection	Portugal, USA, Germany	Gas	Single leak	Industrial pipeline	Kalman filter
2011	Real-time leak isolation based on state estimation in a plastic pipeline	Mexico, France, Venezuela	Water	Single leak	Laboratory	Extended Kalman filter
2011	Leakage detection and location in gas pipelines through an LPV identification approach	Portugal, US, Germany	Gas	Single leak	Industrial pipeline	Kalman filter

Table 2. *Cont.*

Year	Reference	Country	Fluid	Fault	Testing	Filter Type
2011	Calibration of fitting loss coefficients for modelling purpose of a plastic pipeline	Mexico, France	Water	NA	Laboratory	Extended Kalman Filter
2012	Leak isolation with temperature compensation in pipelines	Mexico	Water	Single leak	Laboratory	Extended Kalman filter
2012	Real-time leak isolation based on a fault model approach algorithm in a water pipeline prototype	Mexico	Water	Single leak	Laboratory	Extended Kalman filter
2013	State estimation of pipeline models using the ensemble kalman filter	US	Gas	Single leak	Simulation	Ensemble Kalman filter
2014	Leak detection and location based on improved pipe model and nonlinear observer	Venezuela, France	Water	Single leak	Simulation, Laboratory	Extended Kalman filter
2014	Design and realization of the Kalman filter based on LabVIEW	China	Water	NA	Simulation	Kalman filter
2014	Online burst detection in a water distribution system using the Kalman filter and hydraulic modelling	UK	Water	Single leak	Industrial pipeline	Kalman filter
2015	Modeling and state estimation for gas transmission networks	Iran	Gas	NA	Simulation	Extended Kalman filter
2015	Dynamic model of a new above-ground pipeline using a Kalman estimator-based system	United Arab Emirates		NA	Laboratory	Kalman filter
2016	Research on natural gas pipeline leak detection algorithm and simulation	Mexico, France	Water	Single leak	Laboratory	Adaptative Kalman filter
2017	Water Leak diagnosis in pressurized pipelines: a real case study	Mexico	Water	Single leak	Laboratory	Extended Kalman filter
2017	Real-Time Leak Isolation Based on State Estimation with Fitting Loss Coefficient Calibration in a Plastic Pipeline	Mexico, France	Water	Single leak	Laboratory	Extended Kalman filter
2018	Online leak diagnosis in pipelines using an EKF-based and steady-state mixed approach	Mexico, Spain	Water	Single leak	Laboratory	Extended Kalman filter
2018	EKF-based leak diagnosis schemes for pipeline networks	Mexico, France	Water	Single leak	Laboratory	Extended Kalman filter
2018	Multi-leak diagnosis and isolation in oil pipelines based on Unscented Kalman filter	China	Water	Single leak	Simulation	Unscented Kalman filter

According to API 1155, there are four metrics for evaluating leak detection systems: reliability, sensitivity, accuracy, and robustness. Despite the large amount of academic work based on the Kalman filter, however, only some of them provide a report on such metrics within a field context. In Table 3, these works are listed.

Table 3. Contributions with declared performance metrics.

Paper	Results	Application
A combined Kalman filter—discrete wavelet transform method for leakage detection of crude oil pipelines.	Reliability: 5% of false alarms. Accuracy: 0.26% of error	Main pipelines
Gas pipelines LPV modelling and identfication for leakage detection.	Sensitivity: A leak about 10% of the nominal flow rate was detected 24 minutes after its occurrence	Main pipelines
Identificability of multi-leaks in a pipeline	Accuracy: 12% of error	Main pipelines
Minimal order nonlinear observer for leak detection.	Accuracy: 1.36% of error with respect to the pipeline length in a noisy data scenario.	Main pipelines
Online burst detection in a water distribution system using the Kalman filter and hydraulic modelling.	Reliability: 85% of detected burst.	Pipeline networks
Real-time leak isolation based on a fault model approach algorithm in a water pipeline prototype	Accuracy: 3.6% of error with respect to the ESL.	Main pipelines
Research on natural gas pipeline leak detection algorithm and simulation.	Accuracy: 0.11% of locating error.	Main pipelines

Note that the accuracy of any methodology is strongly determined by the instruments in the physical installation but also by the availability of a proper system model.

5. Future Research

Hitherto, three main approaches for estimating leak locations using the Kalman filter have been reported in the literature. Such methods are based on banks of filters, on pressure and flow rate estimations, and on the direct estimation of leak parameters. Table 2 lists our current state of knowledge regarding leak location studies. It is clear from this table that some studies still do not involve field testing. In addition, one can realize that most of the research has focused on detecting and locating leaks in water pipelines. Very few studies have addressed the problem of leaks in pipelines that transport other types of fluids, such as oil, gas, or heavy oils.

In particular, proposing a leak detection system for heavy oils is important in the petroleum industry because of the enormous increase in oil demand and the progressive exhaustion of low-viscosity oil reservoirs. Moreover, the leak localization in multiphase flow pipelines (typical in oil production) is a pending issue that has not even been deeply addressed with other algorithmic tools. Therefore, there is a clear need for laboratory investigation of leak localization in multiphase flow pipelines.

The detection of single leaks remains an important concern that requires algorithms with better performance metrics; therefore, it is necessary to continue investigating to improve the Kalman filters used for it. Furthermore, most of the proposed Kalman-based approaches for single leaks require measurements from four sensors to perform the leak localization in a pipeline. Therefore, an improvement would be the reduction of this number of sensors. One way to do it is by using the same technique as the inverse transient analysis (ITA) approaches, as proposed in [56], in order to deal with the alteration of the flow through a maneuver such as the opening/closing of a valve; another would be by manipulating the frequency controller of a pump. The goal of this alteration is to obtain more information on the flow status.

At this point, it is necessary to say that ITA approaches and the Kalman-based approaches (proposed until today) have in common that an inverse problem is solved (the identification of the leak parameters) by minimizing the error between the numerical solution of a model and the available recordings. However, the Kalman-based approaches (proposed until today) do not need the generation of transients as ITA approaches need, but they require more measurements to compensate for the lack of information that a transient can give.

Regarding multiple faults, it is necessary that part of the research efforts focus on the development of new tools for the localization of multiple leaks (or faults), since this is an important issue that has not yet been approached meritoriously. It is worth mentioning that the presence of multiple leaks is a very common problem in countries plagued by vandalism and theft of hydrocarbons. Usually, in these places the pipelines are infested by simultaneously activated illegal taps.

Finally, the location of leaks using Kalman filters is a challenging area that will require rigorous experimental validation and addressing some concerns, such as the extension of existing algorithms for complex pipeline configurations, including branched pipelines or pipeline networks.

Author Contributions: Conceptualization, L.T.; methodology, L.T. and J.J.-C.; investigation, J.J.-C. and O.G.; resources, F.-R.L.-E.; data curation, J.J.-C. and O.G.; writing—original draft preparation, L.T., J.J.-C. and L.M.; writing—review and editing, all authors; supervision, F.-R.L.-E.; project administration, F.-R.L.-E.; funding acquisition, L.T., F.-R.L.-E. and L.M. All authors have read and agreed to the published version of the manuscript.

Funding: This research was funded by CONACyT-Convocatoria Atención a Problemas Nacionales 2017—Proyecto 4730, and CONACyT-Secretaría de Energía (Hidrocarburos)—Proyecto 280170.

Conflicts of Interest: The authors declare no conflict of interest. The funders had no role in the design of the study; in the collection, analyses, or interpretation of data; in the writing of the manuscript, or in the decision to publish the results.

Appendix A. Reminders about the Kalman Filter

The Kalman filter was first described and partially developed in technical papers by (among others) the Hungarian émigré Rudolf E. Kálmán [57–59]. It is an algorithm used to solve the so-called *linear quadratic problem*, which consists of estimating the instantaneous state of a linear dynamic system affected by white noise; therefore, it is also known as the linear quadratic estimator (LQE). In fact, the Kalman filter becomes an estimator that is statistically optimal with respect to any quadratic function of the estimation error. The following presentation seeks to briefly summarize relevant concepts presented in several prior works [57,60–63] related to the discrete Kalman filter.

Appendix A.1. The Discrete Kalman Filter

Let us start with the state-space representation (without a direct feedthrough) of a linear dynamic system such as

$$x(k+1) = Ax(k) + Bu(k) + w(k),$$
$$y(k) = Cx(k) + v(k), \tag{A1}$$

where $w(k)$ and $v(k)$ denote uncorrelated white noise processes with zero mean and covariances $Q(k)$ and $R(k)$, respectively. Notice that these noises perturb both the system states and the system outputs.

Since the objective is to find the optimal linear filter, the cost function to be minimized is the expected value of the squared prediction error as follows:

$$J = E\left\{ \| \hat{x}(k+1) - x(k+1) \|_2^2 \right\},$$
$$= E\left\{ (\hat{x}(k+1) - x(k+1))^T (\hat{x}(k+1) - x(k+1)) \right\}. \tag{A2}$$

The Kalman filter can be conceptualized as two distinct phases: "prediction" and "correction". The prediction phase uses the state estimate from the previous time step to produce an estimate of the state one time step ahead into the future at $k + 1$. This predicted state estimate, denoted by $\hat{x}(k+1|k)$, is known as the *a priori* state estimate. Thus, in the correction phase, the *a priori* state estimate is corrected based on the available measurements of the output $y(k+1)$. This improved estimate, denoted by $\hat{x}(k+1|k+1)$, is termed the *a posteriori* state estimate. The covariance matrix of the states, which provides a measure of the estimated accuracy of the state estimate, is

$$P(k) = E\left\{(\hat{x}(k) - x(k))^T (\hat{x}(k) - x(k))\right\}. \tag{A3}$$

Prediction Phase

The estimates of the states (since the noise $w(k)$ is assumed to be zero mean) are updated as

$$\hat{x}(k+1|k) = A\hat{x}(k) + Bu(k), \tag{A4}$$

based on the measurements up to time step k. By taking into account the *a priori* state estimation in (A4), the covariance matrix can be written (after some algebra) as

$$P(k+1|k) = AP(k)A^T + Q(k), \tag{A5}$$

where the fact has been exploited that $\hat{x}(k)$ and $x(k)$ are uncorrelated with $w(k)$.

Correction Phase

Once a new measurement $y(k+1)$ is available, it can be used to correct the estimates as follows:

$$\hat{x}(k+1|k+1) = \hat{x}(k+1|k) + \\ + K(k+1)(y(k+1) - C\hat{x}(k+1|k)), \tag{A6}$$

where clearly the estimates are based on the measurements up to time step $k+1$ and the optimal feedback gain $K(k+1)$ is calculated as

$$K(k+1) = P(k+1|k)C^T \left(CP(k+1|k)C^T + Q\right)^{-1}. \tag{A7}$$

According to Equation (A6), $K(k+1)$ determines which one has more weight in updating the estimated states: the observation error $y(k+1) - C\hat{x}(k+1|k)$ or the prediction of the states based on the internal model $\hat{x}(k+1|k)$. Finally, by taking into account the *a posteriori* state estimation in (A6), the covariance matrix can be updated as

$$P(k+1|k+1) = (I - K(k+1)C)P(k+1|k). \tag{A8}$$

The corresponding block diagram of the Kalman filter is shown in Figure A1.

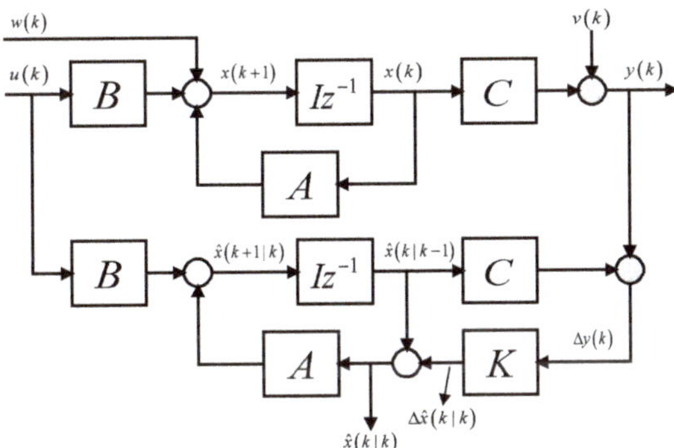

Figure A1. Kalman filter block diagram.

Appendix A.2. Extended Kalman Filter

In the theory of nonlinear state estimation, the *de facto* standard is the extended Kalman filter (EKF), which is the nonlinear version of the Kalman. In many situations, one is confronted with nonlinear system models of the form

$$x\left(k+1\right) = f_k\left(x\left(k\right),u\left(k\right)\right) + w\left(k\right),$$
$$y\left(k\right) = g_k\left(x\left(k\right)\right) + v\left(k\right), \tag{A9}$$

where $w\left(k\right)$ and $v\left(k\right)$ denote uncorrelated white noise processes with zero mean and covariances $Q\left(k\right)$ and $R\left(k\right)$, respectively.

In the EKF, the functions f_k and g_k are used to compute the predicted state (from the previous estimate) and the predicted measurement (from the predicted state), respectively. To update the covariance matrix $P(k)$, however, a first-order Taylor series expansion of (A9) is used. The idea is essentially to linearize the nonlinear system around the current estimate. Thus, at each time step, the Jacobian is evaluated by considering the current predicted states. These matrices are used in the Kalman filter equations.

The extended Kalman filter is then given as follows.

Prediction Phase

$$\hat{x}\left(k+1|k\right) = f_k\left(x\left(k\right),u\left(k\right)\right), \tag{A10}$$

$$F\left(k\right) = \left.\frac{\partial f_k\left(x,u\right)}{\partial x}\right|_{x=\hat{x}(k),u=u(k)}, \tag{A11}$$

$$P\left(k+1|k\right) = F\left(k\right)P\left(k\right)F^T\left(k\right) + Q\left(k\right). \tag{A12}$$

Correction Phase

$$G\left(k+1\right) = \left.\frac{\partial g_{k+1}\left(x\right)}{\partial x}\right|_{x=\hat{x}\left(k+1|k\right)}, \tag{A13}$$

$$K\left(k+1\right) = P\left(k+1|k\right)G^T\left(k+1\right) \times$$
$$\left(G\left(k+1\right)P\left(k+1|k\right)G^T\left(k+1\right) + Q\left(k+1\right)\right)^{-1}, \tag{A14}$$

$$\hat{x}\left(k+1|k+1\right) = \hat{x}\left(k+1|k\right) +$$
$$K\left(k+1\right)\left(y\left(k+1\right) - g_{k+1}\left(\hat{x}\left(k+1|k\right)\right)\right), \tag{A15}$$

$$P\left(k+1|k+1\right) = \left(I - K\left(k+1\right)G\left(k+1\right)\right)P\left(k+1|k\right). \tag{A16}$$

Appendix A.3. Extended Kalman Filter for Parameter Estimation

By augmenting the state vector $x(k)$ with a parameter vector θ, the EKF can be used for state and parameter estimations. The augmented system can be written as follows:

$$\begin{bmatrix} \hat{x}\left(k+1\right) \\ \hat{\theta}\left(k+1\right) \end{bmatrix} = \begin{bmatrix} f\left(x\left(k\right),\theta\left(k\right),u\left(k\right)\right) \\ \theta\left(k\right) \end{bmatrix} + \begin{bmatrix} w\left(k\right) \\ \xi\left(k\right) \end{bmatrix}$$
$$y\left(k\right) = g\left(\hat{x}\left(k\right)\right). \tag{A17}$$

Notice that the parameters are modeled as constant quantities disturbed by white noise.

Appendix A.4. Continuous Extended Kalman Filter With a Prescribed Degree of Stability

Let us consider a continuous nonlinear system that can be represented by the following equations:

$$\dot{x}(t) = f(x(t), u(t)),$$
$$y(t) = h(x(t)), \tag{A18}$$

where $x(t) \in \mathbb{R}^q$ is the state, $u(t) \in \mathbb{R}^p$ the input, and $y(t) \in \mathbb{R}^m$ the output. An observer (A18) can then be designed as follows:

$$\dot{\hat{x}}(t) = f(\hat{x}(t), u(t)) + K(t)[y(t) - h(\hat{x}(t))], \tag{A19}$$

where the state estimate is denoted by $\hat{x}(t)$, and the observer gain $K(t)$ is a time-varying $q \times m$ calculated as

$$K(t) = \mathbf{P}(t)\mathbf{C}^{\mathrm{T}}(t)\mathbf{W}^{-1}, \tag{A20}$$

where $\mathbf{P}(t)$ is a matrix calculated by using the next differential Riccati equation

$$\dot{\mathbf{P}}(t) = (\mathbf{A}(t) + \alpha\mathbf{I})\mathbf{P}(t) + \mathbf{P}(t)(\mathbf{A}^{\mathrm{T}}(t) + \alpha\mathbf{I}) - \mathbf{P}(t)\mathbf{C}^{\mathrm{T}}(t)\mathbf{W}^{-1}\mathbf{C}(t)\mathbf{P}(t) + \mathbf{Q}, \tag{A21}$$

which involves the following Jacobians

$$\mathbf{A}(t) = \frac{\partial f}{\partial x}(\hat{x}(t), u(t)), \quad \mathbf{C}(t) = \frac{\partial h}{\partial x}(\hat{x}(t)),$$

$$\mathbf{P}(0) = \mathbf{P}(0)^{\mathrm{T}} > 0, \mathbf{Q} = \mathbf{Q}^{\mathrm{T}} \geq 0, \mathbf{W} = \mathbf{W}^{\mathrm{T}} > 0.$$

In the Riccati Equation (A21), $\alpha > 0$ is a parameter that provides a stability degree to the estimation. Furthermore, by manipulating this parameter, the estimation rate (convergence time) can be tuned.

References

1. Drumond, G.P.; Pasqualino, I.P.; Pinheiro, B.C.; Estefen, S.F. Pipelines, risers and umbilicals failures: A literature review. *Ocean Eng.* **2018**, *148*, 412–425. [CrossRef]
2. Bermúdez, J.R.; López-Estrada, F.R.; Besançon, G.; Valencia-Palomo, G.; Torres, L.; Hernández, H.R. Modeling and simulation of a hydraulic network for leak diagnosis. *Math. Comput. Appl.* **2018**, *23*, 70. [CrossRef]

3. American Petroleum Institute. *API RP 1130 (2007): Computational Pipeline Monitoring for Liquids*; American Petroleum Institute: Washington, DC, USA, September 2007.

4. Geiger, G.; Werner, T.; Matko, D. Leak detection and locating-a survey. In Proceedings of the PSIG Annual Meeting, Pipeline Simulation Interest Group, Bern, Switzerland, 15–17 October 2003.

5. Besançon, G. *Nonlinear Observers and Applications*; Springer: Berlin, Germany, 2007; p. 224.

6. Besançon, G. Observer tools for pipeline monitoring. In *Modeling and Monitoring of Pipelines and Networks*; Springer: Berlin, Germany, 2017; pp. 83–97.

7. Recent Advances in Pipeline Monitoring and Oil Leakage Detection Technologies: Principles and Approaches. *Sensors* **2019**, *19*, 2548. [CrossRef] [PubMed]

8. Simon, D. Kalman filtering with state constraints: a survey of linear and nonlinear algorithms. *IET Control Theor. Appl.* **2010**, *4*, 1303–1318. [CrossRef]

9. Evensen, G. *Data Assimilation: The Ensemble Kalman Filter*; Springer: New York, NY, USA, 2009.

10. Digernes, T. Real-Time Failure-Detection and Identification Applied to Supervision of Oil Transport in Pipelines. *Model. Identif. Control* **1980**, *1*, 39–49. [CrossRef]

11. Benkherouf, A.; Allidina, A. Leak detection and location in gas pipelines. *IEE Proc. D-Control Theory Appl.* **1988**, *135*, 142–148. [CrossRef]

12. Besançon, G.; Georges, D.; Begovich, O.; Verde, C.; Aldana, C. Direct observer design for leak detection and estimation in pipelines. In Proceedings of the 2007 European Control Conference (ECC), Kos, Greece, 2–5 July 2007; pp. 5666–5670.

13. Geiger, I.G. Principles of leak detection. In *Fundamentals of Leak Detection*; KROHNE Oil and Gas: Reken, Germany, 2005.

14. Emara-Shabaik, H.; Khulief, Y.; Hussaini, I. A non-linear multiple-model state estimation scheme for pipeline leak detection and isolation. *Proc. Instit. Mech. Eng. Part I J. Syst. Control Eng.* **2002**, *216*, 497–512. [CrossRef]

15. Khulief, Y.; Emara-Shabaik, H. Laboratory investigation of a multiple-model state estimation scheme for detection and isolation of leaks in pipelines. *Proc. Instit. Mech. Eng. Part I J. Syst. Control Eng.* **2006**, *220*, 1–13. [CrossRef]

16. Bai, L.; Yue, Q.; Li, H. Sub-sea Pipelines Leak Detection and Location Based on Fluid Transient and FDI. In Proceedings of the the the Fourteenth International Offshore and Polar Engineering Conference, Toulon, France, 23–28 May 2004.

17. Doney, K. Leak Detection in Pipelines Using the Extended Kalman Filter and the Extended Boundary Approach. Ph.D. Thesis, University of Saskatchewan, Saskatoon, SK, Canada, 10 October 2007.

18. Begovich, O.; Navarro, A.; Sanchez, E.N.; Besançon, G. Comparison of two detection algorithms for pipeline leaks. In Proceedings of the IEEE International Conference on Control Applications, CCA 2007, Singapore, 1–3 October 2007; pp. 777–782.

19. Billmann, L.; Isermann, R. Leak detection methods for pipelines. *Automatica* **1987**, *23*, 381–385. [CrossRef]

20. Verde, C. Multi-leak detection and isolation in fluid pipelines. *Control Eng. Pract.* **2001**, *9*, 673–682. [CrossRef]

21. Chaudhry, M.H. *Applied Hydraulic Transients*; Van Nostrand Reinhold: New York, NY, USA, 1979.

22. Verde, C.; Bornard, G.; Gentil, S. Isolability of multi-leaks in a pipeline. In Proceedings of the 4th MATHMOD, Vienna, Austria, 5–7 February 2003.

23. Torres, L.; Besançon, G.; Georges, D. A collocation model for water-hammer dynamics with application to leak detection. In Proceedings of the 47th IEEE Conference on Decision and Control, Cancún, Mexico, 9–11 December 2008; pp. 3890–3894.

24. Torres, L.; Besançon, G.; Georges, D. Collocation modeling with experimental validation for pipeline dynamics and application to transient data estimations. In Proceedings of the European Control Conference, ECC'09, Budapest, Hungary, 23–26 August 2009.

25. Dos Santos, P.L.; Azevedo-Perdicoúlis, T.; Ramos, J.; Jank, G.; de Carvalho, J.M.; Milhinhos, J. Gas pipelines LPV modelling and identification for leakage detection. In Proceedings of the American Control Conference (ACC2010), Baltimore, Maryland, USA, 30 June–2 July 2010; pp. 1211–1216.

26. Navarro, A.; Begovich, O.; Besançon, G.; Dulhoste, J. Real-time leak isolation based on state estimation in a plastic pipeline. In Proceedings of the IEEE International Conference on Control Applications CCA 2011, Denver, CO, USA, 28–30 September 2011; pp. 953–957.

27. Colombo, A.F.; Karney, B.W. Energy and costs of leaky pipes: toward comprehensive picture. *J. Water Resour. Plan. Manag.* **2002**, *128*, 441–450. [CrossRef]

28. Delgado-Aguiñaga, J.; Begovich, O.; Besançon, G. Varying-parameter modeling and extended Kalman filtering for reliable leak diagnosis under temperature variations. In Proceedings of the 20th International Conference on System Theory, Control and Computing (ICSTCC), Sinaia, Romania, 13–15 October 2016; pp. 632–637.

29. Verde, C.; Torres, L.; González, O. Decentralized scheme for leaks' location in a branched pipeline. *J. Loss Prev. Process Ind.* **2016**, *43*, 18–28. [CrossRef]

30. Delgado-Aguiñaga, J.A.; Begovich, O. Water Leak Diagnosis in Pressurized Pipelines: A Real Case Study. In *Modeling and Monitoring of Pipelines and Networks*; Springer: New York, NY, USA, 2017; pp. 235–262.

31. Navarro, A.; Begovich, O.; Sánchez, J.; Besançon, G. Real-Time Leak Isolation Based on State Estimation with Fitting Loss Coefficient Calibration in a Plastic Pipeline. *Asian J. Control* **2017**, *19*, 255–265. [CrossRef]

32. Santos-Ruiz, I.; Bermúdez, J.; López-Estrada, F.; Puig, V.; Torres, L.; Delgado-Aguiñaga, J. Online leak diagnosis in pipelines using an EKF-based and steady-state mixed approach. *Control Eng. Pract.* **2018**, *81*, 55–64. [CrossRef]

33. Delgado-Aguiñaga, J.; Besançon, G. EKF-based leak diagnosis schemes for pipeline networks. *IFAC-PapersOnLine* **2018**, *51*, 723–729. [CrossRef]

34. Liu, P.; Li, S.; Wang, Z. Multi-leak diagnosis and isolation in oil pipelines based on Unscented Kalman filter. In Proceedings of the 30th Chinese Control And Decision Conference (CCDC), Shenyang, China, 9–11 June 2018; pp. 2222–2227.

35. Verde, C. Minimal order nonlinear observer for leak detection. *J. Dyn. Syst. Meas. Control* **2004**, *126*, 467–472. [CrossRef]

36. Verde, C.; Visairo, N. Identificability of multi-leaks in a pipeline. In Proceedings of the American Control Conference (ACC), Boston, MA, USA, 30 June–2 July 2004; Volume 5, pp. 4378–4383.

37. Gong, J.; Cai, J.; Li, X.; Song, S. Research on state estimation of oil pipeline considering adaptive extended Kalman filtering. In Proceedings of the 2007 International Conference on Mechatronics and Automation, Harbin, China, 5–8 August 2007; pp. 1294–1298.

38. Yu, Z.; Jian, L.; Zhoumo, Z.; Jin, S. A combined Kalman filter-Discrete wavelet transform method for leakage detection of crude oil pipelines. In Proceedings of the 9th International Conference on Electronic Measurement & Instruments, Beijing, China, 16–19 August 2009; pp. 3–1086.

39. Vianna, F.; Orlande, H.; Dulikravich, G. Estimation of the temperature field in pipelines by using the Kalman filter. In Proceedings of the 2nd International Congress of Serbian Society of Mechanics (IConSSM 2009), Palic (Subotica), Serbia, 1–5 June 2009; pp. 1–5.

40. Ye, G.; Fenner, R.A. Kalman filtering of hydraulic measurements for burst detection in water distribution systems. *J. Pipeline Syst. Eng. Pract.* **2010**, *2*, 14–22. [CrossRef]

41. Dos Santos, P.L.; Azevedo-Perdicoúlis, T.; Jank, G.; Ramos, J.; de Carvalho, J.M. Leakage detection and location in gas pipelines through an LPV identification approach. *Commun. Nonlinear Sci. Numer. Simul.* **2011**, *16*, 4657–4665. [CrossRef]

42. Navarro, A.; Begovich, O.; Besançon, G. Calibration of fitting loss coefficients for modelling purpose of a plastic pipeline. In Proceedings of the IEEE 16th Conference on Emerging Technologies & Factory Automation (ETFA), Toulouse, France, 5–9 September 2011; pp. 1–6.

43. Pizano-Moreno, A.; Begovich, O. Leak isolation with temperature compensation in pipelines. In Proceedings of the 9th International Conference on Electrical Engineering, Computing Science and Automatic Control (CCE), Mexico City, Mexico, 26–28 September 2012; pp. 1–5.

44. Padilla, E.A.; Begovich, O. Real-time leak isolation based on a fault model approach algorithm in a water pipeline prototype. *IFAC Proc. Vol.* **2012**, *45*, 916–921. [CrossRef]

45. Modisette, J.P. State estimation of pipeline models using the ensemble Kalman filter. In Proceedings of the PSIG Annual Meeting, Pipeline Simulation Interest Group, Prague, Czech Republic, 16–19 April 2013.

46. Guillen, M.; Dulhoste, J.F.; Besançon, G.; Scola, I.R.; Santos, R.; Georges, D. Leak detection and location based on improved pipe model and nonlinear observer. In Proceedings of the European Control Conference (ECC2014), Strasbourg, France, 24–27 June 2014; pp. 958–963.

47. Tian, J.; Wang, S.Q. Design and realization of the Kalman filter based on LabVIEW. Applied Mechanics and Materials. *Trans Tech. Publ.* **2014**, *519*, 1276–1280.

48. Okeya, I.; Kapelan, Z.; Hutton, C.; Naga, D. Online burst detection in a water distribution system using the Kalman filter and hydraulic modelling. *Procedia Eng.* **2014**, *89*, 418–427. [CrossRef]

49. Behrooz, H.A.; Boozarjomehry, R.B. Modeling and state estimation for gas transmission networks. *J. Nat. Gas Sci. Eng.* **2015**, *22*, 551–570. [CrossRef]

50. Al Ghailani, L.; El-Sinawi, A. Dynamic model of a new above-ground pipeline using a Kalman estimator-based system. In Proceedings of the IEEE Conference on Systems, Process and Control (ICSPC), Kuala Lumpur, Malaysia, 19–20 December 2015; pp. 12–15.

51. Zhou, D. Research on Natural Gas Pipeline Leak Detection Algorithm and Simulation. In Proceedings of the 2015 Chinese Intelligent Automation Conference, Fuzhou, China, 2–19 June 2015; pp. 355–361.

52. Verde, C.; Rojas, J. Iterative Scheme for Sequential Leaks Location. *IFAC-PapersOnLine* **2015**, *48*, 726–731. [CrossRef]

53. Choi, D.Y.; Kim, S.W.; Choi, M.A.; Geem, Z.W. Adaptive Kalman filter based on adjustable sampling interval in burst detection for water distribution system. *Water* **2016**, *8*, 142. [CrossRef]

54. Durgut, İ.; Leblebicioğlu, M.K. State estimation of transient flow in gas pipelines by a Kalman filter-based estimator. *J. Nat. Gas Sci. Eng.* **2016**, *35*, 189–196. [CrossRef]

55. Delgado-Aguiñaga, J.; Besançon, G.; Begovich, O.; Carvajal, J. Multi-leak diagnosis in pipelines based on Extended Kalman Filter. *Control Eng. Pract.* **2016**, *49*, 139–148. [CrossRef]

56. Brunone, B. Transient test-based technique for leak detection in outfall pipes. *J. Water Resour. Plan. Manag.* **1999**, *125*, 302–306. [CrossRef]

57. Kalman, R.E. A new approach to linear filtering and prediction Problems. *J. Basic Eng.* **1960**, *82*, 35–45. [CrossRef]

58. Lauritzen, S.L. Time series analysis in 1880: A discussion of contributions made by TN Thiele. *Int. Stat. Rev./Revue Int. Stat.* **1981**, *49*, 319–331. [CrossRef]

59. Lauritzen, S.L. *Thiele: Pioneer in Statistics*; Clarendon Press: Oxford, UK, 2002.

60. Ljung, L. Asymptotic behavior of the extended Kalman filter as a parameter estimator for linear systems. *IEEE Trans. Autom. Control* **1979**, *24*, 36–50. [CrossRef]

61. Song, Y.; Grizzle, J.W. The extended Kalman filter as a local asymptotic observer for nonlinear discrete-time systems. In Proceedings of the American Control Conference, Chicago, IL, USA, 24–26 June 1992; pp. 3365–3369.

62. Andrews, A. *Kalman Filtering: Theory and Practice Using MATLAB*; Wiley: New York, NY, USA, January 2001.

63. Isermann, R.; Münchhof, M. *Identification of Dynamic Systems: An Introduction with Applications*; Springer: New York, NY, USA, 2010.

MDPI
St. Alban-Anlage 66
4052 Basel
Switzerland
Tel. +41 61 683 77 34
Fax +41 61 302 89 18
www.mdpi.com

Journal of Marine Science and Engineering Editorial Office
E-mail: jmse@mdpi.com
www.mdpi.com/journal/jmse